Gorillas in the Mist

Gorillas in the Mist

Dian Fossey

A MARINER BOOK

HOUGHTON MIFFLIN COMPANY

BOSTON NEW YORK

Frontispiece: A portrait of Uncle Bert, Flossie, and their two-week-old son, Titus, made after field observations. Jay Matternes has captured the essential nature of the gorilla — the protectiveness of the silverback, the maternal concern of the mother, the vulnerability and dependence of the infant.

First Mariner Books edition 2000

Visit our Web site: www.hmhco.com

Library of Congress Cataloging in Publication Data
Fossey, Dian.
Gorillas in the Mist
Bibliography: p.
Includes index.
1. Gorillas — Behavior. 2. Mammals — Behavior.
QL737.P.96F7 1983 599.88'460451 82-23332
ISBN 0-618-08360-x (pbk.)
ISBN 978-0-618-08360-2
Printed in the United States of America

DOC 20 19 18 17

To the memories of Digit,
Uncle Bert, Macho, and Kweli

The mist of light
from which they take their grace
Hides what they are.

—RICHARD MONCKTON MILNES
(LORD HOUGHTON) 1809–1885

Acknowledgments

M ANY OF US have dreams or ambitions we hope to fulfill someday. My own, to go to Africa to study the mountain gorillas, might never have succeeded had it not been for the Henry family of Louisville, Kentucky, who loaned me the collateral for my first safari to Africa in 1963. It was at that time I met the gorillas of the Virunga Volcanoes in the then Democratic Republic of the Congo and Dr. Louis S. B. Leakey at Olduvai Gorge in Tanzania.

Three years later, Dr. Leakey chose me to undertake a long-term field study of the mountain gorillas. From that day until his death in 1972, he was a constant source of encouragement and optimism.

I shall always remember Dr. Leakey as I last saw him standing on the observation deck of the Nairobi airport to see me off to Rwanda and the gorillas. While his white hair streamed in the breeze, he cheerfully waved his aluminum crutches in the air. Even as the plane taxied down the runway, the silver glitter of Dr. Leakey's waving crutches was still visible.

Dr. Leakey's fervent hope that the mountain gorilla research proj; ect would be as successful as Dr. Jane Goodall's great study of free living chimpanzees convinced one of his close friends, Mr. Leighton A. Wilkie, to provide the funds to launch my project. I remain deeply indebted to The Wilkie Brothers' Foundation, not only for their initial support but also for their generosity in financing a renewal of the research after its first phase was terminated by a rebellion and for their continued financial assistance toward data compilation in America.

The National Geographic Society's Committee for Research and

Exploration began their magnanimous support of the Karisoke Research Centre in 1968, and their generous contributions continue to the present day. Not only has the National Geographic Society made possible the first long-term study of the mountain gorilla, but the Society has gone out of its way to provide technical assistance, material, and equipment to myself and numerous students. Among those of the Society who have given so unfailingly of their time, efforts, and support are: Dr. Melvin M. Payne, Chairman of the Board of Trustees; Edwin W. Snider, Secretary of the Committee for Research and Exploration; Mary G. Smith, Senior Assistant Editor of the Research Grant Projects; Robert E. Gilka, Senior Assistant Editor of Photography; Joanne M. Hess, Chief of the Audiovisual Department; Ronald S. Altemus, of Audiovisual; W. Allan Royce, Assistant Director of Illustrations; and Andrew H. Brown, Assistant Editor. To these and others of the National Geographic family who have contributed to the output and accomplishments of the Karisoke Research Centre, I shall always remain indebted.

Within recent years, as the study expanded, financial support was generously given by the L. S. B. Leakey Foundation for specific projects. I wish to express my deepest appreciation to the many members of the Foundation who have contributed toward the mountain gorilla study. Among these are the late Allen O'Brien, founder of the L. S. B. Leakey Foundation, and Mr. Jeffrey R. Short, Jr., who provided both guidance and financial support in addition to friendship. My continued gratitude must be extended to Mary Pechanec and Joan Travis, who carried on with the initial aims of the Leakey Foundation for nearly a decade following Dr. Leakey's death.

I owe a great deal to Professor Robert Hinde of Cambridge University, who so patiently and thoughtfully supervised the analysis and completion of my doctorate and several scientific papers. Dr. Hinde's constant encouragement meant much, particularly during the long period when the thesis was far from being a reality.

Special thanks are due to Mr. Jay Matternes for his exceptional rendition of the true nature of the gorilla in the portrait of Uncle Bert, Flossie, and Titus in this volume, and also for his invaluable analyses of gorilla skeletal material accumulated during the course of the research.

For excellent film documentation and deep, ensuing friendship, I wish to express my indebtedness to Robert M. Campbell, Alan Root, and Warren and Genny Garst, all photographers of rarity in that they respected the nature of their subject, the gorilla, over and above their personal desires to obtain footage. Additionally, I wish to thank Joan and Alan Root for a comradeship that began in 1963 at Kabara, where, because of their tolerance, I was first privileged to meet the mountain gorillas. In later years both Joan and Alan contributed enormously toward turning the dream into a reality by sharing some of their knowledge of Africa with me.

During the first two years of the study, congenial hospitality and warm friendship were always available from Mr. Walter Baumgärtel, owner of the Travellers Rest Hotel in Kisoro, Uganda, a pioneer who cared about the Virungas and the gorillas long before most of the world grew concerned over their future.

There have been numerous people in Rwanda who have provided loyalty and friendship during bleak or lonely periods. I shall always be grateful to Mrs. Alyette DeMunck, who gave so much assistance, good sense, and companionship when helping me set up the Karisoke Research Centre after termination of the study in the Democratic Republic of the Congo. Likewise, I remain greatly indebted to Mrs. Rosamond Carr of Gisenyi, Rwanda, whose warm and gracious personality were so welcomed throughout the entire study period. I also wish to thank Dr. Lolly Preciado, whose boundless enthusiasm, coupled with her medical skills, contributed much to Karisoke in spite of the demands of her own strenuous work with the lepers of Rwanda. Lolly, Mrs. DeMunck, and Mrs. Carr are just three of the many individuals who were always there when needed in a land where the bonds of friendship and giving of self seem to be a way of life.

Additionally, there have been numerous members of the American Embassy staff based in Kigali, Rwanda, who have gone out of their way to assist myself and Karisoke students over seemingly insuperable hurdles that arose from time to time. Among the many who have contributed to Karisoke, I would like to thank Mr. and Mrs. R. E. Kramer, who selflessly volunteered their personal time and efforts toward solving problems that I could not attend to from the top of a mountain. To Ambassador and Mrs. Frank Crigler, I owe a large portion of my heart and Karisoke's survival, particularly during very low times. Their constant active and moral support will never be forgotten.

My gratitude needs also to be extended to the many students who have participated in the accumulation of long-term records as well as toward the maintenance of the Karisoke Research Centre during my absences. These include: T. Caro, R. Elliott, J. Fowler, A. Goodall, A. Harcourt, S. Perlmeter, A. Pierce, I. Redmond, R. Rombach, C. Sholley, K. Stewart, A. Vedder, P. Veit, D. Watts, W. Weber, and J. Yamagiwa.

I want to pay special tribute to the memory of one young woman whose ambition to work with the mountain gorillas was never fulfilled. Debre Hamburger had been given the name Mwelu, meaning "a touch of brightness and light," by Africans with whom she worked as an archeological research assistant in another area of Africa. Debbie died of cancer shortly before her scheduled period of study at Karisoke. Her ashes have been scattered along the trails which she might have walked within the Virungas. Her wish "for a little more time" was not granted, but her words have become symbolic to all of those who have been able to work with the gorillas at Karisoke.

The long-term study in the isolation of the Virunga Mountains would have been impossible without the assistance of my dedicated African staff, most of whom have remained with me since the study began: Gwehandagaza, the head porter, who has provided Karisoke's only link with the outside world by dauntlessly trudging up and down a long muddy trail twice a week to bring food and mail from the nearest town of Ruhengeri; Nemeye and Rwelekana, two indefatigable trackers whose skills have enabled Karisoke researchers to fulfill their goals; Kanyaragana and Basili, the housemen who have greatly contributed toward making the Karisoke camp somewhat of an oasis in the midst of a wet wilderness.

I thank the Government of Rwanda for their generosity in allowing me to establish the Karisoke Research Centre within the Parc des Volcans. I am deeply grateful to the President of Rwanda, Général-Major Juvénal Habyarimana, whose interest and support toward conservation of endangered animal species has been both consistent and sincere. I would like to thank the members of the Rwandan Ministry of Foreign Affairs for the authorization of foreign students at Karisoke. Also, my gratitude must be extended to the Rwandese Office of Tourism and National Parks (ORTPN) for the privilege of being able to continue the research of the mountain gorillas.

Paulin Nkubili, formerly the Chef des Brigades of Ruhengeri, deserves a medal for his integrity, inflexibility, and unyielding exercise of his genuine efforts toward the preservation of wildlife and the prosecution of poachers who threaten the sanctity of the

Rwandese sector of the Virunga Volcanoes. Mr. Nkubili's demonstrations of courage against the odds will never be forgotten.

There are numerous others who have contributed, and continue to contribute, toward the active conservation of gorillas. I am profoundly grateful to these, the hundreds of concerned people who have donated to the Digit Fund. The aim of this nonprofit, tax-exempt corporation, established in memory of the slain gorilla Digit and staffed by volunteers, is to finance antipoacher patrols within the Virungas. Donations, which are tax deductible, may be made to The Digit Fund, Karisoke Research Centre, c/o Rane Randolph, C.P.A., P. O. Box 25, Ithaca, New York 14851. All contributors receive a newsletter reporting on events in the field.

I would like to acknowledge gratefully the efforts of Dr. Glenn Hausfater, who made it possible for me to come to Cornell University in March 1980, and who dealt so patiently with withdrawal symptoms on my return to civilization. Glenn Hausfater also conceived the idea of forming the Karisoke Board of Scientific Directors, a committee designed for the continuance of scientific research and the maintenance of the Karisoke Research Centre during my interim in the United States.

This book grew through its embryonic stages to final publication because of the tremendous patience and constructive editing skills of Anita McClellan, an editor who went beyond the call of duty in dealing with a "bushy" author. Further, Anita became a devoted disciple of the gorillas and understood why certain traumas in Africa plagued the initial publication date. I am deeply indebted to Anita. Additionally, I wish to thank Stacey Coil of Cornell University, who clean-typed the chapter drafts and virtually knows the text by heart, and David Minard, the illustrator, who has succeeded in blending artistry and accuracy in his splendid drawings throughout the text.

Lastly, I wish to express my deepest gratitude to the gorillas of the mountains for having permitted me to come to know them as the uniquely noble individuals that they are.

DIAN FOSSEY
November 1982

Contents

Preface

*G*ORILLAS IN THE MIST recounts some of the events of the thirteen years that I have spent with the mountain gorillas in their natural habitat and includes data from the fifteen years of continuing field study. Mountain gorillas live only on six extinct mountains within the Virunga Volcanoes and do not frequent the two active volcanoes of the chain. The region inhabited by the gorillas is some twenty-five miles long and varies in width from six to twelve miles. Two thirds of the conservation area lies in Zaire (formerly known as the Democratic Republic of the Congo) in the Parc National des Virungas; about 30,000 acres of conservation area lie in Rwanda and are known as the Parc National des Volcans. The small remaining northeastern portion of the mountain gorillas' habitat lies in Uganda and is known as the Kigezi Gorilla Sanctuary.

My research studies of this majestic and dignified great ape — a gentle yet maligned nonhuman primate — have provided insight to the essentially harmonious means by which gorillas organize and maintain their familial groups and also have provided understanding of some of the intricacies of various behavioral patterns never previously suspected to exist.

In 1758 Carl Linnaeus, the first serious student of classification, officially recognized the close relationship between humans, monkeys, and apes. He devised the order name Primates to encompass them all and to denote their high ranking in the animal kingdom. Man and the three great apes — orangutan, chimpanzee, and gorilla — are the only primates without tails and, like most primates, have five digits on each hand and foot, the first of which is opposable. Anatomical features shared by all primates are two mammae (nipples), orbits directed forward to permit binocular vision, and, usually, a total of thirty-two teeth.

Because of the scanty record of ape fossils, there is no universal agreement on the origin of the two families, Pongidae (the apes) and Hominidae (mankind), which have been separated for millions of years. None of the three great apes is considered ancestral to modern man, *Homo sapiens,* but they remain the only other type of extant primate with which human beings share such close physical characteristics. From them we may learn much concerning the behavior of our earliest primate prototypes, because behavior, unlike bones, teeth, or tools, does not fossilize.

Several million years ago the chimpanzee and gorilla lines had already separated from one another, and the orangutan line even earlier than that. Throughout the eighteenth century there remained a considerable amount of confusion in distinguishing between orangutans, chimpanzees, and gorillas. The orangutan was the first to be recognized as a distinct genus — only because of its remote habitat in Asia. It was not until 1847, on the basis of a single skull from Gabon, that the gorilla was confirmed as a separate genus from the chimpanzee.

Just as there are separate subspecies among orangutan and chimpanzee, there are separate subspecies of gorilla, also with morphological variations related primarily to habitat. In western Africa there remain some 9000 to 10,000 lowland gorilla (*Gorilla gorilla gorilla*) in the wild. It is this subspecies most frequently seen in captivity and mounted in museum collections. Some 1000 miles to the east within the Virunga Volcanoes of Zaire, Uganda, and Rwanda live the last surviving mountain gorillas (*Gorilla gorilla beringei*), the subjects of my field study. Only about 240 mountain gorillas remain in the wild. None are found in captivity. The third subspecies is known as the eastern lowland gorilla (*Gorilla gorilla graueri*). Only about 4000 *graueri* remain in the wild, mainly in eastern Zaire, and less than two dozen live in captivity.

There are some twenty-nine morphological differences between the lowland and the mountain gorilla, adaptations related to altitudinal variations. The mountain gorilla, the more terrestrial of the two and living at the highest altitude in the gorillas' range, has longer body hair, more expanded nostrils, a broader chest girth, a more pronounced sagittal crest, shorter arm limbs, a longer palate, and shorter, broader hands and feet.

In the wild only some 4000 gorillas (including all three subspecies) now live in reputedly protected areas. Advocates for establishing captive gorilla populations thus feel justified in attempting to preserve this most endangered of the great apes in zoos or similar institutions. Because of the strong kinship bonds of gorilla families, the capture of one young gorilla may involve the slaying of many of its familial group, and certainly not every animal collected from the wild reaches its destination alive. Moreover, three times more gorillas have been taken from the wild than have been born in captivity, and gorilla deaths in confinement continue to outnumber gorilla births. I cannot concur with those who advocate saving gorillas from extinction by killing and capturing more free-living individuals only to exhibit them in confinement.

Conservation of any endangered species must begin with stringent efforts to protect its natural habitat by the enforcement of rigid legislation against human encroachment into parks and other game sanctuaries. Next, confinement facilities should be encouraged to extend new and promising programs that replace solitary wire-and-cement cages with more natural group settings rather than to expend energy on acquiring additional exotic species for display.

For captive gorillas, trees should be available to climb and material such as straw, branches, or bamboo supplied for nest building. Food could be allotted in small portions throughout the day and require some degree of preparation such as peeling and stripping of stalks or even searching for randomly distributed items supplied at various locations within the enclosure. Access to the outdoors should be provided; contrary to popular opinion, gorillas greatly enjoy basking in the sun. Of prime importance to the reclusive gorilla are obscured niches where captive animals may withdraw, as desired, not only from the presence of people but also from one another, as is the species' habit in the wild.

Those bearing the heavy responsibility of caring for captive gorillas should be encouraged to exchange so-called nonbreeders between populations, an inherent process among free-living gorillas and one that avoids inbreeding and also stimulates productivity. Once facilities improve the physical conditions under which many gorillas are kept, propagation success should follow, not automatically but certainly more often than now holds true in an unnecessary number of sterile environments for isolated colonies of captive gorillas.

The late Dr. Louis Leakey almost prophetically realized that the mountain gorilla, the subspecies scientifically recognized and described in 1902, might possibly be doomed to extinction in the same century in which it had been discovered. It was for this reason that Dr. Leakey wanted a long-term field study done on the mountain gorilla, which had, by 1960, been studied in the wild only by George Schaller.

Dr. Leakey's planning was indeed fortuitous. In the six and a half years between Schaller's excellent study and the beginning of my own, the ratio of adult gorilla males to females within the Kabara area of the Virunga Volcanoes had dropped from 1:2.5 to 1:1.2, accompanied by a halving of the population. Moreover 40 percent of the mountain gorillas' protected habitat was in the process of being appropriated for cultivation purposes. The human encroachment pressure on the Virunga parks subjects gorilla groups to increased overlapping of their home ranges, and causes higher frequencies of aggression between groups. If mountain gorillas are to survive and propagate, far more active conservation measures urgently need to be undertaken. The question remains, is it already too late?

Among all researchers who have worked in the African field, I consider myself one of the most fortunate because of the privilege of having been able to study the mountain gorilla. I deeply hope that I have done justice to the memories and observations accumulated over my years of research on what I consider to be the greatest of the great apes.

Family Lines

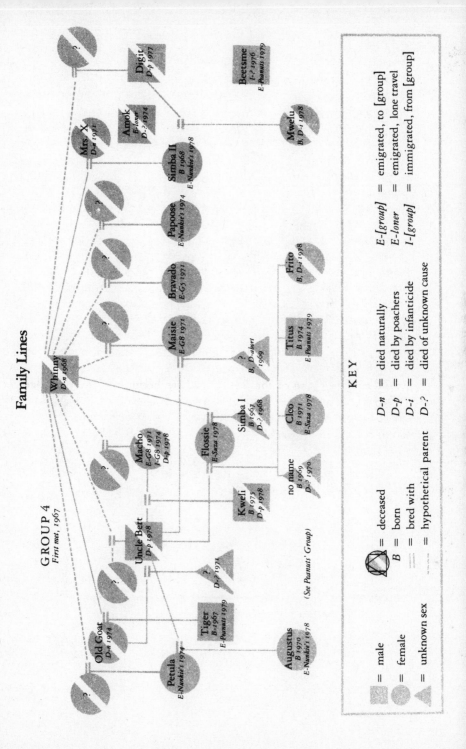

GROUP 4
First met, 1967

(See Peanuts' Group)

KEY

▨ = male	D-n = died naturally	E-*[group]* = emigrated, to [group]
◗ = female	D-p = died by poachers	E-*loner* = emigrated, lone travel
◮ = unknown sex	D-i = died by infanticide	I-*[group]* = immigrated, from [group]
	D-$?$ = died of unknown cause	

⊘ = deceased	
B = born	
⎯⎯ = bred with	
- - - - = hypothetical parent	

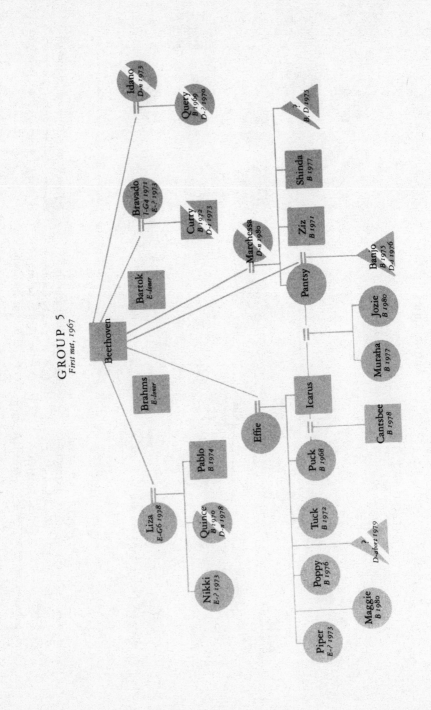

GROUP 5
First met, 1967

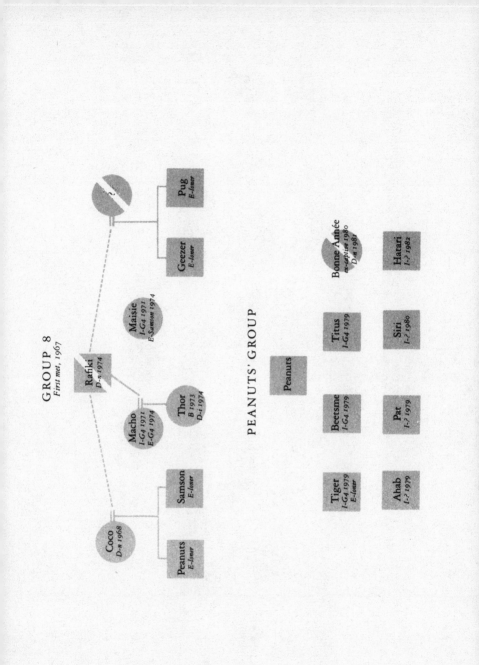

GROUP 8
First met, 1967

Coco
D-n 1968

Peanuts
E-loner

Samson
E-loner

Macho
I-G4 1971
E-G4 1974

Thor
B 1973
D-i 1974

Rafiki
D-n 1974

Maisie
I-G4 1971
E-Samson 1974

?

Geezer
E-loner

Pug
E-loner

PEANUTS' GROUP

Peanuts

Tiger
I-G4 1979
E-loner

Beetsme
I-G4 1979

Titus
I-G4 1979

Bonne Année
ex-captive 1980
D-n 1981

Ahab
I-? 1979

Pat
I-? 1979

Siri
I-? 1980

Harari
I-? 1982

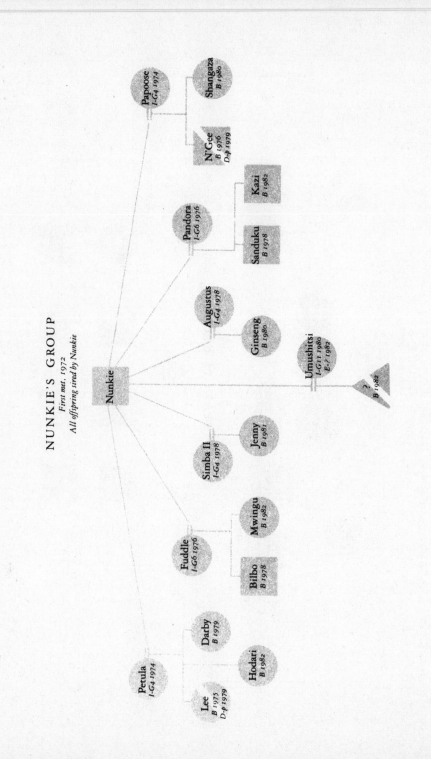

NUNKIE'S GROUP
First met, 1972
All offspring sired by Nunkie

Nunkie

Petula
I-G4 1974

Papoose
I-G4 1974

Pandora
I-G6 1976

Augustus
I-G4 1978

Simba II
I-G4 1978

Fuddle
I-G6 1976

Darby
B 1979

Lee
B 1975
D-? 1979

Hodari
B 1982

Bilbo
B 1978

Mwingu
B 1982

Jenny
B 1981

Ginseng
B 1980

Umushitsi
I-G11 1980
E-? 1982

?
B 1982

Sanduku
B 1978

Kazi
B 1982

N'Gee
B 1976
D-? 1979

Shangaza
B 1980

Gorillas in the Mist

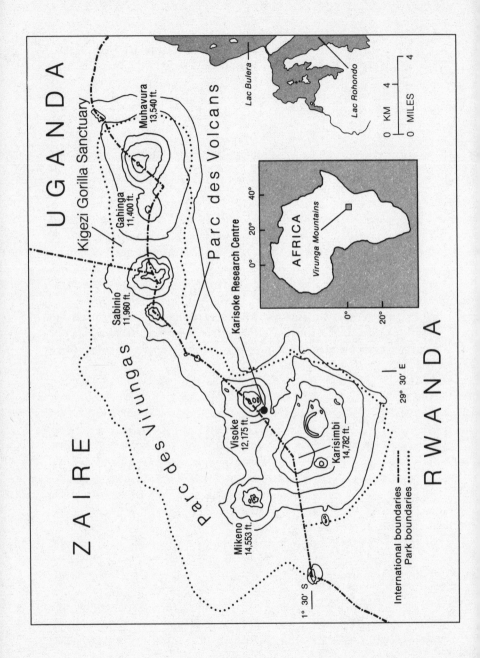

1 | In the Mountain Meadow of Carl Akeley and George Schaller

J SPENT MANY YEARS longing to go to Africa, because of what that continent offered in its wilderness and great diversity of free-living animals. Finally I realized that dreams seldom materialize on their own. To avoid further procrastination I committed myself to a three-year bank debt in order to finance a seven-week safari throughout those parts of Africa that most appealed to me. After months spent planning my itinerary, most of which was far off the normal tourist routes, I hired a driver, by mail, from a Nairobi safari company and flew to the land of my dreams in September 1963.

Two of the main goals of my first African trip were to visit the mountain gorillas of Mt. Mikeno in the Congo and to meet Louis and Mary Leakey at Olduvai Gorge in Tanzania. Both wishes came true. How vividly I still can recall Dr. Leakey's sparkling interest in hearing that I was on my way to visit briefly the gorillas at Kabara in the Congolese sector of the Virunga Mountains, where George Schaller had worked a few years previously. Dr. Leakey spoke to me most enthusiastically about Jane Goodall's excellent field work with the chimpanzees at the Gombe Stream Research Centre in Tanzania, then only in its third year, and he stressed the importance of long-term field studies with the great apes. I believe it was at this time the seed was planted in my head, even if unconsciously, that I would someday return to Africa to study the gorillas of the mountains.

Dr. Leakey gave me permission to walk around some newly excavated sites at Olduvai, one of which contained a recently discovered giraffe fossil. As I ran down a steep slope, my exultation at being free under African skies was abruptly shattered, along with my right ankle, when I fell into a dig containing the new find. As the ankle cracked, the sudden pain induced me to vomit unceremoniously all over the treasured fossil. As if this wasn't humiliating enough, I had to be ignominiously hauled out of the gorge, piggyback style, by disgusted members of the Leakeys' staff. Mary Leakey then very kindly served me cool lemon squash while we watched the swelling ankle turn from various shades of blue to black. Both she and my driver felt that the intended climb into the Virungas to search for gorillas would have to be forfeited. Neither of them realized that the accident only strengthened my determination to get to the gorillas I had come to meet in Africa.

Two weeks after leaving the Leakeys and aided by a walking stick carved by a sympathetic African encountered along the road, I, the hired driver, and a dozen porters carrying the basics of camping gear and food began the arduous five-hour climb to the remote Kabara meadow. Kabara lies at 10,200 feet immediately adjacent to the 14,553-foot Mt. Mikeno in the Parc des Virungas in Zaire, formerly known as the Democratic Republic of the Congo. Some three years before my visit of 1963, Kabara had been the study site of George B. Schaller. An eminent American scientist, he was the first person to conduct a reliable field study of the mountain gorillas, amassing 458 hours of observation within that area. The Kabara meadow also contained the grave of Carl Akeley, an American naturalist who had been responsible for urging the Belgian government to create the Albert National Park for the protection of mountain gorillas and their 400,000-year-old volcanic habitat.

In 1890 the mountains had been the object of a twenty-year dispute between Belgium (representing the present Zairoise portion), Germany (the Rwandan area), and Britain (the Ugandan side). It was only in 1910 that the boundaries were finally settled. By 1925 some 190 square miles were set aside and the park was established. Carl Akeley had convinced King Albert of Belgium to expand the protected area so that by 1929 most of the Virunga chain was included. It was then called the Albert National Park. In 1967 the Zairoise named their section the Parc National des Virungas and the Rwandese called theirs the Parc National des Volcans. In Uganda the Virunga gorilla habitat was designated the Kigezi Gorilla Sanctuary in 1930. Akeley died when revisiting Kabara in 1926 and was buried on the meadow's edge in accordance with his wishes. He had considered

Kabara one of the loveliest and most tranquil spots in the world.*

On my first visit to Kabara in 1963 I was fortunate in meeting Joan and Alan Root, photographers from Kenya who were camped at the meadow while working on a photographic documentary of the mountain gorillas. Both Joan and Alan kindly overlooked the intrusion of a somewhat hobbly and inquisitive American tourist into their secluded mountain workshop and allowed me to accompany them on some of their extraordinary contacts with the relatively unhabituated gorillas of Kabara. It was only because of their generosity, coupled with the skill of Sanwekwe, a Congolese park guard and tracker, that I was able to contact and photograph the animals during that brief visit. Sanwekwe had worked as a boy tracking gorilla for Carl Akeley; as a man he worked for George Schaller. Nearly twenty years later he became my friend and skilled tracker.

I shall never forget my first encounter with gorillas. Sound preceded sight. Odor preceded sound in the form of an overwhelming musky-barnyard, humanlike scent. The air was suddenly rent by a high-pitched series of screams followed by the rhythmic rondo of sharp *pok-pok* chestbeats from a great silverback male obscured behind what seemed an impenetrable wall of vegetation. Joan and Alan Root, some ten yards ahead on the forest trail, motioned me to remain still. The three of us froze until the echoes of the screams and chestbeats faded. Only then did we slowly creep forward under the cover of dense shrubbery to about fifty feet from the group. Peeking through the vegetation, we could distinguish an equally curious phalanx of black, leather-countenanced, furry-headed primates peering back at us. Their bright eyes darted nervously from under heavy brows as though trying to identify us as familiar friends or possible foes. Immediately I was struck by the physical magnificence of the huge jet-black bodies blended against the green palette wash of the thick forest foliage.

Most of the females had fled with their infants to the rear of the group, leaving the silverback leader and some younger males in the foreground, standing tense with compressed lips. Occasionally the dominant male would rise to chestbeat in an attempt to intimidate us. The sound reverberated throughout the forest and evoked similar displays, though of lesser magnitude, from gorillas clustered around

---

* His body lay there undisturbed for fifty-three years before grave robbers, Zairoise poachers, violated the site and carried off his skeleton in 1979.

him. Slowly, Alan set up his movie camera and proceeded to film.
The openness of his motions and the sound of the camera piqued
curiosity from other group members, who then treed to see us more
clearly. As if competing for attention, some animals went through
a series of actions that included yawning, symbolic-feeding, branch-
breaking, or chestbeating. After each display, the gorillas would look
at us quizzically as if trying to determine the effect of their show.
It was their individuality combined with the shyness of their behavior
that remained the most captivating impression of this first encounter
with the greatest of the great apes. I left Kabara with reluctance
but with never a doubt that I would, somehow, return to learn more
about the gorillas of the misted mountains.

* * *

My reunion with Kabara, Sanwekwe, and the gorillas came about
as a direct result of a visit by Dr. Leakey to Louisville, Kentucky,
where I was continuing my work as an occupational therapist in
order to pay off the huge bank loan amassed for the first safari.
Vaguely remembering me as the clumsy tourist of three years earlier,
Dr. Leakey's attention was drawn to some photographs and articles
I had published about gorillas since my return from Africa. After a
brief interview, he suggested that I become the "gorilla girl" he
had been seeking to conduct a long-term field study. Our conversation
ended with his assertion that it was mandatory I should have my
appendix removed before venturing into the remote wilderness of
the gorillas' high altitude habitat in central Africa. I would have
agreed to almost anything at that point and promptly made plans
for an appendectomy.

Some six weeks later on returning home from the hospital sans
appendix, I found a letter from Dr. Leakey. It began, "Actually
there really isn't any dire need for you to have your appendix removed. That is only my way of testing applicants' determination!"
This was my first introduction to Dr. Leakey's unique sense of humor.

Eight more months passed before Dr. Leakey was able to obtain
funds to launch the study. During the interim I finished paying for
my 1963 safari while virtually memorizing George Schaller's two
superlative books about his 1959–60 field studies with the mountain
gorillas, as well as a "Teach Yourself Swahili" grammar book. Quit-
ting my job as an occupational therapist and saying goodbye to the
children who had been my patients for eleven years was difficult,

as were the farewells to Kentucky friends and my three dogs. The dogs seemed to sense that this was going to be a permanent separation. I can still recall them — Mitzi, Shep, and Brownie — running after my overladen car as I drove away from my Kentucky home to head for California to say farewell to my parents. There was no way that I could explain to dogs, friends, or parents my compelling need to return to Africa to launch a long-term study of the gorillas. Some may call it destiny and others may call it dismaying. I call the sudden turn of events in my life fortuitous.

At the end of 1966, Leighton Wilkie, the man responsible for financially backing Jane Goodall into her long-term chimpanzee study, told Dr. Leakey that he also would be willing to initiate another long-term great ape study. Like Louis Leakey, Leighton Wilkie felt that by studying man's closest living relatives, the great apes, new light could be shed on how our ancestors might have behaved. With his support the finances of my project were assured.

Thus in December 1966, I was again Africa-bound. This time only gorillas were my goal. By an incredible stroke of luck, I chanced to run into Joan Root at the Heathrow airport in London while waiting for a Nairobi-bound plane. Both she and Alan were stunned that I intended to drive from Nairobi to the Congo, a distance of some 700 miles, then seek government permission to work at Kabara, and, lastly, carry out the study by myself. They shared the opinion of many friends that lone females, especially those fresh from America, should not be expected to try any one of the above "impossibilities," much less all three combined.

Once we were in Nairobi, Joan accompanied me on numerous shopping sprees. Because of her long experience on safaris, she saved many hours, and undoubtedly many mistakes, by helping me select functional camp supplies and equipment such as tents, lights, stoves, and bedding. Dr. Leakey purchased, after some perilous test drives through Nairobi's crowded streets, an antiquated, canvas-topped Land-Rover that I later named "Lily." Little did I know that in seven months' time Lily was going to be responsible for saving my life.

When all of the gear was finally assembled, Jane Goodall kindly invited me to visit the Gombe Stream Research Centre for two days to show me her methods of camp organization, data collecting, and, as well, to introduce me to her lovable chimpanzees. I fear that I was not an appreciative guest, for I was desperately keen to reach Kabara and the mountain gorillas.

Finally Alan Root, still doubting the sanity of myself and Dr. Leakey, said that he would accompany me in his Land-Rover during the long drive from Kenya to the Congo, nearly halfway across Africa. Without Alan I don't know if I would have succeeded in coaxing Lily over some of the escarpment goat-trail roads prevalent in Africa at that time. Nor might I have handled with Alan's ease many of the complexities involved in obtaining the government permits necessary for working at Kabara within the Parc des Virungas.

On the morning of January 6, 1967, Alan and I, accompanied by some Congolese park guards and two Africans willing to stay on as camp staff, arrived at the small village of Kibumba situated at the base of Mt. Mikeno. There, exactly as I had done with my driver three years earlier, we selected several-dozen porters to carry up the camping gear to the remote Kabara meadow. Neither the porters' village nor the forest's huge, ancient, moss-laden *Hagenia abyssinica* trees seemed to have changed during my three-year absence. Elated, I climbed the nearly four thousand feet between Kibumba and Kabara, where I established my camp within the heartland of the ancient dormant volcanoes. I was thrilled to find Kabara so unchanged even to the presence of two delightfully mischievous ravens (*Corvultur albicollis*). They absconded with every scrap of unguarded food before eventually learning to unzip the tent flap in search of concealed food.

Alan could stay at Kabara only for two days, but during that time he worked around the clock. At camp he supervised mundane necessities such as the digging of a latrine secluded by potato-sacking walls, placement of barrels to store water, and planning drainage ditches around my tent. To our mutual regret, we never had a visual contact with the gorillas during his two-day stay, though we did hear two groups exchanging "hootseries" from high on Mt. Mikeno's slopes. We also found fresh tracks of a gorilla group in the relatively flat saddle area adjacent to the mountain. In my excitement I promptly took off on the trail swath left by the gorillas through dense herbaceous foliage in the certainty that I would encounter the group at any moment. Some five minutes of "tracking" passed before I was aware that Alan was not behind me. Perplexed, I retraced my steps and found him patiently sitting at the very point where we had first encountered the trail.

With the utmost British tolerance and politeness, Alan said, "Dian, if you are ever going to contact gorillas, you must follow their tracks to where they are going rather than backtrack trails to where they've been." That was my first lesson in tracking, and one that I've never forgotten.

The next day I felt a sense of panic while watching Alan fade into the foliage near the descending edge of the Kabara meadow. He was my last link with civilization as I had always known it, and the only other English-speaking person on the mountain. I clung on to my tent pole simply to avoid running after him.

A few moments after Alan's departure one of the two Africans in camp, trying to be helpful, asked, *"Unapenda maji moto?"* Forgetting every word of Swahili memorized over the past year, I burst into tears and zipped myself into the tent to escape imagined "threats." About an hour later, feeling the fool, I asked the Congolese to repeat his statement slowly. Did I want hot water? Whether for tea or bath he didn't specify, but this seemed to be the panacea necessary for all *wazungus'* (white people's) rough times. I accepted a couple of gallons of hot water as graciously as possible with many *asantes* (thank you's), hoping to convince the Africans that their concern was deeply appreciated.

The following morning it was field work, or the actual searching for gorillas, that took precedence over the endless list of camp chores such as setting up clothes-drying lines, placing the catchment barrels in optimal places to collect rainwater, and teaching the staff how to care for the kerosene lamps and stove purchased in Nairobi. Like a bored housewife, I relegated these and many more tasks to the evening hours. Daylight belonged to the gorillas.

* * *

On my first full day in the field I had scarcely walked more than ten minutes away from camp before seeing a lone male gorilla sunbathing on a horizontal tree trunk that projected over a small lake nestled in a corner of the Kabara meadow. Even before I could get my binoculars out of their case, the startled animal leapt from the tree and disappeared into the dense vegetation of the adjacent mountain slope. I spent the entire day trying to catch up with him but, of course, my climbing ability was no match for that of a single frightened gorilla. Oddly enough that brief observation was the first and only one of its kind in which I encountered a gorilla resting in such an exposed area. Only later was I to learn that gorillas tend to avoid both open meadows and relatively large bodies of water, most likely because these were the places where they frequently encountered people.

By the second day, a Congolese park guard arrived to serve as a temporary "tracker" until Sanwekwe, the wonderful old guide whom I had first met with Joan and Alan Root, could come to camp. The substitute had no previous experience in tracking, which became increasingly evident as the long and tiring day of circuitous meanderings revealed absolutely nothing of gorilla sign. The third day proved equally unrewarding as far as gorillas were concerned but retrospectively provided me with many laughs. For several hours the man and I had been following a trail through dense herbaceous foliage when suddenly we spotted a gorilla-sized black object that appeared to be sunbathing some one hundred feet away from us on the opposite side of a deep ravine. Slowly, I uncased my binoculars, readied my notebook, pen, and stopwatch before finding an obscured vantage point from which to observe the animal, which seemed to be sunning contentedly on the open slope. For more than an hour the study object did not move. The guide quietly snored behind me while my stopwatch ticked away the time. Although I realized that gorilla observations required patience, this great "first" seemed unduly testing because the first page of "data" remained absolutely blank after an hour. Finally, I awoke my guide and asked him to remain where he was so that I could creep closer to the sunbathing animal. I'll never forget the chagrin felt upon realizing that the "gorilla" I had been observing for over an hour was actually a giant forest hog (*Hylochoerus meinertzhageni*). Upon seeing me the animal crept into thicker foliage and disappeared. Two days later I found his aged body lying on the forest's floor under a large *Hagenia* tree. He had apparently died of natural causes.

All of the surprises didn't occur necessarily during the daytime. On the fourth night at Kabara I was rudely awakened when tossed out of my cot and rolled to the other side of the tent still encased in my sleeping bag. The entire tent was shaking as if the pent-up furies of the long-extinct volcanoes were unleashing themselves for a great eruption. Hearing deep rumbling sounds, I felt anger rather than fear to think that the study, only just begun, was coming to such an abrupt end. About a minute of shaking and rumbling passed before the air was filled with both auditory and odoriferous clues to explain the disruption. Three elephants had discovered that tent poles made good side scratchers and one chose to leave his colossal calling card just outside the tent's veranda. These three, and later others, were to become frequent visitors to camp, and I was always thrilled by their curiosity and lack of fear. I could not, however, keep them out of my promising vegetable garden. After about the third *tembo* (elephant) raid, I decided that I would have to sacrifice salads for the duration.

Because of the near daily encounters with elephant, buffalo, giant forest hog, and, of course, gorilla, the time spent in the field was by far more exciting than the hours that, of necessity, had to be spent in camp. Almost immediately, I became bogged down in paperwork and have remained that way ever since. Copious notes concerning everything from weather to bird and plant life, poacher activities, and, naturally, all details of the gorilla contacts needed to be typed nightly.

The 7-foot by 10-foot tent became a combined bedroom, office, bath, and drying area for my clothes, constantly wet in the rain forest climate. Within the tent wooden crates I had covered with exotically colored native cloth served as desks, chairs, cupboards, and file cabinets. I prepared and ate meals in the second room of the men's hut, a small wooden building some thirty to thirty-five years old which had been vandalized extensively since I had first seen it. The staff, whose number had grown to three with the arrival of Sanwekwe, prepared their meals on a fire in the middle of their room. They never seemed daunted by the constant smoke that permeated the entire cabin and often left me tearing and gasping for breath.

The men's meals consisted mainly of mammoth portions of sweet and white potatoes, multicolored beans, corn, and occasionally fresh vegetables they carried up from their village, Kibumba, at the mountain's base. Any misgivings I first had about eating my more varied diet around the Africans were soon discarded, since they tended to scoff politely at the contents of the tinned food that formed the mainstay of my diet. Once a month in Kisoro, Uganda, a small town about two hours' drive from the base of Mt. Mikeno, I would stock up on cans of hot dogs, Spam, powdered milk, margarine, corned beef, tuna, hash, and various vegetables, as well as boxes of noodles, spaghetti, oatmeal, and bags of sweets. Bread, cheese, and other fresh foods only lasted about two weeks on the mountain. A month thus tended to be divided into two parts: feast for the first half and fast throughout the remaining days. At least eggs were plentiful, thanks to a prolific hen named Lucy. Lucy and her mate, Dezi, were given to me by Sanwekwe, who assumed I would fatten them up for the pot. Instead, they were to become my first pets in Africa, and I grew extremely fond of them over the years.

Whenever food supplies ran low I was quite content with potatoes: mashed potatoes, fried potatoes, baked potatoes, boiled potatoes. It was indeed fortunate that I really like potatoes. Occasionally though, at the end of a month, I would run out of cigarettes just about the same time that Sanwekwe would run out of pipe tobacco. For us both this was a real deprivation. Fastidiously, we would ration the dwindling supply; he by mixing in dead leaves with his remaining tobacco, and I by allowing myself only two or three puffs at a time on a treasured cigarette. The absurdity of these "breaks" inevitably reduced us both to giggles much like two delinquent schoolchildren.

Sanwekwe possessed a marvelous sense of humor in addition to being an untiring tracker and a man who cared deeply for the gorillas and other animals of the forest. He taught me just about all that I was to learn about tracking and proved a reliable companion during the many days spent hiking throughout the rugged terrain, usually in pouring rain. Because of Sanwekwe's help I eventually was able to find three groups of gorillas within the study area of some two square miles that lay adjacent to and along the slopes of Mt. Mikeno.

Gorillas live in relatively stable, cohesive social units known as groups, whose compositions are altered by births, deaths, and occasional movements of individuals in or out of a group. Group sizes vary from two to twenty animals, and average about ten individuals. A typical group contains: one silverback, a sexually mature male over the age of fifteen years, who is the group's undisputed leader and weighs roughly 375 pounds, or about twice the size of a female; one blackback, a sexually immature male between eight and thirteen years weighing some 253 pounds; three to four sexually mature females over eight years, each about 200 pounds, who are ordinarily bonded to the dominant silverback for life; and, lastly, from three to six immature members, those under eight years. Immatures were divided into young adults between six and eight years, weighing about 170 pounds each, juveniles between three and six years, weighing some 120 pounds, and infants, from birth to three years, weighing between 2 and 30 pounds.

The prolonged period of association of the young with their parents, peers, and siblings offers the gorilla a unique and secure type of familial organization bonded by strong kin ties. As the male and female offspring approach sexual maturity they often leave their natal groups. The dispersal of mating individuals is perhaps an evolved pattern to reduce the effects of inbreeding, though it seems that maturing individuals are more likely to migrate when there are no breeding opportunities within the group into which they are born.

During the early days of the study at Kabara, it was difficult to establish contacts because the gorillas were not habituated or accustomed to my presence and usually fled on seeing me. I could often choose between two different kinds of contacts: obscured, when the gorillas didn't know I was watching them, or open contacts, when they were aware of my presence.

Obscured contacts were especially valuable in revealing behavior that otherwise would have been inhibited by my presence. The drawback to this method was that it contributed nothing toward the habituation process. Open contacts, however, slowly helped me win the animals' acceptance. This was especially true when I learned that imitation of some of their ordinary activities such as scratching and feeding or copying their contentment vocalizations tended to put the animals at ease more rapidly than if I simply looked at them through binoculars while taking notes. I always wrapped vines around the binoculars in an attempt to disguise the potentially threatening glass eyes from the shy animals. With gorillas, as is often the case with humans, direct staring constitutes a threat.

Not only was it necessary to get the gorillas accustomed to the bluejeaned creature who had become a part of their daily lives, it was also very necessary for me to know and recognize the particular animals of each group as the amazing individuals they were. Just as George Schaller had done some seven and a half years before me, I relied heavily upon "noseprints" for identification purposes. There is a tendency for the gorillas of each group to resemble one another, especially within matrilineal lines. As no two humans have exactly the same fingerprints, no two gorillas have the same "noseprint" — the shape of the nostrils and the outstanding troughs seen on the bridges of their noses. Since the gorillas initially were unhabituated, I had to use binoculars, but even from a distance I could quickly make sketches of noseprints seen on the more curious group members peeking back at me from partially hidden positions in the dense vegetation. These sketches proved invaluable at a time when close-up photography was out of the question. Also, I would have needed a third hand in order to manage a camera, binoculars, and note taking, not to mention carrying on with the imitative routine of feeding, scratching, and vocalizing needed to relax the gorillas as well as to arouse their curiosity.

Occasionally I did take out my camera, especially when the sun was shining. Probably one of the most publicized pictures of gorillas in the wild was taken at Kabara during the second month of my study when the gorillas were beginning to trust me. It shows a line-up of sixteen gorillas posing like so many Aunt Matildas on a back porch. The group had been day-nesting and sunbathing when I contacted them, but upon my approach they nervously retreated to obscure themselves behind thick foliage. Frustrated but determined to see them better, I decided to climb a tree, not one of my better talents. The tree was particularly slithery and, try as I might, no amount of puffing, pulling, gripping, or clawing succeeded in getting me more than a few feet aboveground. Disgustedly, I was about to give up when Sanwekwe came to my aid by giving one mighty boost to my protruding rump; tears were running from his eyes as he was convulsed in silent laughter. I felt as inept as a baby taking its first step. Finally able to grab on to a conveniently placed branch, I hauled myself up into a respectful semislouch position in the tree about twenty feet from the ground. By this time I naturally assumed that the combined noises of panting, cursing, and branch-breaking made during the initial climbing attempts must have frightened the group on to the next mountain. I was amazed to look around and find that the entire group had returned and were sitting like front-row spectators at a sideshow. All that was needed to make the image complete were a few gorilla-sized bags of popcorn and some cotton candy! This was the first live audience I had ever had in my life and certainly the least expected.

That day's observation was a perfect example of how the gorillas' sense of curiosity could be utilized toward their habituation. Nearly all members of the group had totally exposed themselves, forgetting about hiding coyly behind foliage screens because it was obvious to them that the observer had been distracted by tree-climbing problems, an activity they could understand.

Stimulating the gorillas' curiosity was but one aspect of the habituation process learned over time. It was a while though before I realized that standing upright or walking within their view increased the animals' apprehension. That discovery marked the beginning of my knuckle-walking days. Crawling toward groups on knuckles and knees and maintaining contacts in a seated position not only kept me at the gorillas' eye level but also conveyed the impression that I was settled and not about to barge into their midst. After a contact was established I learned that if I partially concealed my celery-chomping self, their curiosity would inevitably draw them from behind thick clumps of foliage or induce them to climb trees in order to see me better. Previously when I had been completely visible throughout a contact, the gorillas were content to remain obscured and peek at me through vegetation — which did not contribute much toward my observations of their behavior. I therefore changed my tactics from climbing trees to view the gorillas to leaving the trees for the gorillas to climb to view me.

Initially I often had to wait for up to a half an hour, pretending to feed on foliage, before the gorillas gave in to their inquisitiveness and climbed trees surrounding me. Once their curiosity was satisfied, they would resume their usual activities, forgetting that I was there. This is what I had come to observe.

For a number of months I imitated the gorillas' chestbeats by slapping my hands against my thighs in studious mimicry of their rhythm. The sound was instantly successful in gaining the gorillas' attention, especially when they were at distances over one hundred feet. I thought I was very clever but did not realize that I was conveying the wrong information. Chestbeating is the gorillas' signal for excitement or alarm, certainly the wrong message for me to have sent as an appeasement signal. I stopped mimicking chestbeating and only use it now when trying to hold newly encountered groups, whose curiosity upon hearing chestbeats from a human being nearly always overcomes their instinct to flee.

Whenever approaching a group for a contact I always tried to select an observational point containing a good solid tree for the gorillas to climb. There were many times, however, when logistics gave way to fatigue. This was especially true after I had climbed for several hours up forty-five-degree slopes, had waded through a morass of muddy trails, had to hack my way through pillared vegetation, or had been crawling on my hands and knees for a length of time through punishing foliage like nettles. My nose, a most protrusive one, suffered more nettle stings than the rest of my body, which was protected by heavy gloves, long underwear, heavy jeans, socks, and high boots. Most people, when thinking of Africa, envision dry plains sweltering under a never-ending sun. When I think of Africa

I think only of the montane rain forest of the Virungas — cold and misty, with an average annual rainfall of seventy-two inches.

Frequently the mornings were sunny, but I soon learned that these teasing starts were a hoax. For this reason my knapsack always included raingear in addition to the daily necessities of camera, lenses, film, notebooks, and the luxury of a thermos containing hot tea. The usual weight of the knapsack, between fifteen and twenty pounds, became almost unbearable on extended treks when the long directional microphone of the twenty-pound Nagra tape recorder was added. All too vividly I can recall my temptation to ditch the lot when nearing the end of particularly arduous tracking sessions. At such times, the knowledge that the gorillas were somewhere ahead of me was the only impetus to keep on going.

The Kabara groups taught me much regarding gorilla behavior. From them I learned to accept the animals on their own terms and never to push them beyond the varying levels of tolerance they were willing to give. Any observer is an intruder in the domain of a wild animal and must remember that the rights of that animal supersede human interests. An observer must also keep in mind that an animal's memories of one day's contact might well be reflected in the following day's behavior.

* * *

I ran out of "following days" with the Kabara gorillas at 3:30 P.M. on July 9, 1967, when Sanwekwe and I returned to the Kabara meadow after one of our usual rewarding days with the gorillas. Camp was surrounded by armed soldiers who informed me that a rebellion was going on within the Kivu Province of Zaire, the new name for the Belgian Congo. I must be "evacuated for my own safety."

The next morning I was "escorted" down the mountain by soldiers and porters carrying all my camp equipment, personal belongings as well as my beloved chickens Lucy and Dezi. Overhead flew the two semitame ravens, seemingly as confused and perplexed as I about the abrupt loss of our mountain home. At the base of Mt. Mikeno, after the three-hour descent, the ravens left me to return to the meadow and the barren, empty square where my tent had been for six and a half months.

I spent two weeks confined in Rumangabo, the outlying post for both the park headquarters and the military in that particular area of the Kivu Province. This extremely unpleasant period was aggravated by my viewing the towering slopes of Mt. Mikeno from my room, constantly wondering if I would ever be able to return to the gorillas of the mountains.

By the end of the first week no one at the park headquarters seemed willing or able to explain why I was being detained. The park staff's apprehension increased noticeably when soldiers from the adjacent military camp set up road blockades around the entrances and exits of the park headquarters. From bits and pieces of conversation, I learned that the barricades were being set up for the protection of a general who would soon be arriving at Rumangabo from the besieged town of Bukavu, where he was leading an uprising. It was only after a "visit" to the army camp that I realized, on reading a military cable, that I was earmarked for the general. With chances for my release lessening each hour I remained in captivity, I decided to escape, using Lily's registration plates as a ruse.

At that point, Lily was still a Kenyan vehicle, and the exchange fees from Kenyan to Zairoise registration were roughly $400. I managed to convince the soldiers that all my cash was kept in Kisoro, Uganda, and that we only had to go to Kisoro to pick up my money in order to register Lily properly in Zaire. The enticement of that much cash, along with the anticipated acquisition of the car as well as a cooperative hostage, was too much for the soldiers to resist. They agreed to "escort" me to Uganda under armed guard.

Working throughout the night before our two-hour drive, I managed surreptitiously to load Lily with my data, photographic equipment, and Lucy and Dezi. I had kept a small .32 automatic pistol locked up at Kabara but of course had never used it. Upon my arrival at Rumangabo I had given the pistol to a sympathetic park guard for safekeeping. The man befriended me during my detainment by sneaking in fresh food and also by keeping me posted on the political situation. The night preceding my intended escape into Uganda, the guard stealthily returned the pistol and advised that it be kept handy during the two-hour drive, particularly at the border between Zaire and Uganda. He said that Bunagana, the border post, was heavily manned by soldiers who might not take kindly to my exit into Uganda, even if only temporarily. The logistical problem then arose of how to keep a pistol, even a small one, handy and at the same time concealed from the half-a-dozen armed soldiers who were to serve as my "escorts." I finally decided to risk hiding the pistol in the bottom of a half-filled box of Kleenex tucked unobtrusively in the open glove compartment of the dashboard. I wedged rusty bolts and small car tools around the carton in the hope of stabilizing it for the rough ride over the unpaved lava-rock road to the border. The last thing I needed was to have the gun bounce out onto the lap of the soldier next to me!

The men were in high spirits when we started off the following morning, spirits that rose considerably along the way as a result of numerous forced "pub" stops at local *pombe* (native beer) bars. They certainly never noticed my undue fascination with the bouncing Kleenex box.

The border stop was all that my park guard friend had anticipated — a lengthy and garrulous battle of wills. The military stationed at the border said I could walk the five-odd miles to Kisoro and leave the Land-Rover with them; the soldiers from Rumangabo refused to walk and would not let me go on alone. The tissue-thin authorization papers issued at the Rumangabo military camp allowing my "temporary" entry into Uganda were being snatched from hand to hand by drunken soldiers and equally drunken custom officials. It was obvious, however, that the general's name impressed even the most belligerent of the border's soldiers. After several hours of heated argument, during which I had remained absolutely silent, Lucy laid an egg. I jumped up and down, clapped my hands, and otherwise played the absolute fool extolling Lucy's great talent. A silence fell over the soldiers as they gazed at me unbelievingly. Finally all, border and military guards alike, agreed that I was a first-class *bumbavu* (idiot) and had to be considered harmless. The barricade was opened.

Twelve years before these events, a wonderful man by the name of Walter Baumgärtel had established a delightful home-away-from-home in Kisoro for both gorilla researchers and tourists. His Travellers Rest Hotel had been an oasis to many scientists preceding me, George Schaller among them. I had met Walter on my first safari in 1963, and during the six-and-a-half-month study in 1967 had grown to think of him as one of the kindest and most endearing friends I had made in Africa. Ten minutes after crossing the border into Uganda, I spun my car into the driveway of Walter's hotel, grabbed the car keys, ran through the front door, where a gathering of wide-eyed, open-mouthed refugees from Zaire had suddenly collected. I continued running the length of the hotel to the farthest room. Diving through cobwebs, I buried myself under a bed where, cowardly, I remained until the uproar created by Walter's swift action in calling the Ugandan military to arrest the Zairoise soldiers had ceased. However, the first thing I did upon emerging was to congratulate Lucy properly for her well-timed egglaying. The egg had broken during the melee.

After several days of interrogation in Kisoro, where the word had been relayed that I would be shot on sight if I tried to go back to Zaire, I drove to Kigali, the capital of Rwanda, for further questioning. I then flew back to Nairobi for my first reunion with Dr. Leakey in seven months — not exactly the kind of reunion we had anticipated.

He was waiting at the Nairobi airport wearing one of his ear to ear grins that implied "Well, we fooled them, didn't we?" After a brief discussion, we both decided I should return to the Virungas rather than work with the lowland gorillas in West Africa or the orangutans in Asia. In Nairobi I learned that I had been declared missing and assumed dead by the United States Department of State. Dr. Leakey and I therefore had to check in with the local American Embassy. The chargé d'affaires at that time flatly declared it was impossible for me to return to Rwanda. In his words, I would be "immediately extradited to Zaire as an escaped prisoner."

This was just the kind of an encounter Dr. Leakey loved. He and Embassy representatives asked me to leave the room and closed the doors. For nearly an hour, their voices could be heard bellowing throughout the halls of the Embassy. Dr. Leakey finally emerged with his usual zest and a sparkling twinkle in his eyes that impishly hinted this had been a particularly enjoyable and successful debate.

Only because of the continued generosity of Leighton Wilkie was I able to assemble in Nairobi the basic equipment needed for a second start. Within two weeks I flew off to the Rwandan side of the Virungas. There still were gorillas to find and mountains to climb. It was like being reborn.

My settlement in Rwanda was made fairly easily because of the help of an extraordinary Belgian woman. Alyette DeMunck had been born in the Kivu Province of Zaire and was richly endowed with commonsense knowledge of the country and its traditions. Thanks to her, I was able to begin a census count of the gorillas on the Rwandan side of the volcanoes almost immediately after arrival.

Searching for a new campsite offering as much as Kabara had

was a challenging adventure. I began on Mt. Karisimbi, the 14,782-
foot volcano lying southeast of Mt. Mikeno. I was disappointed to
find Karisimbi's slopes crowded with vast herds of cattle and had
to climb to 12,000 feet before being able to set up a bivouac camp
in an uninhabited area about a half-hour from the Zairoise border.
Nineteen weeks had passed since I had last seen a gorilla, but luck
was with me when I found fresh gorilla trail leading right to one
of the three study groups I had known at Kabara. The contact was
one of the most wonderful welcome-home gifts I ever could have
received. The gorillas recognized me and held their ground at about
fifty feet. I was able to see that an infant had been born since I
left.

An eleven-day search of most of the Rwandan side of Karisimbi
was disheartening because of the abundance of cattle and poachers
throughout the park and the total absence of gorilla sign. However,
one morning on a sparkling clear day, I climbed to the barren, moon-
like alpine meadows of Karisimbi to a vantage point that enabled
me to see the entire twenty-five-mile-long Virunga chain of extinct
volcanoes. With binoculars I saw very promising gorilla country in
the gently rolling saddle terrain between Mts. Karisimbi and Visoke.
As I sat on the high windy meadow contemplating all that the future
held, two ravens came flying up from the vast ocean of green forest
lying below. Cawing insistently, they glided in to take up begging
positions for my leftover lunch scraps. Their relative shyness suggested that they probably were not the pair from Kabara, but their
presence at that time and place seemed propitious.

More than a decade later as I now sit writing these words at camp,
the same stretch of alpine meadow is visible from my desk window.
The sense of exhilaration I felt when viewing the heartland of the
Virungas for the first time from those distant heights is as vivid
now as though it had occurred only a short time ago. I have made
my home among the mountain gorillas.

Second Beginning: Karisoke Research Centre, Rwanda

RWANDA IS ONE of the most densely populated countries in the world. Consisting of only 10,000 square miles, about one-eighth the size of Kenya and smaller than the state of Maryland, Rwanda contains 4.7 million people, a population expected to double by the end of the century. Rwanda, known as the "little Switzerland of Africa," is also one of the world's five poorest countries, with about 95 percent of the population barely managing to survive on small farm lots, called *shambas,* of about two and one half acres each. Terracing techniques are relied upon to utilize nearly all of the available land for cultivation. However, even with such stringent escarpment devices, the population is living above the carrying capacity of the terrain. Each year another 23,000 families will need new land plots to grow their food and support their livestock.

In 1969, 22,000 acres were removed from the Parc National des Volcans for the cultivation of pyrethrum, daisylike flowers made into natural insecticide and sold in European markets for foreign exchange. The remainder of the park consists of only 30,000 acres, or ½ of 1 percent of Rwanda's land total. However, the Rwandan Ministry of Agriculture is considering taking another 40 percent, or 12,000 acres, of the remaining parkland for cattle-grazing schemes for some of the nation's 680,000 cattle — livestock maintained in spite of the country's extreme population pressure. There is no buffer zone between the cultivation and the gorilla parkland. The fertile soil adjacent to the park contains 780 inhabitants per square mile. The people freely cross back and forth into the park to collect wood, set illegal traps for antelope, collect honey from wild bee hives, graze cattle, and plant plots of potatoes and tobacco. Encroachment upon this terrain may be responsible for the mountain gorilla becoming one of the seven or so other rare species both discovered and extinct within the same century.

*   *   *

Alyette DeMunck helped me make preparations to begin a second safari into the utopian saddle region I had seen from Mt. Karisimbi's high alpine meadow. In Lily the Land-Rover and Alyette DeMunck's Volkswagen bus, both loaded with camping equipment, we drove in a northeasterly direction around the foothills of Karisimbi and Visoke on extremely rough, unpaved boulder-strewn tracks traversed by countless herds of cattle and goats. Three hours later the tracks ended in a densely cultivated area lying about 8000 feet amid *shambas* and fields of pyrethrum. There we hired several dozen porters to carry all of my gear for the five-hour climb to the 10,000-foot saddle terrain that lay deep in the montane rain forest adjacent to Mt. Visoke, obscured in the fog far ahead of us.

The porters were mainly Bahutu people of the Bantu race, the main agriculturists of the area. More than four centuries previously the Watutsi people of the Hamitic race came down from the north and subjugated the Bahutu who were living in the region that came to be known as Rwanda. A type of feudalism developed as the Watutsi, who owned cattle, took over the land. The Bahutu then had to pay in services or goods for the right to use the cattle and pastureland. In time the Bahutu became the serfs of the Watutsi kings. The two castes remained distinct throughout most of the German and Belgian colonial period until 1959, when the Bahutu overthrew their Watutsi masters. Rwanda became independent from Belgium in 1962 with the Bahutu in power. The revolution and its aftermath lasted well into 1973 and caused the slaughter of thousands of Watutsi and the exodus of many thousands more. To this day some bitterness remains between the two races.

Many of the Watutsi who remained in Rwanda tended cattle and, because of land scarcity, were grazing vast herds illegally within the park when I arrived in 1967. Over the thirteen years of my research there, I came to know one Mututsi (singular of Watutsi) family very well. Also, mainly within the Parc des Volcans, I encountered a third race, known as Batwa. These were members of a semipygmoid tribe, the lowest in the Rwandan caste system. Traditionally they are the poachers, hunters, and honey gatherers. Their notorious activities within the park were to have great repercussions, both in my life and in the lives of the gorillas I would meet.

The barefoot Bahutu porters were cheerful as they deftly adjusted the loads Alyette DeMunck and I divided among them. Next, each man pulled long hanks of grass to fashion into a compact circular pad to protect his head from the burden before picking up his walkingstick, or *fimbo*. The sticks, as I soon discovered, were needed for balance on the mud-slicked parts of the trail, and also proved tremendously helpful in pulling one's weight from deep bogs where elephants had walked. Boots were not available in Rwanda at the time, and the plastic shoes that could be bought in local markets would have been useless in the suction of the muddy trail, where mud frequently comes up to the knees.

Leaving Lily surrounded by curious villagers and a *zamu* guard to watch over her, Mrs. DeMunck and I fell in behind the line of porters who eagerly had begun shoving their way through the hordes of children who had gathered around. The women were left behind, since it was their daily job to tend to the fields, cut and gather firewood, collect water, and care for the young ones. Many of the women were visibly pregnant, yet carried infants in skin slings on their backs while often keeping an eye on toddlers near their feet.

As the long scraggly train of porters passed through a narrow maze of trails across the cultivated fields, spirited greetings were exchanged between them and the working women. People seemed to be everywhere, a marked contrast to heading up to my first camp from the small Congo village of Kibumba and immediately plunging into the utter stillness of the dark forest below Kabara. However, the natives of Kibumba and this Rwandan village, Kinigi, were curious and friendly. Both men and women were swathed in yards of cloth rather than in the discarded European clothing that later became more favored apparel. Like the adults, the children went barefooted and seemed unaware of their own various stages of undress, ranging from nothing at all to tattered bits of sacking cloth. Heavy rain pelted down on all of us. Despite my plastic raingear, I found myself shuddering while watching the laughing children gaily skip along the side of our procession, seemingly without a care in the world.

As we walked through the newly planted pyrethrum fields, thick fog concealed much of the land-clearance devastation. However, closer to the trail, one could see smoldering trunk stumps of age-old *Hagenia* trees, the only remnants of what once had been a magnificent forest. I longed for the joyful sense of exhilaration I had always felt when beginning the climb to Kabara. The ascending route here was more like walking through a bombsite in the aftermath of a war.

About a half an hour before reaching Visoke's immediate slopes we came to a bamboo zone that previously also had been part of the park but was now doomed to be culled for pyrethrum and people. Today, in that same area, some six tin rondavels and a large parking lot exist for the convenience of tourists. I am grateful that I knew this region even as late as 1967, for it will never be the same.

Once within the thick bamboo, I felt a small bit of the magic of the wilderness when fresh elephant droppings were sighted, as well as signs of gorilla. From the bamboo belt the trail led through a cool rock tunnel, some five to six feet in width and about thirty feet in length. The crumbling sides of the lava tunnel bore the impressions of years of scrapings created by the rough hides of elephants using the tunnel as a passageway between the forest above and the bamboo below. The firm floor of the tunnel bore the smooth, undulated patterns of elephant footprints and its misty air was filled with their odor. Ten years later, when most of the elephants of the parklands had been killed by poachers, the sides of the tunnel were to become covered with thick layers of moss, obliterating for ever signs of one of the many animal species among the original inhabitants of the Virunga Volcanoes.

The tunnel created a dramatic entrance into the world of the gorilla. It served as a passageway between civilization and the silent world of the forest, for it opened upon a vista of densely vegetated slopes where massive, moss-clad *Hagenia* trees formed a partial canopy along the sides of the trail.

The *Hagenia abyssinica* tree is the most common tree found in the Virunga saddle area and becomes progressively less abundant on higher mountain slopes probably because the steeper slopes cannot support its bulk. There appears to be little regeneration of the *Hagenia* in the saddle or lower slope region between altitudes of 8600 and 10,800 feet, yet in the subalpine and alpine zones numerous young *Hagenia* saplings have begun to grow. George Schaller wonderfully described the *Hagenia* as having "the appearance of a kindly, unkempt old man." Their trunks, which may grow to some eight feet in diameter, and their huge, armchairlike padded limbs support a profuse variety of mosses, lichen, ferns, orchids, and other epiphytes. The tree seldom reaches more than 70 feet, and in the saddle area of the parkland *Hagenia* canopy covers only about 50 percent of the region; thus a rich undergrowth of herbaceous foliage may flourish. Gorillas favor many of the epiphytes supported by the limbs of the *Hagenia* rather than the trees' long pinnate leaves or lilac-colored pendulous flower clusters. One favored gorilla food item harbored by the *Hagenia* is a narrowleaf fern (*Pleopeltis excavatus*), which hangs down individually suspended from thick moss pads on the tree's nearly horizontal lower limbs. Gorillas frequently settle themselves comfortably on a soft cushion of moss, disengage a big wedge of moss, and sit with it on their laps, idly picking out the fern, leaf by leaf. Many aged *Hagenia* tree trunks are partly hollow and provide homesteading sites for a variety of animals ranging from hyrax (*Dendrohyrax arboreus*), genet (*Genetta tigrina*), mongoose (possibly *Crossarchus obscurus*), and dormouse (*Graphiurus murinus*) to squirrel (possibly *Protoxerus stangeri*).

Sharing the saddle terrain with the *Hagenia* is the *Hypericum lanceolatum,* known as St. John's Wort in Europe. The *Hypericum* is a more delicate tree than the *Hagenia* and has a wider altitudinal variation growing from the park boundaries around 8600 feet to about 12,000 feet in the alpine zone, where it becomes more stunted in size. Throughout the saddle the *Hypericum* reaches 40 to 60 feet, but its relatively spindly trunks and limbs cannot support the heavy moss cushions so characteristic of the *Hagenia*. Amid the latticework of the *Hypericum*'s small, pointed, lanceolate leaves and brilliant yellow, waxy-petaled flowers cling multiple long wispy strands of *Usnea* lichens that resemble Spanish moss. The tree also serves as a host for the red-flowering parasitical plant *Loranthus luteo-aurantiacus,* which belongs to the mistletoe family and is greatly favored as a food by gorillas. The elasticity and the thinness of the branches of the *Hypericum* are probably the reasons why they are so frequently utilized by the gorillas as nesting material both on the ground and (less often) in the tree itself.

The third most common tree sharing restricted portions of the saddle and lower Visoke slopes with the *Hagenia* and the *Hypericum* is the *Vernonia adolfi-friderici*. The *Vernonia* may attain a height of some 25 to 30 feet, growing in stands where its dense canopy often prohibits the growth of herbaceous ground foliage. The tree has wide, softly textured leaves, extremely tough branches, and stems from which sprout either buds or small clusters of white-petaled, lavender-tipped flowers. Gorillas seem to prefer the nutty tasting *Vernonia* buds and will either sit in the tree or simply bend a branch down to the ground to pluck the buds off one by one just as humans eat a strand of grapes. The wood of the tree is also a favored gorilla food item, either when sodden or rotten. Indeed, within numerous areas of the Virungas, the *Vernonia* has been so extensively used by gorillas for feeding, playing, and nesting activities that only broken stumps remain to give an idea of previous gorilla population concentrations.

The path that we followed between the herbaceous slopes on our right and the saddle terrain slightly below on our left was more defined than it had been in the village because it was kept open by elephant and buffalo as well as by an overflow from numerous streams draining off the mountain.

The first hour and a half of the climb was the steepest part and, as the altitude increased, breathing — for me at least — became somewhat of a raspy affair. I was delighted when the porters wanted to stop for a rest and a smoke. The spot they chose was a small meadow clearing where elephant and buffalo droppings had accumulated around a stream running through the middle of the clearing. The air was pure elixir and the running water refreshingly sweet and cold. The heavy fog and drizzle were just beginning to give way to the promise of welcomed sun. For the first time I could fully appreciate the extent of the herbaceous foliage that abounded on Visoke's steeper slopes to the north side of our trail. The terrain seemed to be very promising gorilla country. I grew tremendously eager to discover what lay ahead west of us, deeper within the heartland of the Virungas.

Considerably apprehensive, the porters were now quieter than they had been far below in their village. Nevertheless, they were willing to go on even though it seemed that only a few had ever been this far into the mountains before. We continued climbing at an easier gradient for more than an hour before coming to the beginning of a long meadow corridor densely carpeted with a variety of grasses, clovers, and wildflowers. Distributed throughout the meadows, like so many powerful sentries, stood magnificent *Hagenia* trees, bearded by long lacy strands of lichen flowing from their orchid-laden limbs. The entire scene was backlit by sunlight, giving all a specular dimension no camera could record or eye believe. I have yet to see a more impressive spot in all of the Virungas or a more ideal location for gorilla research.

Exactly at 4:30 P.M. on September 24, 1967, I established the Karisoke Research Centre — "Kari" for the first four letters of Mt. Karisimbi that overlooked my camp from the south; "soke" for the last four letters of Mt. Visoke, whose slopes rose north some 12,172 feet immediately behind the 10,000-foot campsite.

With the site chosen, the next logical step was the selection of a Rwandese camp staff from the line-up of porters. A number of men were eager for permanent work, and in no time at all I had the beginnings of the Karisoke staff helping to set up tents, boil water, collect firewood, and unpack essential supplies and equipment. My tent was set up alongside a swiftly flowing creek. About one hundred yards farther back in the meadow nestled closely to Mt. Visoke's slopes, another tent was set up for the newly chosen African staff recruited from the porter line.

Since that day I never have had the slightest difficulty in recalling the elation felt upon being able to renew my research with the mountain gorilla. Little did I know then that by setting up two small tents in the wilderness of the Virungas I had launched the beginnings of what was to become an internationally renowned research station eventually to be utilized by students and scientists from many countries. As a pioneer I sometimes did endure loneliness, but I have reaped a tremendous satisfaction that followers will never be able to know.

A distinct language barrier existed between the Rwandese and myself during the early days following Karisoke's establishment. Alyette DeMunck, who had an excellent command of languages, had to leave following a mere few days at camp. I spoke only Swahili and the Rwandese only Kinyarwanda. Thus, much of our communication was carried out via hand gestures, head nods, or facial expressions. Africans have a great facility for learning languages quickly because they do not tend to rely upon the crutch of books. It was therefore easier for me to teach the men Swahili than for them to try to teach me Kinyarwanda.

Most of the Rwandese porters I hired that day have remained with me as loyal and dedicated assistants. Some of the men enjoyed being in the forest, so I trained them, just as Sanwekwe had trained me, in tracking skills. Others preferred to stay in camp and so learned the basic fundamentals of tent-cleaning, washing of clothes and dishes, and some very simple cooking. Woodmen of necessity had to be strong, hardy, and uncompromisingly obedient to one golden rule: No conveniently located standing trees or even any fallen trees supporting plant or animal life were to be cut. Human nature, being what it is, resulted in a more rapid turnover of woodmen than of trackers or cabin staff.

* * *

In 1967 the Parc des Volcans of Rwanda had only about a dozen park guards and an unmotivated Conservator. Most of the men were totally unfamiliar with and afraid of the forest and preferred remaining in their villages with families and friends. The park itself came under the jurisdiction of the *Directeur des Eaux et Forêts* (Director of Water and Forests) in the Ministry of Agriculture. There was no central organization, as is now the case, for management of the park. Honey gatherers, cattle grazers, and poachers, most of whom were friends or relatives of the park guards, were free to come and go across the park boundaries as they wished. With the exception of a handful of European residents who occasionally climbed or camped overnight in the mountains, the park held virtually no interest for anyone but its illegal encroachers. Indeed, upon my arrival in Rwanda, I was told by numerous Europeans that few, if any, gorillas resided on the Rwandan side of the Virungas and that I would be wasting my time searching for them. I was inclined to think otherwise.

The word *poacher* implies "to put into a bag" and is derived from the French word *poche* meaning "bag" or "pocket." The main poacher prey within the Virungas are two antelope species: the bushbuck (*Tragelaphus scriptus*) and the duiker (*Cephalophus nigrifrons*). These graceful animals are either killed directly by spears or arrows or they suffer lingering deaths after being caught in spring traps concealing wire or hemp nooses triggered upon the slightest pressure to catch and snare any animal's foot.

When on a hunting foray within the park, poachers and cattle grazers live in very simple structures, known as *ikiboogas,* built around large hollow boles of ancient *Hagenia* trees. The poachers usually spend several nights in the forest — depending on their hunting luck — and they commonly smoke hashish around their campfires at night. The larger the game sought for the following day, the larger the amount of hashish required to bolster the hunters' courage. When away from the *ikiboogas* and out on hunts, the men, usually

Batwa, hide their hashish pipes, extra wire snares, smoked antelope meat, or food brought from their villages in deep recesses of *Hagenia* tree trunks. Cattle grazers also hide their *ibianzies* (milk jugs) in thick vegetation not far from their *ikiboogas*. It did not take long to learn where to search for the illegal possessions of cattle grazers or poachers in order to discourage the encroachers.

When out for small game such as antelope, poachers usually travel singly or in small bands and are often accompanied by dogs who wear handmade metal clappers attached to their necks by antelope-skin collars. The clappers are stuffed with leaves while the hunters are searching for game trail. When the dogs encounter fresh spoor, there is no further need for quiet, so the poachers remove the mufflers and allow the dogs to lead them to their prey.

It is impossible to recall how many times I have been tracking gorillas on Visoke's higher slopes and suddenly heard the cries of poachers accompanied by the howls of their dogs and the distinct clamor of the dogs' clappers as they pursued their prey, usually to its death. Occasionally the chases took place in the open meadows some 500 to 900 feet below in the saddle area. There were times I was able to applaud silently when a near-exhausted duiker or bush-buck managed to escape its pursuers by cleverly switching directions back and forth across small grassy glades until reaching the protective thick foliage of Visoke's slopes. The escapes left both poachers and dogs confused, running in circles in the meadow while the prey rested amid their refuge of brambles and thistles. I wondered when, if ever, the presence of two tents, myself, and the small Karisoke staff would be able to daunt the activities of poachers. During the early days of the research, the illegal hunters, waving spears or brandishing bows and arrows, even leapt over the pegs of my tent as blithely as the antelope they and their dogs chased across the camp's meadow.

On one occasion while surreptitiously following such a party, I saw a boy crouching behind a tree ahead of me intently poising an arrow in the direction of a bushbuck the other poachers were trying to flush out of thick brush at the base of Visoke's slopes. I managed to catch the ten-year-old, son of the leading poacher of the Virungas, Munyarukiko, and carried the boy and his weapons back to my tent in the hope that his capture might allow me to arrange a palaver with his father and other high-ranking poachers. I needed to talk with them face to face about the urgency of leaving at least the mountain slopes free of trapping and hunting pursuits for the sake of the remaining gorillas. The hostage enjoyed his two-day stay at camp. He served as an appealing mediator, along with the Rwandese of my staff, when he was exchanged for Munyarukiko's promise that Visoke's slopes were to be considered sacrosanct against further trap-setting or hunting. For some time, to the best of my knowledge, Munyarukiko kept his word. However, the saddle terrain in 1967 contained large herds of elephant, buffalo, and antelope, and it be-came a poacher's hunting ground, particularly because years of poach-ing had decimated game adjacent to the lower park boundaries. As the game at lower altitudes within the Parc des Volcans grew more meager, antipoacher work became a greater part of Karisoke's day-to-day fight for the survival of the gorillas, as well as for other animals that occupied the saddle area of the forest.

Encountering a trapline and being able to cut the traps down before they have snared a victim is always rewarding. It is equally rewarding to be able to release unharmed antelope shortly after they have been snared and watch them leap away from their intended doom. The flexible poles needed to spring the traps are usually made of bamboo. They can easily be seen in the forested, green areas of herbaceous foliage, but finding them in the bamboo zone itself is more difficult. The spring traps were often so cleverly concealed that after several hours of searching in dense bamboo mazes I felt surrounded by mirages of traps. The trackers and I always found it comical, albeit humiliating, to be crawling or walking through thick bamboo growth in search of traps and suddenly have our wrists or ankles jerked up from under us when inadvertently caught in nooses concealed under light soil covering. We always carried the nooses back to camp, either to burn or to throw into the latrine as a guarantee they would not be used again. For the same reason, the poles to which the nooses were attached were cut when found.

Pit traps, about eight to twelve feet deep, are floored with sharply pointed bamboo stakes that impale any unsuspecting creature having the misfortune to fall into them. I was one such unsuspecting creature on a day when out alone chopping my way through a tall dense nettle field with my *panga,* a machete-like cutting tool. It was quite a shock to be aboveground one second and eight feet below the next, feeling the electric sting from the countless welts of the nettles that had overgrown the pit. Fortunately, it was an old and long-neglected pit whose stakes had rotted and fallen down. I felt a sense of panic as I looked up at the bright blue sky high above me and realized that it was early in the morning. The staff would not be looking for me until dusk, many hours away. Luckily, the *panga* had fallen in with me so I began chipping away a series of hand and foot holds in the crumbling sides of the pit until it was possible to reach the vine-like roots growing nearer the top. This was one of the very few times in my life when I was grateful for my six-foot height. Late that afternoon, I returned to the pit to cover it securely with stout tree limbs so no other creature could ever fall into it.

A neckrope noose structure is usually found in dense blackberry patches. It could be triggered immediately by the head of a browsing antelope as its nose leads it toward the tender shoots and fruits growing in the thick brush. The victims caught in such traps slowly strangle to death, their futile struggles only serving to tighten the binding nooses.

An increasingly rare kind of ambush, probably because of the reduced size of herds, was the tree-trunk stockade built along buffalo trails but always leading to precipices where the enclosures narrowed to small openings at the cliff edge. Much like those built by early American Indians, the narrowing corrals confined buffalo to the dropoff point. Poachers and dogs drove the animals from behind while poachers with spears awaited the victims below after their fall. Whenever the trackers and I encountered the remnants of these ambushes in the forest, the men recalled stories they had heard from their fathers of these massacres. The buffalo cemeteries sometimes found at the base of the cliffs told the rest.

A visitor to camp unintentionally discovered another type of trap, but fortunately only a few of these have since been found. The newcomer was crawling on all fours through dense vegetation and was just about to put one hand forward to lean upon when, instinctively, he froze. Looking down at the exact spot where his hand would have touched the ground, he saw a cleverly concealed wire-spring noose partially covered by dirt. Following the ascent of the single strand of wire with his eyes, he saw that it was connected to three large tree trunks, each about two feet in diameter, six feet in length, and perched about three and one-half feet above him. Immediately he realized that the slightest pressure by his hand upon the small dirt area encircled by the wire noose would have released the trunks directly above to crush him to death. With great presence of mind, the student slowly backed out of the ambush and, once standing upright outside, tripped the taut wire. Harmlessly and victimless, the heavy logs smashed to the ground with an earthshaking *thud*.

I remain puzzled about the specific purpose of that type of trap. The several I have seen lacked the height necessary to snare buffalo and were also unduly heavy to kill antelope, usually caught by the far less complex method of wire or hemp nooses attached to poles. It seemed most likely that this type of log ambush was intended for wild bush pigs (*Potamochoerus porcus*), game that the younger African members of the Karisoke staff recalled as numerous in 1967 when parkland was being taken for cultivation.

Even during my own time it was becoming obvious that the supply of wildlife, particularly at lower altitudes, was not keeping up with the number of animals poachers were extracting from the park. For this reason their hunting forays and trapping took them higher into the forest and more regularly into gorilla habitat.

Although not usually the intended victims of traps, gorillas do get snared. A gorilla's tremendous strength enables it to break free from the trap site and escape, wearing the tight wire noose around its ankle or wrist. I observed three gorillas who had each been caught by a wire noose around the wrist. They learned to substitute their feet for the preparation and stabilization of food items; however, each became noticeably weaker before disappearing from their groups. None were ever seen again.

A fourth well-documented case was that of a forty-four-month-old juvenile gorilla who had been followed from the day of her birth to the day of her death. She had always been a lively, playful, and captivating youngster until one morning when she unknowingly romped into a concealed wire noose. Instantly the wire tightened around her ankle, holding her a prisoner to the pole. The rest of the group's members hysterically displayed around her — running, breaking branches, chestbeating, and screaming in their helplessness, not knowing what to do to release her. Later the same day, in desperation she broke away from the trap pole with the wire still around her ankle. For sixty agonizing days the wire embedded itself deeper and deeper into her flesh, until the emaciated little juvenile mercifully died of gangrene coupled with pneumonia. During this period she had become progressively weaker. The group adjusted their travel pace to allow her to keep up with them, but the young gorilla was doomed from the start.

In two other cases where juveniles had been snared around their wrists, it appeared that the silverback leader of the group was able to release them immediately by using his canines to slide the wire free over their hands while they were still attached, by the wire, to the trap pole. Most likely this particular silverback had had far more experience with trap encounters than the leader of the doomed juvenile's group. Another individual, an adult female, was possibly a trap victim in her youth, because she was missing some of the fingers of both hands when first seen as an adult. When she eventually gave birth, I was thrilled to observe the dexterity with which she was able to care for her new baby in spite of her tremendous handicap.

Some gorilla groups seemed to be more "trap-wise" than others, perhaps because of having had more experiences with the havoc caused by traps. One day I watched a group purposely deflect their travel route from a fairly visible line of arched bamboo poles, each tautly attached to lethal wire nooses. Although the trapline had been set recently, one had already snared a duiker, which had died after a futile struggle to free itself from the wire's grasp. The trackers, a camp guest, and myself instantly spread out to destroy the dozen or so remaining traps before they could inflict further damage. I was worried about what effects the noise we were making while cutting the poles would have on the gorillas, who were moving down into much thicker foliage below us. They seemed unconcerned, however, probably because they had seen and recognized us and did not connect us with poachers.

Just as we had finished demolishing all of the traps on the slope, a tremendous screaming outbreak came from the gorillas, who were then nearly obscured in tall vegetation at the base of the slope. Horrified, we ran down toward a pole that was bobbing up and down as its victim fought the grip of the noose. Despite the presence of the group, I unintentionally broke one of my cardinal rules about remaining silent when near them and began screaming "No! No! No!" at the thought that yet another gorilla had been snared. The animals, understandably startled by my behavior, had begun moving away by the time we reached the trap. There is no way to explain our relief upon finding a young duiker, rather than a gorilla, gamely fighting the wire noose.

Long experience dictated our standard routine for releasing antelope from traps. The first step is to get the animal's head covered, usually with a jacket, for this serves to restrain somewhat the violence of an antelope's desperate struggles. Next, the fragile legs need to be immobilized to prevent the victim from breaking bones or tearing ligaments and muscles. Confining the legs requires strength and agility because the desperation of a trapped animal fighting for its life seems to make it incredibly strong. Only then may the noose be removed followed by a check to ascertain that the victim has not been unduly harmed. If the antelope appears able to fend for itself when released, the head covering is then removed.

Just before we began working the noose off this plucky little duiker's leg, I chanced to look in the direction the gorillas had taken and burst into uncontrolled laughter. Sitting in a row on a large *Hagenia* branch about twenty feet away were the four adolescent males of the group. They seemed totally fascinated by our activities. The intensity of their concentration gave the impression of their lending moral support to our efforts. Farther in the background, the rest of the group could be seen peering over the foliage and staring at us curiously. With a rooting section like this, we felt that nothing could go wrong. Sure enough, the instant we uncovered the freed antelope's head, it hopped up and with one mighty leap disappeared into the surrounding vegetation. The four gorillas watched it briefly before chestbeating and casually descending the tree now that the show was over. Once again I marveled at the sense of curiosity gorillas possess.

*   *   *

It took some four years after establishing Karisoke to clear the cattle out of the saddle area between Mts. Visoke, Karisimbi, and Mikeno and reduce poacher activity in the same region. The gorillas were then able to leave the overcrowded mountain slopes and expand their ranges into the saddle terrain. However, when a group traveled far from camp and encountered traplines or poachers in unfamiliar ranges, it was necessary to resort to a technique I called herding.

The drastic decision to herd a gorilla group was made only when animals were in potentially dangerous poacher or trap areas. I always made that decision with the greatest reluctance, because it meant deliberately disturbing a group and influencing their usual ranging pattern. Under compelling circumstances I felt that this was justifiable.

Preparation for herding consisted of supplying everyone at camp, both staff and willing students, with poacher dog bells I had acquired over the years of raiding *ikiboogas*. The threatened gorilla group is then located but not contacted. Silently we space ourselves out in a large arc roughly 150 feet behind the group and remain out of sight. Once in position, we launch the near-equivalent of an actual poacher attack by ringing dog bells and imitating poachers' hunting calls. The feigned "attack" is planned so that the gorillas may be herded in a direction of safety, usually toward Visoke's slopes. The noises are not made continuously but frequently enough to get the gorillas started, if they stop for prolonged periods, or to prevent them from circling back to their previous locations. As the gorillas flee from their unseen herders, they leave diarrhetic dung deposits along their flee route and an overpowering fear odor in the air. If there are two silverbacks in a group, the dominant male takes the lead to direct the route and speed of travel for the females and young immediately behind him. The subordinate silverback brings up the rear as a watchdog, additional defense for the vulnerable animals in the middle. Usually after about fifteen minutes of flight, their pace slows down for brief rest periods. Once a group is out of a danger area and is headed back into familiar terrain, the gorillas are left alone, with no attempt made to contact them that day. This drastic but effective technique was used sparingly. I consider it by far the lesser of two evils when compared with what could happen if a group surrounded by poachers or traps is left to its own defense devices.

In the process of patrolling the saddle, I soon learned that poachers did not take lightly to having their traps destroyed. One means by which they expressed their displeasure was by use of *sumu,* an African word meaning "poison" but used as a general term for black magic in the central portion of Africa. Sometimes poachers would cut two sticks from a tree, form them into a cross, and stick them into the ground along the trail leading to their trapline. This Christian-like symbol was meant to convey a death threat to anyone crossing beyond the point where the sticks were placed. Some of the men who worked in the forest on antipoacher patrols expressed genuine fear of the cross sign and refused to enter the protected area. I was able to talk others out of their apprehensions, but *sumu* exerts a powerful influence in the daily lives of many Rwandese, especially those living in the more remote areas near the Kivu Province of Zaire, where the most prodigious *umushitsi* (witch doctors) practice.

Pombe (banana beer) loaded with potion was the most common way of dispensing *sumu.* There were other methods, however. Burying an animal's rib bone along a trail that an intended victim would be walking was reputedly effective even if only the person's shadow fell across the buried bone site. A more costly *sumu* procedure involved the slaughtering of a goat or chicken by a high-ranking *umushitsi* while chanting magical words and the name of the intended victim. Wherever he was, the victim would reputedly become fatally ill the instant the goat's or chicken's throat was slit. I do know of one man who died as a result of this procedure.

At varying times, all the Africans who worked at camp have been exposed to the virulence of *sumu*. If they think they have been poisoned, usually by some foreign element added to their *pombe,* they are absolutely convinced that nothing but the power of an expert *umushitsi* can save them. They begin preparing for their own funeral by wearing their best clothes daily to be ready when their doom arrives. In this way they may be sure that the clothes will be buried with them and not fall into anyone else's hands. Countertreatment by a highly qualified *umushitsi* is extremely expensive — about the equivalent of one month's salary. At first when asked for treatment money, I felt the requests were hoaxes. However, as I watched some of the staff literally begin to waste away before my eyes, I had to believe that *sumu* exerted far too strong a force among the Africans for an outsider to understand. Eventually I accepted their belief in the power of black magic. I paid for the thirty days of remedies obtained only in the hut of a witch doctor and tried not to show the astonishment I always felt when the men returned to work fully recovered and wearing their ordinary clothes.

Not all *sumus* were meant to kill. Seregera was an older African who had come from the deeply superstitious Kivu Province of Zaire and asked to work at camp as a *zamu* (guard). I found him somewhat threatening in both appearance and manner. The camp's three younger staff members became intimidated by him. One of the men, Kanyaragana, had the courage to bring me evidence of a *sumu* procedure being carried out at camp by Seregera. Late one afternoon, thoroughly terrified at his own daring, the Rwandese came into my tent and from his pocket pulled out an object resembling a miniature mummified head partially covered with hair. On closer examination I found that the "head" was roughly carved from heavy wood and bore a remote resemblance to my Roman-nosed countenance. I was told that the hair was mine, plucked from my hairbrush by Seregera over a period of weeks. According to Kanyaragana, once the head was completely covered by the hair of its image, it was to be pulverized by a witch doctor and placed into the food and tea of the victim, in this case walnut-head Fossey. Shortly thereafter I was expected to become totally subservient to the whims of the hair-collector, providing, of course, that I didn't notice any strange ingredients in my food or tea. I tossed the thing back to Kanyaragana so that he could return it before Seregera found it missing. I then promptly proceeded to clean my hairbrush, a habit assiduously retained many years later even when in America.

Unknown to me at that time, Seregera was also a poacher. He was to become one of the biggest elephant killers within the Virungas when he acquired a hunting rifle about the same time that Munyarukiko also obtained one.

Poachers were frequent users of *sumu,* probably because many of their black magic ingredients came from the animals and vegetation of the forest. With courage bolstered by hashish, they killed silverback gorillas for their ears, tongues, testicles, and small fingers. The parts, along with some ingredients from an *umushitsi,* were brewed into a concoction reputed to endow the recipients with the virility and strength of the silverback. Some of the younger men at camp reluctantly admitted that their fathers believed in the power of the silverback potion, but that they themselves scorned the tradition. Fortunately, this practice now seems to be on the decline. Gorillas, silverbacks in particular, were also killed for their skulls and hands. The grisly trophies were sold either to tourists or resident Europeans in the nearby towns of Ruhengeri and Gisenyi for the equivalent of about $20. That practice was short-lived, but only after at least a dozen silverbacks had been slaughtered.

Because of the brutality of the poachers' crimes, I found it far easier to tolerate the encroachment of the cattle-grazing Watutsi despite the extensive damage their cattle inflicted upon the park's vegetation. The tradition of cattle grazing within the Virunga Volcanoes goes back at least four hundred years, and the names of most meadows and hills have been passed down through the centuries by Watutsi herdsmen. Cattle are grazed by the male members of a Watutsi family, and as many as three generations may share in the responsibilities of tending their family herds. While the elder members are out with the cows, the youngest son is left at the *ikibooga* to guard it and the calves and to keep the perpetual campfire going. *Ibianzies,* the hollowed wooden containers into which the cows are milked, are passed down from father to son and kept hidden around the forest *ikibooga*.

Even the cattle and the actual grazing regions within the park are passed down from generation to generation. The descendants of one family of Watutsi had used the meadows around Karisoke for countless years before my arrival and considered this area their own though they knew they were trespassing within a park sanctuary. The leader of this particular family was a regal old man of indeterminate years by the name of Rutshema. Two of his sons, Mutarutkwa and Ruvenga, and even their young sons helped Rutshema with the responsibilities of herding and caring for their 300 head of cattle, one of the largest herds within the park.

Considering that this family, one among many Watutsi clans, had known the meadows between Visoke and Karisimbi for years, I found it extremely difficult to insist that they take their cattle outside the park boundaries. The reader may justifiably ask, "Well then, why did you do it?" My answer now, as it was a decade and a half ago, is quite simple. One cannot compromise on conservation goals within established park areas. I could in turn ask, "Are parks that have been established to protect flora and fauna going to remain intact or should they be exploited by encroachers for personal profit?"

For several years I spent countless days forfeiting observations with the gorillas because of the necessity to drive cattle out of the park. A less time-consuming chore, though one equally repugnant, was mixing up the herds belonging to several separate Watutsi clans. That maneuver created chaos among bulls while obviously destroying long-cherished bloodlines between distinct familial herds. After several years of Fossey versus cattle, the Watutsi finally left the park permanently to graze their herds elsewhere. Oddly enough, Mutarutkwa of Rutshema's clan never harbored any resentment and later became one of my best friends as well as the leader of antipoacher patrols I established to drive poachers from the park.

*   *   *

One way to grasp the immensity of the cattle problem throughout the Virungas was by air. Early one morning in 1968 I flew over the two active volcanoes in Zaire and the six dormant ones shared by Zaire, Rwanda, and Uganda. That flight can only be described as an ethereal experience.

We began with the most easterly volcano, 13,540-foot Mt. Muhavura, meaning "he who shows the way or guides." Traditional as a holy mountain where only good spirits may rest, Muhavura has the longest oral and written history about cattle and human encroachment. Approximately one third of Muhavura lies in Uganda where the Kigezi Gorilla Sanctuary was formed in 1930. Initially the reserve encompassed some eighteen square miles, but because of the pressure for cultivation was reduced to nine square miles in 1950 and decreased even further since that time. I found the mountain's high barren slopes frosted with hail, a harsh contrast to the thick bamboo growth and cultivation that fringed its base. The mountain was far bleaker than I had imagined because of the extent of its barren rock surfaces covered only with scrubby lichen growth.

A flat meadow strip separated Mt. Muhavura from the least spectacular of all the volcanoes, 11,400-foot Mt. Gahinga. *Gahinga* means "the Hill of Cultivation" because its surrounding cols have traditionally been used as passing places for Rwandese peasants who traveled to Ugandan blacksmiths in order to obtain hoes for their agricultural needs. The small mountain's relatively uniform slopes are thickly covered with bamboo and *Hypericum* woodlands all the way to the summit, which contains a steeply sided swampy crater. Gahinga appeared to offer slightly more potential for gorillas than Muhavura but is limited by size and by a preponderance of bamboo, strictly a seasonal gorilla food. The meadows between Gahinga and Muhavura to the east and Sabinio to the southwest appeared narrow enough so that gorillas could travel from Gahinga to the neighboring mountains. But when these corridors were used by humans, usually smugglers, gorillas would not risk crossing the clearings between the mountains.

We next circled over the third mountain in the chain, 11,960-foot Mt. Sabinio, meaning "Father of the Teeth" because of its five jagged and serrated summit ridges. Sabinio is considered the oldest mountain within the Virungas and was as impressive from the air as it is in profile from the ground. Its upper heights are bleak and inhospitable-looking, yet below the alpine zone grow thick *Hypericum* woodlands interspersed among a diversity of other trees along steep-sided ridges separated by gullies containing ample herbaceous foliage. Sabinio's base is fringed by a broad bamboo belt that, as on all the mountains, directly abuts cultivated areas. Its narrow ridges greatly confine the travel of antelope and thus attract trap-setters because of the relative ease with which such game may be snared. However, for the same reason, the mountain was less attractive to cattle grazers than others within the Virunga range.

A very narrow meadow corridor separates Sabinio from a small bamboo-covered volcano, Muside, which in turn is separated from Mt. Visoke by seven and one half miles of small bamboo hills. In 1969, the hill chain still served as a passageway between the three eastern volcanoes just described and the three western ones. Therefore, the gorilla populations of the two halves of the six dormant Virungas were not yet isolated from one another. But, even then, the narrow chain was slowly being nibbled by cultivation, which soon resulted in the permanent severance of the two regions and their animal inhabitants.

Once over Mt. Visoke the pilot dipped the plane down so close above camp that the staff actually thought we were going to drop in for a cup of tea! Visoke, 12,172 feet in height, means "a place where the herds are watered." The term does not refer to Visoke's large crater lake on the summit but to a traditionally used cattle drinking spot, Ngezi Lake, adjacent to the northeastern slopes of the mountain. This was the first time I had seen the magnificent crater lake, some 400 feet in diameter, with steep sides of flowering alpine vegetation. I also had never fully realized the enormity of the mountain's surface, which as yet showed little erosion. Herbaceous foliage covers a large portion of the slopes, making them near-ideal gorilla habitat. Except for the eastern side, the remainder of the mountain is surrounded extensively by saddle terrain leading toward Mts. Karisimbi and Mikeno. This is the area I refer to as the heartland, or the core of the Virunga Volcanoes. It will probably become the last stronghold for the mountain gorilla.

From the air it was possible to comprehend fully just how much of the park had been usurped for cultivation. The remnants of the old line of evergreens that had marked the original 1929 boundaries along the Rwandan side of the six dormant volcanoes stood like weary soldiers at a ravaged post. Above the evergreens, the devastated forest was pocked with smoke from burning *Hagenia* trees where small land plots were being cleared for pyrethrum cultivation. The pillage extended up to 8860 feet on Visoke and to 9680 feet on Karisimbi.

The Common Market's pyrethrum scheme had vastly affected the gorillas' range as well as those of elephant and buffalo by the excision of 22,000 acres from the Parc National des Volcans. The land removed consisted primarily of bamboo but included some *Hagenia* forest. In 1967, a year before my flight, at the base of Mt. Visoke I had contacted one of the gorilla groups whose range was to be nearly depleted by cultivation. The fringe group, which came to be known as Group 6, were then forced to move higher onto Visoke's slopes, where their range both abutted and overlapped with those of the main study groups. My first meeting with the group occurred in an untouched *Hagenia* woodland that later was to become a camping ground and parking lot for tourists. As late as 1971, six years following the settlement of *shambas,* the fourteen animals were still seeking to use a few strips of unplowed land lying between the vast fields of pyrethrum. Group 6 would follow these small extensions of forest, none of which was more than fifty feet wide, feeding on indigenous vegetation. Ignoring crops of peas, beans, and potatoes, the gorillas clung only to the remnants of forest, going so far as a fifth of a mile from the mountain as though searching for the whereabouts of what had been their range. On several occasions, the villagers would climb to camp and ask me to head the group back onto Visoke's slopes, a task those of us at camp readily did so that no harm might befall the gorillas. At other times I contacted Group 6 on Visoke's lower eastern slopes adjacent to the *shambas.* Accustomed to working well within the forest, I could not get used to the incongruity of observing gorillas to the accompaniment of the sounds of villagers' voices and the mooing, bleating, and crowing of their livestock and fowl. Group 6, however, seemed to ignore the noises of civilization only some fifty yards below them, though the same animals would nearly always immediately flee at the sound of human voices deeper in the forest.

By 1975, the new border between the park and the farmland would be marked by eucalyptus saplings and young evergreens. Later, in a futile effort to render the border more impressive, twelve tin rondavels were purchased and placed at intervals of 3 miles around the entire Rwandan side of the park. Ostensibly the tin huts were for the purpose of housing park guards to facilitate patrol work within the interior of the park. This plan could have been effective if the guards had been supervised, but, as it happened, the huts were seldom used. They either disintegrated or were moved to the parking lot at Mt. Visoke's base for the use of tourists.

At the time of my flight another assault on the forest had yet to be inflicted. Three years later, reputedly for the purpose of "conservation," a 2½ mile swath, varying in width from 30 to 40 feet, was cut and burned within the interior of the park along the frontier between Zaire and Rwanda. The new international border adhered to the map only where land contours made it convenient to do so. The long scar through the montane rain forest looked like a tornado's aftermath and was just about as useful. A European technical aide, enthusiastically lauding the scheme, actually believed that neither poachers nor game would cross the barren demarcation zone between Rwanda and Zaire because it had been definitively marked. I felt compelled to ask him where I might go to apply for visas for gorilla, elephant, buffalo, or antelope wishing to visit their relatives on the other side.

We next flew over my backyard, a five-mile stretch of saddle area between Visoke and the 14,553-foot Mt. Mikeno. Mikeno, one of the two oldest Virunga Volcanoes, means "poor" and signifies a place of inhospitable slopes that prevent human habitation. For myself, this was a very nostalgic part of the aerial survey — nearly one year had passed since I had parted from my meadow camp at Kabara. It was difficult to control my excitement as the plane skimmed over the trees of the dense forest toward the small grassy clearing ahead of us. Like most seeking reunions with the past, I was bitterly disappointed. The utopian meadow of my memories was filled with cattle, and the hut that had served as a shelter for the guards had been demolished.

Leaving Kabara behind, the plane finally soared upward more than 4000 feet toward the desolate, unworldly summit pinnacles of Mikeno, where the sheer rock surfaces glimmered with hail deposits. During the ascent I tried to forget what I had seen at Kabara, but I was plagued by the thought that if there were cattle grazers then there certainly were poachers. What had become of all the gorillas I had first known? My despondency was partially eased by the extraordinary close-up views we obtained of the near-perpendicular rock walls, canyons, and citadels of Mikeno's uppermost heights. Their desolation seemed to make them all the more powerful, almost supernatural.

Reluctantly, we descended to fly farther west into Zaire toward the two active volcanoes — the 11,381-foot Mt. Nyiragongo, named after the spirit of a woman, and the 10,023-foot Mt. Nyamuragira, meaning "Commander" or "Overseer." While working at Kabara I had always considered these two unpredictable volcanoes as Mike-

Above: The Virunga mountain range in central Africa is formed of two active and six dormant volcanoes shared by Zaire, Rwanda, and Uganda. The only remaining mountain gorillas (*G. gorilla beringei*) live on these dormant volcanoes. In the foreground are the three easterly mountains—Muhavura, Gahinga, and Sabinio; in the background the three westerly mountains—Visoke, Karisimbi, and Mikeno. The author's Karisoke Research Centre is located at 10,000 feet adjacent to Mt. Visoke. (*Dian Fossey*)

Overleaf: This portrait of a gorilla family was taken in the Kivu Province of the former Democratic Republic of the Congo (now Zaire) where Dian Fossey began her study of the rare mountain gorilla. Sixteen group members huddle around their regal silverback leader, at left. Curiosity at the spectacle of the author trying to climb a tree, an activity they could understand, drew them into the open. (*Dian Fossey © National Geographic Society*)

Above left: Uncle Bert, silverback leader of Group 4, feeds on *Galium* vine growing in a tall *Hagenia* tree while his offspring play near the base of the tree. The youngsters of a gorilla group do not require as much food as adults, so often spend their time playing while their elders, who can weigh up to four hundred pounds, feed. (*Dian Fossey*)

Above right: Pantsy and her baby enjoy a rest together. Infants spend the first months of life in close body contact with their mothers; they nest with them at night and are carried for travel by the mothers in a protected, ventral position during the daytime. (*Dian Fossey*)

OPPOSITE

Top: Heavy mists and fogs often roll into the saddle area like so many ocean waves before climbing to obscure the mountain peaks for days on end. At such times, gorillas trying to expand their ranges in the saddle terrain seem to become disoriented as they attempt to return to known ranges on Mt. Visoke's slopes. (*Dian Fossey*)

Bottom: Geezer, a blackback or sexually immature male of Group 8, peers at the observer through dense vegetation on the misty slopes of Mt. Visoke, where rainfall averages seventy-two inches a year. Herbaceous ground foliage provides plentiful food for gorillas, though much of their favored bamboo zones at lower altitudes adjacent to the mountains have been taken over for human cultivation of pyrethrum. (*Dian Fossey*)

Overleaf: Under the watchful eye of her mother, Pantsy, and near her sunning father, Icarus, four-month-old Muraha wobbles over to greet a member of her family group during a resting period. Within its natal group, a gorilla infant is surrounded by constant security, protection, and care to assure its survival in the family. (*Dian Fossey*)

ABOVE

Top: Only when the adults of Group 5 trusted Dian Fossey did they allow her to settle down among them and their offspring and accept her presence without question. (*Peter G. Veit*)

Bottom: Puck, a young female of Group 5, greets the author at the beginning of a day's contact. Because of dense vegetation, Dian Fossey informs the gorillas of her approach by giving contentment vocalizations, one of many signals that gorillas use when communicating among themselves. (*H. van Rompaey*)

Opposite: During a rainy day-nesting period, the author's special friend from Group 4, the silverback Digit, gave a harmless wide yawn, exposing his canines, which are not as large in female gorillas. (*Dian Fossey*)

Above: Poachers hunting within the Parc des Volcans of Rwanda or the Parc des Virungas in Zaire sleep at night in *ikiboogas,* temporary shelters made in the hollow boles of ancient *Hagenia* trees. Here the illegal hunters smoke antelope meat obtained from snare victims or from animals killed by arrows and spears. (*Dian Fossey*)

Opposite: The leading poacher prey in the Virungas are antelope, either killed on sight or snared by trap nooses concealed in a small hollow hole underneath a layer of dirt and leaves. The noose is tautly connected to flexible bamboo or sapling poles, which spring upward on pressure from an animal passing above the noose. This duiker never had a chance. Like most, it had fought for its life by futilely thrashing against the grip of the noose around its leg and died before it could be rescued. Larger antelope like the bushbuck sometimes gnaw off their feet to escape from the lethal hold of a wire noose. All too often gorillas are caught in snares set for antelope. (*Dian Fossey*)

Right: A trap set by poachers snared forty-six-month-old Lee, the silverback Nunkie's first offspring. For sixty days the wire embedded itself deeper and deeper into the bone of the young female's ankle until she succumbed to gangrene and pneumonia. Poachers account for two thirds of gorilla deaths. (*Dian Fossey* ©*National Geographic Society*)

Above: An antipoacher patrol, organized at Karisoke by the author, bivouacs in the forest overnight. The men search the protected park area to destroy traps, confiscate poacher weapons, and free animals from traps whenever possible. Such active conservation is crucial to the survival of the remaining animals of the Virunga Mountains. (*Dian Fossey*)

Opposite: Adult male gorillas used to be killed by poachers for purposes of *sumu*, an African word implying black magic. Poachers cut off the gorillas' ears, tongues, testicles, and small fingers, then brewed the parts into a concoction designed to endow the human recipient with the virility of a silverback. Today poachers capture gorilla infants for sale to foreign zoos and snare gorillas in traps meant for other animals. (*Dian Fossey*)

Although the fingers of this adult female's hands were suspected to have been maimed by poachers' traps, Pandora of Nunkie's Group not only managed to survive but proved herself extremely capable of caring for her firstborn. (*Ann Pierce*)

· Munster, an adult male, ranges with his group on Visoke's eastern slopes above
cultivated fields that support a human population density of 780 persons per square
mile adjacent to the boundaries of the Parc des Volcans. In 1969 Rwanda's portion of
the Virungas almost was halved when 22,000 acres were excised for pyrethrum development. For several years thereafter gorillas had to be herded back onto Visoke's slopes
whenever they confusedly sought remnants of the park range their predecessors had
once roamed. (*Ann Pierce*)

Gorilla foot and knuckle imprints are easily spotted along muddy trails, and the size of the imprint gives a good idea of the age of the animal — in this case a silverback. In dense foliage gorilla trails can be followed by attention to the feeding remnants or dung deposits that survive even the passage of buffalo or elephant herds. (*Dian Fossey*)

no's lively little sisters. Indeed, still conspicuous from the air were several long black fingertip lava flows running through miles of dense green forest. These were the results of several minor eruptions that had occurred the previous year, eruptions that had turned the night sky over Kabara brilliantly crimson. It was quite exciting to dive into the sulfurous and somewhat satanic-appearing craters of the embryonic volcanoes, which were still belching their way into maturity in probably much the same manner the dormant volcanoes now accommodating gorillas had been formed during the past million years.

We flew back toward Rwanda to explore the 14,782-foot Mt. Karisimbi. It is a tradition that the souls of the good will spend their eternity on the summit of Karisimbi, a name derived from the word *nsimbi* for white cowry shells. How perfectly *nsimbi* describes the mountain's top because its peak is frequently covered with hail. The uppermost cone of Karisimbi is encircled by vast meadow rings that lie about 12,000 feet in height and contain numerous small lakes and water conduits. As I had expected, the meadows were filled with cattle, some 3000 head in all, amoebic-structured into herds by their grazers.

Five minutes after leaving Karisimbi's meadows we circled and landed on Ruhengeri's grass strip runway. Once the plane's two engines came to a deafening silence, I had the impression of having flown through a million years of time in only ninety minutes.

3 | Karisoke Field Impressions

M Y KNOWLEDGE OF KABARA'S FATE, as gained from the air, made the research at Karisoke seem more imperative than ever. However, even the prospect of unknown gorillas to identify and habituate did not ease my mind about the destiny of the Kabara population.

At Kabara I had studied three groups totaling 50 individuals. During the first year at Karisoke I concentrated observations on four main groups that totaled 51 individuals living within the 9½ square-mile study area around camp. These groups, identified by number according to the order in which contacted, were Groups 4, 5, 8, and 9. Other groups encountered were considered fringe groups, whose ranges either abutted or overlapped those of the main study groups, or were totally unhabituated groups met during census work on other mountains.

Since I tried to distribute observation hours evenly among the four main study groups, lapses of several days could occur between successive contacts with any one of them. My tracking ability of necessity improved, because the trails were older and longer than if each group had been tracked daily, and the Rwandese on my staff were yet to become skilled trackers.

A good six months were to pass before the men felt confident enough to go out into the forest and track by themselves. Even then, they clearly preferred not going more than an hour from camp and were reluctant to follow trails older than two or three days because of the distances involved. With old trails two trackers, rather than one, went out together. Much of the terrain was still unfamiliar to them and they retained a natural apprehension of possible encounters with wildlife or poachers.

Teaching Rwandese how to track was far easier than instructing the students who eventually came to Karisoke. The locals' senses, especially their eyesight, were more acute. When training anyone, I always led the way for a couple of days, explaining the factors that determined the route taken. Sometimes I purposely strayed from an actual gorilla trail (occasionally unintentionally) to see how long it would take those behind me to realize the error. Another beneficial teaching ruse was furtively to press a series of my own knuckleprints along a section of damp earth going in the opposite direction to the knuckleprints of the gorillas being followed. How Sanwekwe would have loved this bit of chicanery! Those being trained would excitedly discover my knuckleprints and confidently follow them only to find no gorilla spoor ahead. This method proved to be the best way to teach people not to blunder about when on difficult trails — trails on grassy meadows or rocky slopes in particular, where even one bootprint can destroy a vital tracking clue.

Following gorilla track in thick herbaceous foliage is in fact child's play. Most vegetation bends in the direction of a group's travel, knuckleprints may be found impressed upon intermittent dirt patches or trails, and chains of gorilla dung deposits provide other clues as to the direction of the animals' passage. The individuals of a calmly moving group do not travel one after another. There may be nearly as many trails as group members, so I attempt always to follow the most central trail. Numerous cul-de-sacs occur wherever individuals depart from the main route to go off and feed by themselves. I learned eventually that the false leads could be identified by the presence of two layers of foliage. The top one is bent in the direction of the group's travel and the lower is bent in the opposite direction where an individual has gone off on its own before returning to follow the group.

In extremely dense, tall foliage, much circuitous tracking time could be saved by looking ahead of a group's trail for signs of disturbance of vegetation or of branches in distant trees where gorillas have climbed to feed. This technique was especially helpful in the saddle areas, where gorilla spoor could be nearly eradicated by passage of elephant or large herds of buffalo. The ground signs that might survive between the miniature craters left by the elephants' feet are the gorillas' typical trilocular dung deposits or their feeding remnants, such as the unmistakable peelings of thistle and celery stalks. Often, gorilla trail merges briefly with or zigzags in and out of buffalo trail. Whenever this happens and visual clues are obscured by vegetation I feel with my fingertips for the deep imprints left by the cloven hooves of the buffalo to realize that I am on the wrong path. Because gorillas always seek fresh untrampled vegetation for feeding purposes, they seldom travel along buffalo trails for any distance.

Unfortunately, the reverse is not true. Characteristically bovine in nature, buffalo are very trail-oriented, particularly in thick vegetation. Upon encountering gorilla trails, they often follow them like so many cows heading for the barn. On several occasions, without intention I found myself following gorillas who were in turn being followed by buffalo. Twice the gorillas, either in vexation or perhaps with a sense of joie-de-vivre, turned and charged directly toward the buffalo, which speedily turned tail and retreated unknowingly toward me. In retrospect, the subsequent confrontations had all the comical ingredients of a Laurel and Hardy movie. I had the option of climbing any available tree or diving headfirst into vegetation — too often nettles — that fringed the trail of the oncoming herd. I was always more than willing to let buffalo have the right-of-way. This is one of the first rules any person must learn when working in the domain of wild animals and is one that some learn the hard way.

Tracking is an enjoyable challenge, though there were times when trackers became convinced that their four-legged quarry had sprouted wings, so faint were the clues. This was especially true when trying to follow the trail of a lone silverback gorilla rather than a group, trails more than a week old, trails crossing relatively barren regions such as meadows or lava rockslides, and trails traversed by ungulates sharing the gorillas' terrain.

One morning along the trail of a lone silverback I was belly-crawling under a long dank tunnel roofed by a fallen *Hagenia* tree and sided by dense vines. With relief I saw a sunlit opening about fifteen feet ahead and wormed toward it enthusiastically while dragging my knapsack behind me. Upon reaching it I grabbed on to what appeared to be the base of a sapling in order to pull myself out of the gloomy tunnel confines. The intended support not only hauled me out of the tunnel but dragged me through several feet of nettles before I had the sense to let go of the left leg of a very surprised buffalo. The odoriferous deposits of his justifiable fright took several days to wash out of my hair and clothing.

Much can be gained by crawling, rather than walking, along gorilla trail, a fact I discovered one day by accident. Traces of a silverback's pungent body odor, resembling human nondeodorized sweat smell, permeated vegetation the gorilla had traveled through some twenty-four hours previously. Had I been walking after the lone silverback that day rather than crawling, I never would have realized the importance of olfactory clues existing at ground level. There are two types of sweat glands existing in gorilla skin. The axillary region of the adult male contains four to seven layers of large apocrine glands responsible for the powerful fear odor of the silverback, an odor only weakly transmitted by the adult female. The palms and soles of males and females contain apocrine glands and a high concentration of eccrine glands that have an important lubrication function. Both types of gland would appear to be evolutionary adaptations for terrestrial travel and olfactory communication, particularly for adult male gorillas.

The most outstanding odor found along fresh gorilla trail emanates from the dung deposits. Healthy gorillas leave chains of dung lobes similar in texture and smell to those of horses. When gorillas travel at an unhurried pace, the three-lobed sections may be deposited in a chain with the lobes attached to one another by strands of fibrous vegetation. If the animals have been feeding on fruit such as wild blackberries (*Rubus runssorensis*) or the plum-sized *Pygeum africanum*, the seeds, or even the whole fruit, can be found intact in the dung and can provide clues as to where the group had been ranging. The relative age of dung can be determined by the number of flies swarming around it, as well as the amount of eggs the flies have laid on the dung's surface. Countless hundreds of small white eggs are laid within minutes following defecation and begin hatching within eight to twelve hours, the variation dependent on the weather. Weather always has to be considered when determining the age of a trail. Sunny warm days make fresh spoor, such as dung or foliage discards, appear old by drying them out after only a few hours of exposure, whereas rain or heavy mist have exactly the opposite effect.

I found it helpful during the early days of the study to return to camp with fresh dung specimens and vegetation discards and then record their aging process under various weather conditions. Repetition of this simple procedure soon improved my ability to gauge the age of trails accurately. To evaluate distances more precisely I set up stakes outside the tent, 50 to 250 feet apart, so that actual rather than approximate measures became familiar.

The dung of lactating females is often covered with a whitish sheath, possibly a result of the tendency gorilla mothers have to eat the feces of their offspring during the infant's first four to six months of life. Diarrhetic dung, either with or without a mucoid sheath or flecks of blood, when deposited by only one individual of a group, often signifies that the individual is ill. When numerous animals of a group leave diarrhetic dung along a trail, it is an indication that the gorillas have been alarmed by another group or, more likely, by poachers. These types of deposits are always found on flee trails created when a group has rapidly run, almost single file, from a potential threat. The time I spend following a flee trail seems horridly prolonged because of growing apprehensions about what may be found at the end of it.

Occasionally, various groups acquire a communal cestode parasite (*Anoplocephala gorillae*), an infection that could not be correlated with either seasonal or range patterns. Large flatworm segments, about 1 inch long, are most frequently found in feces deposited in night nests and, when examined early in the morning, the dung contents of the nests seem virtually alive, crawling with activity.

All age and sex classes of gorillas have been observed eating their own dung and, to a lesser extent, that of other gorillas. Coprophagy is most likely to occur after prolonged day-resting periods during the rainy season, when both feeding and travel time are minimized. The animals simply shift their buttocks slightly to catch the dung lobe in one hand before it contacts the earth. They then bite into the lobe and while chewing smack their lips with apparent relish. The eating of excrement occurs among most vertebrates, including humans, who have certain nutritional deficiencies. Among gorillas coprophagy is thought to have possible dietary functions because it may allow vitamins, particularly Vitamin B_{12}, synthesized in the hind gut, to be assimilated in the foregut. Since the activity is usually observed during periods of cold wet weather, I am inclined to relate the "meals" to instant warmed TV dinners!

Between age and sex classes dung sizes vary tremendously, ranging from around 3 inches for silverbacks, to $3/8$–1 inch for infants. Analyzing the dung contents of the night nests makes it possible to determine the composition of fringe or census groups, and is also a reliable means of learning if births or transfers have occurred within study groups. (Most births occur during the night and night nests contain nearly half of the dung deposited by an individual over a twenty-four-hour period.)

Gorillas are diurnal and build their nests in different locations each evening. Ninety-eight percent of gorillas' night nests are built from nonfood vegetation, since food items such as thistles, nettles, and celery are not suitable nesting material. Adult night nests are sturdy, compact structures, sometimes resembling oval, leafy bathtubs made from bulky plants such as *Lobelia* (*Lobelia giberroa*) and *Senecio* (*Senecio erici-rosenii*). Construction is concentrated on the rim of the nest, which is composed of multiple bent stalks, the leafy ends of which are tucked around and under the animal's body for a more "cushiony" central bottom. Nests can be built in trees as well as on the ground, but because of adult gorillas' great weight nests are more commonly found on the ground. Favored nesting locations during the rainy season are in the sheltered hollows of tree trunks and nests may be made only of moss or loose soil. These types of nests not only offer protection from the elements but also provide early morning snacks in the form of decayed tree bark and roots.

Nests built by immatures are often only flimsy clusters of leaves until practice enables the construction of a solid, serviceable nest. The youngest animal observed consistently building and sleeping within his own night nest was thirty-four months old. Ordinarily a youngster remains sleeping in the mother's nest until the female again gives birth.

Some degree of predetermination is shown in the choice of night-nesting sites when gorillas are in areas adjacent to the park boundaries or near routes frequently used by poachers. The animals then tend to select knolls or open slopes offering good vantage points from which to view the surrounding terrain. This same type of choice also occurs when other gorilla groups are nearby. Less selectivity is demonstrated in the choice of day-nesting sites, although on sunny days areas with optimal sun exposure are far more frequently used than shaded or heavily treed regions.

For many years the slopes immediately behind camp were a part of the ranges of Groups 4 and 5. On dozens of occasions I found that the females and younger group members built their night nests about one hundred feet up on the slope near camp, whereas the silverbacks nested at the hill's base. This arrangement made it almost impossible for anyone to approach the gorillas undetected. When either Group 4 or 5 nested behind camp, I would approach them cautiously the following dawn in the hope of observing the animals before they awoke. Without fail I would almost step on a sleeping silverback sentry obscured in the tall foliage at the base of the slope. It was difficult to know which of us was the more shocked as the rudely awakened animal instantly jumped to his feet screaming in alarm before running uphill to "defend" his family, all now thoroughly awakened.

Vestiges of tree nests last as long as four years, far longer than those constructed in ground foliage, which last some five months, depending upon weather conditions or location. Clusters of night nests made from tall lobelia plants often yield interesting information concerning the frequency and length of gorillas' use of certain areas. Lobelias continue to grow in height even after their top leafy crowns have been broken off for nests. I have estimated that these plants grow about two or three inches a year. An area containing circles of lobelia stalks, some 10 feet tall, suggests that nesting sites were perhaps built there about thirty years earlier.

There is some speculation that night nests either offer protection from the weather or may be an innate activity remaining from gorillas' ancestral tree-living prototypes. Both points of view are plausible. I have observed numerous zoo gorillas born in captivity who apparently innately, rather than imitatively, utilized any remotely suitable object to shape around or under their bodies, much in the same manner that free-living gorillas use vegetation. Once I watched a lady's large straw hat blow into a zoo enclosure and be immediately retrieved by an adult female gorilla. The animal painstakingly ripped the hat into shreds to "build" a flimsy nest around herself while staunchly defending her nesting material against the other individuals in the cage.

Normally gorilla groups spend about 40 percent of their days resting, 30 percent feeding, and 30 percent traveling or travel-feeding — times when both movement and eating occur simultaneously. Around the Karisoke Research Centre's study area of 9½ square miles there are seven major vegetation zones, each attractive to gorillas at various times of the year according to weather and season.

The saddle zone is relatively flat terrain lying between the three westerly volcanoes (Mts. Visoke, Karisimbi, and Mikeno) and interspersed with hills and ridges no more than 98 feet high. The saddle contains the richest variety of vines and herbaceous ground foliage, in addition to having the highest frequency of *Hagenia* and *Hypericum* trees.

The *Vernonia* zone is found in small areas of the saddle as well as on the lower slopes of Visoke. The flowers, bark, and pulp of *Vernonia* trees are favored gorilla food. This tree species is so frequently selected for nesting and play activities that it is becoming increasingly rare in some areas of previous abundance.

The nettle zone is found in small sections of the saddle and on the lower Visoke slopes, but the main nettle area lies at the western base of Visoke in a dense belt varying in width from one to two fifths of a mile.

The bamboo zone is a limited region found primarily along the eastern boundary of the park and is responsible for seasonal movements of Group 5. Only a few isolated clumps of bamboo grow in the saddle of Group 4's range, but when the bamboo begins shooting, the group leaves the mountain slopes and unerringly travels straight to the bamboo clumps, indicating their keen recollection of both season and location of food sources.

The brush zone is found mainly along ridges of Visoke's slopes and, to a lesser extent, on hills in the saddle. I consider it a separate zone because it contains a high density of favored fruit shrubs and trees, such as blackberry and *Pygeum,* and rarer trees and brush whose bark is avidly sought by the gorillas.

The giant lobelia zone is found 11,480 to 12,465 feet on Visoke's upper slopes. This area is frequented by gorillas during drier months when the high mountain vegetation retains moisture from nightly mists. For this reason succulence can be obtained from the brush, trees, and foliage characteristic of the region.

The Afro-alpine zone encompasses the highest portion of the mountain summits and consists mainly of open grass or lichen-covered meadows. This is a sparse, bleak area containing little gorilla vegetation.

Gorillas travel more rapidly in areas where food resources are limited, and also when they are undertaking "exploratory sallies" — treks into unfamiliar terrain. Such ventures appear to be the means by which either a lone silverback or a group can expand its saddle-zone range. Range expansion into the saddle avoids extensive overlapping with other groups, as was the case on Visoke's slopes in the late 1960s. Often when tracking gorillas on these long crosscoun-

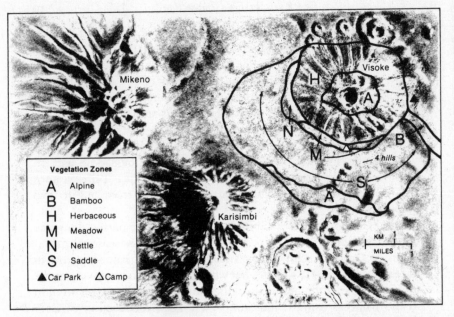

Major Gorilla Food Vegetation Zones within the Karisoke Research Centre's Study Area

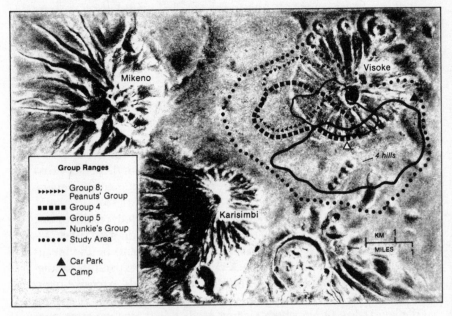

Gorilla Group Ranges within the Karisoke Research Centre Study Area

try treks, I whimsically pictured the silverbacks urging on their group members by saying, "Okay guys, let's just see what's on the other side of this next little hill!" Frequently the animals ended up in totally unsuitable gorilla habitat and had to traverse back and forth in order to find small oases containing food vegetation before renewing their quest for satisfactory terrain. Sometimes their travel routes were so erratic that I became certain, especially on foggy days when the mountaintops were hidden from view, that the animals were either lost or extremely disoriented.

Acquisition of new range area is more often achieved within the saddle zone than on the slopes, because the saddle's expansive land surface offers a greater abundance and variety of preferred vegetation. Gorillas feed upon some fifty-eight plant species from the seven zones in the study area. Leaves, shoots, and stems form about 86 percent of their diet and fruits only 2 percent. Dung, dirt, bark, roots, grubs, and snails are also eaten, but to a far lesser extent than foliage. The most common herbaceous plants consumed are thistles, nettles, and celery — which could grow up to eight feet. The scraggly *Galium* vine forms the bulk of the gorillas' diet, most likely because it, unlike other vegetation, grows at nearly all levels of the forest from amid thick ground foliage to the tops of tree branches, where it is more easily obtained by agile immatures than by adults.

There is the possibility that gorillas improve their habitat within tall herbaceous vegetation both in the saddle and on the mountain slopes. Cattle and buffalo, with their corneous, sharp hooves, sever plant stems underfoot; but gorillas' hands and feet, with their padded soles, press herbaceous foliage into the earth, and thereby cause more rapid regeneration because of the increased number of shoots sprouting from the nodes of the semiburied stems. By marking off small plots of foliage traversed only by gorillas, some plots frequented only by bovines, and the remaining used by neither, I was able to see, within a six-week period, that the sections covered by gorillas had a far denser growth of vegetation, particularly nettles and thistles.

Competition over food resources is seldom observed among gorillas unless the sources of favored food are restricted by short seasonal growth or clumped in small areas. One such example is the *Pygeum* fruit tree that grows oaklike about 60 feet tall and is found only on a few mountain ridges. Because of the relative scarcity of the trees and their brief fruiting season — only two to three months a year — the ridges that support them attract concentrations of gorilla groups all at one time. It is a spectacular sight to watch massive silverbacks gingerly climbing to the highest branches in search of the small delicacies. Because of status, silverbacks have first culling choice while animals of lesser rank wait their turns at the bottom until the patriarchs descend. After gathering mouthfuls and handfuls of the fruit, the gorillas skillfully maneuver themselves to the nearest sturdy perch upon which to sit and enjoy their meager harvest.

Another scarce and keenly sought food is related to mistletoe. At altitudes around 10,000 feet it grows on spindly trees such as *Hypericum*. Thus immature animals are able to collect the leafy flowered stalks more proficiently than weighty adults who frequently have to sit under trees waiting for *Loranthus* tidbits to fall their way. Youngsters who make the mistake of painstakingly descending to the forest floor to eat their collection more comfortably are usually bothered by pilfering adults who have no trouble "bullying" the young out of their acquisitions.

Still another special food is bracket fungus (*Ganoderma applanatum*), a parasitical tree growth resembling a large solidified mushroom. The shelflike projection is difficult to break free from a tree, so younger animals often have to wrap their arms and legs awkwardly around a trunk and content themselves by only gnawing at the delicacy. Older animals who succeed in breaking the fungus loose have been observed carrying it several hundred feet from its source, all the while guarding it possessively from more dominant individuals' attempts to take it away. Both the scarcity of the fungus and the gorillas' liking of it cause many intragroup squabbles, a number of which are settled by the silverback, who simply takes the item of contention for himself.

Group disputes also arise when restricted feeding sites containing prized foods create crowded conditions. The most common example occurs whenever an entire group seeks access to limited bamboo patches such as are found in the saddle zone. This also happens in the dry months when gorillas go on soil-eating binges on Visoke's ridges, where some earth is particularly rich in calcium and potassium. For many years one cavernous "dig" was favored by Group 5. The ridge supporting numerous trees had been so dug out by the gorillas that the tree roots formed exposed gnarled supports for the vast caves created by the animals' repeated soil digging.

Upon approaching this region, the leader of Group 5 went first as a matter of course, while other group members resigned themselves to waiting outside the favored cave. It was eerie to watch the huge silverback magically disappear beneath a web of tree roots into total blackness. When he emerged, covered with the sandy crumbs of his feast, he moved off, leaving the cavern to the other group members. In order of rank, they disappeared into its depths. Their subsequent screams and pig-grunts reflected the overcrowded conditions.

Group 4 chose their dirt mainly from sandy slides. Year after year the slides also attracted swallows to bathe and nest in the loose dirt. Much like Group 5, Group 4 headed for these barren areas during dry seasons to scoop up the soil with their hands and ingest handfuls of dirt. Even after hours of observation at these spots, I never saw gorillas attempt to catch adult swallows, their young, or their eggs.

Since gorillas mainly eat vegetation, food preparation involves manual and oral dexterity, attributes with which gorillas are well endowed. Perhaps for this reason gorillas have not yet been observed fashioning objects within their environment as tools. By contrast, free-living chimpanzees are renowned for their clever adaptations of twigs and leaves to serve as tools for obtaining both food and water.

Possibly gorillas have never been observed improvising tools to obtain food because the resources of their habitat meet their needs. Once, following a four-month dry spell in 1969, swarms of termites passed through the study area. I expected that the gorillas would, chimpanzee style, improvise twigs to extract the termites from the decayed tree stumps. However, they totally ignored the termites and waded their way past the infested areas to feed on surrounding vegetation.

On warm sunny days when group contentment is at its highest, feeding and resting periods are frequently accompanied by soft purring sounds resembling stomach rumbling; thus I named them "belch vocalizations." Typically, one animal expresses its feeling of well-being by giving a series of disyllabic belch vocalizations, *naoom, naoom, naoom*. This brings a chain of similar responses from other animals nearby, thus establishing both the location and the identification of the individuals participating in the exchange. The sound serves as the perfect communication for humans to imitate when initiating contacts with gorilla groups either partially or totally obscured in vegetation. By its use I can inform the animals of my presence and allay any apprehensions they might have on hearing the noise of vegetation being broken near them. It is an extraordinary feeling to be able to sit in the middle of a resting group of gorillas and contribute to a contented chorus of belch vocalizers.

The belch vocalization is the most common form of intragroup communication. In its prolonged form it expresses contentment, though a slightly shortened version may serve as a mild disciplinary rebuke toward young animals. A stronger disciplinary vocalization is the "pig-grunt," a series of harsh, staccato grunts resembling the sounds of pigs feeding at a sty, and frequently used by silverbacks when settling squabbles among other members of their groups. Females direct the vocalization toward other adults when conflicts over food arise or when right-of-way on trails occurs, and also toward their infants, particularly during the last stages of the weaning process. Young individuals will pig-grunt among themselves when complaining during rough play with their siblings or peers.

Popular literature generally describes roars, screams, or *wraaghs* as the main components of the gorilla vocabulary. Indeed, during the initial part of my study, these were the most frequent sounds I heard from the as yet unhabituated gorillas whenever my presence posed an element of threat to them. Gorilla vocalizations have always interested me, and I have spent many months recording sounds in the field and later analyzing them spectrographically at Cambridge University. The work proved most rewarding when the high frequency of alarm calls was slowly replaced by undisturbed intragroup vocalizations, sounds I used to gain further acceptance by the gorillas.

In late 1972, when student observers began working at Karisoke, instruction in the art of belch vocalizing was one of the first lessons taught. Several newcomers never quite got on to imitating the sound properly. One person's rendition of the belch vocalization sounded exactly like a goat's bleat, but within several weeks, the gorillas even became accustomed to his individual greeting call.

* * *

At times, students as well as I have unexpectedly encountered gorillas before we were aware of the animals' nearness. Such occasions could provoke charges, especially if interactions were occurring between groups, when the animals were traveling in a precarious range area (like one frequented by poachers), or if an infant had recently been born.

Understandably, such circumstances compelled highly protective strategies from a group's silverback leader. Once I was charged when climbing through tall vegetation up a steep hill to meet Group 8, thought to be several hours away. Suddenly, like a pane of broken glass, the air around me was shattered by the screams of the five males of the group as they bulldozed their way down through the foliage toward me. It is very difficult to describe the charge of a gorilla group. As in the other charges I have experienced, the intensity of the gorillas' screams was so deafening, I could not locate the source of the noise. I only knew that the group was charging from above, when the tall vegetation gave way as though an out-of-control tractor were headed directly for me.

Upon recognizing me, the group's dominant silverback swiftly braked to a stop three feet away, causing the four males behind him, momentarily and ungracefully, to pile up on top of him. At this instant I slowly sank to the ground to assume as submissive a pose as possible. The hair on each male's headcrest stood erect (pilo-erection), canines were fully exposed, the irises of ordinarily soft brown eyes glinted yellow — more like those of cats than of goril-las — and an overpowering fear odor permeated the air. For a good half-hour all five males screamed if I made even the slightest move-ment. After a thirty-minute period, the group allowed me to pretend to feed meekly on vegetation before they finally moved rigidly out of sight uphill.

Only then could I stand up to check out the cause of human shout-ing that I had heard coming from the base of the slope about four hundred feet below. There, standing along a trail used extensively for cattle at this early stage of my work, stood a group of Watutsi herdsmen. They had been drawn by the gorillas' screams from various parts of the adjacent forest where they were grazing cattle. I later learned the men were certain I had been torn to shreds, and upon seeing me stand upright were convinced that I was protected by a very special kind of *sumu* against the wrath of the gorillas, whom they feared deeply.

Once the men moved out of sight I continued to follow Group 8 — at a distance — to discover that they had been interacting with Group 9 when I had attempted to contact them. Trail sign indicated Group 9 had also taken part in the charge but had halted before reaching me. It was only when descending the slope that I discovered a lone silverback directly below me. His presence made Group 8's charge far more understandable. Upon hearing the sounds of my approach through the thick vegetation, the gorillas probably thought I was the lone male whose presence neither group would have tolerated.

Though you know the charging gorillas are simply acting defensively and do not wish to inflict physical harm, you instinctively want to flee, an impulse that automatically invites a chase. I have always been convinced of the intrinsically gentle nature of gorillas and felt their charges were basically bluff in nature, so never hesitated to hold my ground. However, because of the intensity of their screams and the speed of their approaches, I found it possible to face charging gorillas only by clinging to surrounding vegetation for dear life. Without that support, I surely would have turned tail and run.

Like all charges, this one was really my fault for having climbed the steep slope to approach directly beneath the animals without first identifying myself. Other charges have occurred when students, also accidentally, made the same error. Some census workers who encountered unfamiliar gorilla groups outside the study area had to return to their camps several times to change clothes because of reflexive reactions prompted by charges. People who hold their ground usually are not hurt unless they are unknown to the gorillas, but even then they only occasionally receive a moderate slap from a passing animal. People who run are not so fortunate.

A very capable student once made the same mistake as I had when approaching Group 8 from directly below. He was climbing through extremely dense foliage in a poacher area and noisily hacking at vegetation with his *panga,* not knowing the group was near. The faulty approach provoked a charge from the dominant silverback, who could not see who was coming. When the young man instinctively turned and ran, the male lunged toward the fleeing form. The gorilla knocked him down, tore into his knapsack, and was just beginning to sink his teeth into the student's arm when he recognized a familiar observer. The silverback immediately backed off, wearing what I was told was an "apologetic facial expression" before scurrying back to the rest of Group 8 without even a backward glance.

Another person who ran away from the charge of an unfamiliar group was someone who had always scoffed at the idea of pacifying gorillas with introductory vocalizations on approaching them. His actions around gorillas were often jerky and almost aggressive in nature. He was able to spend nearly a year working with habituated animals before his luck ran out. In the lead of a large boisterous group of tourists he approached two interacting groups from directly below and was instantly charged by a silverback, who rolled with him for some thirty feet, breaking three of his ribs, and then bit deeply into the dorsal surface of the man's neck. The bite would have been fatal had it pierced the jugular vein on the neck's ventral surface. This person survived to brag about his "close shave" without acknowledging his violation of basic gorilla protocol.

In another incident a young tourist tried to pick up an infant from Group 5 "to cuddle" in spite of the alarmed screams given by the group. Before he got his hands on the youngster, the infant's mother and the group's silverback defensively charged, causing the boy to turn and run. He fell and both gorilla parents were instantly on his back, biting him and tearing at his clothing. Many months later in Ruhengeri I saw that he still bore deep scars from the encounter on his legs and arms.

Charge anecdotes do the gorilla an injustice. Were it not for human encroachment into their terrain, the animals undoubtedly would have to charge only when defending their familial groups from intrusion by other gorillas. I remain deeply concerned about having habituated gorillas to human beings. This is one reason I do not habituate them to members of my African staff. Gorillas have known Africans only as poachers in the past. The second that it takes a gorilla to determine if an African is friend or foe is the second that might cost the animal its life from a spear, arrow, or bullet.

How ironic it is that probably less than a hundred men, armed with bows and arrows, spears or guns, have been allowed to plague the wildlife in the parklands that form the last stronghold for the mountain gorilla. The strongest counterstrategy against the abuse encroachers bestow upon the wildlife of the Virungas may be that of active conservation.

* * *

Active conservation is a straightforward issue. It begins with providing personal incentive on a one-to-one basis with individual Africans, not only to take pride in their park but also to assume personally some of the responsibility toward the protection of their heritage. Given the incentive, active conservation is accomplished by very fundamental needs such as boots for the rangers' feet, decent clothing and raingear, ample food, and adequate wages. Thus equipped, hundreds of antipoacher patrols have set out from Karisoke into the heartland of the Virungas to cut traps, confiscate encroachers' weapons, and release newly trapped animals from snares. Active conservation within a steadily shrinking internationally designated sanctuary filled with poachers, traps, herdsmen, farmers, and beekeepers needs to be supplemented by Rwandese and Zairoise enforcement of anti-encroacher laws as well as severe penalties for the illegal sale of poached animals for their meat, skins, tusks, or for financial profits. Active conservation does not rule out any other long-term conservation approaches.

Theoretical conservation as a sole conservation effort is in marked contrast to active conservation. To an impoverished country such as Rwanda, an abstract rather than practical approach is more appealing. Theoretical conservation seeks to encourage growth in tourism by improving existing roads that circle the mountains of the Parc des Volcans, by renovating the park headquarters and tourists' lodging, and by the habituation of gorillas near the park boundaries for tourists to visit and photograph. Theoretical conservation is lauded highly by Rwandese government and park officials, who are understandably eager to see the Parc des Volcans gain international acclaim and to justify its economic existence in a land-scarce country. These efforts attract increasing numbers of sightseers to the Parc des Volcans. In 1980 alone, the park's revenue from tourism more than doubled over that received in 1979.

There is a failure to realize that the immediate needs of some 200 remaining mountain gorilla, and also of other Virunga wildlife now struggling for survival on a day-to-day basis, are not met by the long-term goals of theoretical conservation. Gorillas and the other park animals do not have time to wait. It takes only one trap, one bullet to kill a gorilla. For this reason it is mandatory that conservation efforts be actively concentrated on the immediate perils existing within the park. Next to these efforts, all others become theoretical. Educating the local populace to respect gorillas and working to attract tourism do not help the 242 remaining gorillas of the Virungas survive for future generations of tourists to enjoy. Theoretical conservation has good long-term goals that needlessly ignore desperate immediate needs.

Far from the public's eye active conservation continues in the Parc des Volcans with a handful of dedicated people who work tirelessly behind the scenes to protect the park and its wildlife. One outstanding person, who risked his position for what he believed, is Paulin Nkubili. As Rwandese Chef des Brigades, he inflicted strong penalties upon both buyers' and sellers' game illegally poached from the Parc des Volcans. By his actions, he also essentially eliminated the trophy market involving the sale of gorillas' heads and hands for souvenirs. There are some members of the Watutsi clan of Rutshema, a people who for generations grazed cattle illegally in the park, who themselves became active conservationists by leading antipoacher patrols in the Virungas. Paulin Nkubili, loyal members of the Karisoke Research Centre staff, and those of the patrols are each personally motivated in their unheralded efforts and rewarded only in the knowledge of their accomplishments. The hope for the future of the Virungas lies in the hands of just such individuals.

4 | Three Generations of One Gorilla Family: Group 5

*I*RONICALLY ENOUGH it was poachers who introduced me to the first group encountered on Mt. Visoke, Group 4. Two Batwa had been hunting for duiker with bows and arrows, and, upon hearing a screaming outbreak from the slopes of Visoke, came to camp to tell me of the gorillas' whereabouts.

I followed the poachers to the group and returned to camp elated upon having contacted gorillas the first day after having established the Karisoke camp. While typing up my field notes that evening I heard chestbeats and gorilla vocalizations from Visoke's slopes just behind my tent. The sounds were approximately a mile from where I had left Group 4 earlier that day. Because gorilla groups ordinarily travel only some 400 yards a day, I realized that this had to be a second Karisoke study group, Group 5.

The next morning I climbed toward the source of the previous evening's vocalizations, picked up the gorillas' trail, and tracked them to a ridge of thick trees high above the tented camp. Upon seeing me, all the animals instantly hid, with the exception of a young juvenile who treed to chestbeat and flamboyantly swing through the branches before leaping with a crash into the foliage below. Instantly I named him Icarus. The other group members, fifteen in all as I was later to learn, retreated about twenty feet farther from where they had been feeding and peeked at me shyly through dense vegetation. However, the imp, Icarus, boldly climbed a tree again either to show off his acrobatic ability or to stare curiously at the first human being he had ever seen munching wild celery stalks.

Within the first half-hour of the contact with Group 5 I realized that there were two silverbacks in the group who maintained protective flank positions around the females and young. The two males were easily located and recognized because of their disharmonic vocalizations. The elder, dominant male gave deep alarm *wraaghs* and was named Beethoven; the younger silverback had higher-pitched calls and was named Bartok. I later identified an older blackback male in the group and couldn't resist naming him Brahms. Four females were also sighted carrying bug-eyed infants of varying ages. One of the adults calmly sat under the tree where Icarus was vigorously displaying. She protectively hugged an infant to her breast and showed some concern at Icarus' antics. I felt certain that she was the mother of the young acrobat because of their strong facial resemblance and his periodic need to go to her for reassurance. For no particular reason I named the female Effie and the bright-eyed infant she clutched to her breast Piper. After nearly an hour's contact, the gorillas began moving off to feed. Since one of my basic rules is never to follow a group when they choose to leave, I also left, though Icarus briefly remained treed amid his batons of branches.

The habituation of Group 5 progressed smoothly because of the regularity of my contacts. I was able to approach the animals to within twenty feet during the first year of the work at Karisoke. Beethoven was tolerant of the other two adult males, Bartok and Brahms, since he appeared to rely upon them as watchdogs for the protection of the group's females and young. The highest-ranking female, Effie, along with her daughter Piper, about two years old, and Icarus, between five and six years, maintained the closest proximity to Beethoven, who was consistently indulgent and good-natured whenever his offspring tumbled in play around his huge, silvered bulk. The second-highest-ranking female in the group was Marchessa, who seemed apprehensive of Effie, although Marchessa's daughter, Pantsy, about a year and a half old, never hesitated to mingle among Effie's clan to play with Piper and Icarus. Pantsy had been so named because of a chronic asthmalike condition that affected her vocalizations. More often than not, Pantsy's eyes and nose drained severely, but Marchessa never was observed attempting to clean the infant's face. Two of the remaining four adult females in the group were never given names; I was not absolutely certain which was which because of their tendencies to remain obscured in dense vegetation.

The other two females, Liza and Idano, were the last individuals of Group 5 to be named once they lost their shyness of me and could be clearly identified.

Icarus enhanced contacts with the group because of his insatiable curiosity and boldness, which often prompted risky displays in all sizes of trees from spindly saplings to sturdy old *Hagenia*. One day while trying out a new routine on a tree limb not solid enough for such antics, the little elf-eared fellow unintentionally crashed to the ground along with the branch from which he had been swinging. The splintering noises had barely died away before the air was filled with the indignant roars and screams of Beethoven and Bartok. The two males bluff-charged toward me with the group's females bringing up the rear as though they all held me responsible for the fall. The animals stopped about ten feet away when they saw Icarus, still intact, climb another tree impervious to the furor he had created. The mischievous imp appeared all angelic innocence, but the two silverbacks remained quite tense. The air was filled with their pungent fear odor.

I released my clammy hold on the vegetation I had clutched, when, much to my dismay, Icarus' sister, Piper, climbed into the broken sapling he had just discarded. The little juvenile began an uncoordinated series of spins, twirls, kicks, and chest pats. She exuded blasé self-importance as the attention of myself and the gorillas' was riveted on her. No high-wire artist ever had such a rapt audience. The glances of the silverbacks darted back and forth between Piper and myself as if they expected me to leap forward and grab her at any moment. When our eyes met, they roared their disapproval. Suddenly, Icarus broke the nervousness building up among the group members. He climbed playfully into Piper's tree and began a chasing game that led them both back to the watchful group. All three silverbacks then released their tension by chestbeating and running through the tall foliage before leading the group uphill and away.

On a slope gorillas always feel more secure when positioned above humans, or even approaching gorillas. I never relished climbing up to a group from directly below, but there were times when the thickness of the vegetation compelled me to do so. I vividly remember one such contact when crawling up to Group 5 and carrying a heavy Nagra tape recorder. Just about twenty feet below the gorillas, who could be heard feeding above, I softly vocalized to make my presence known. I set up the microphone in a nearby tree and stabilized the tape recorder on the ground. A number of curious infants and juveniles climbed into trees above to stare intently down at the unaccustomed equipment. Upon recognizing me, they began playing pretentiously in flimsy *Vernonia* saplings. The feeding sounds among the adults, still out of sight farther up the slope, stopped as the youngsters' dare-deviltry spurred them into wilder and noisier acrobatics. Just as expected, the silverbacks instantly led the females, all hysterically screaming, in a bluff-charge to within ten feet. Because of the incredible intensity of the screams, the needle of the modulation meter on the tape recorder went berserk, bouncing far above the proper intake level. I tried to bend down to adjust the machine's volume, but the slightest movement incited renewed charges from the overwrought animals. Forgetting all about the microphone, I dramatically whispered to myself, "I'll never get out of this alive!" When the tape ran out I could only stand helplessly by, glancing alternately at the worried silverbacks directly above and the frantically spinning empty sound reel in the machine at my feet. Only when the group eventually climbed out of sight was it possible to turn off the recorder. That night when listening to the tape in the cabin, my theatrical words, sandwiched between two screaming charges, came as a complete surprise and reduced me to gales of laughter, for I had forgotten ever having whispered them during all the excitement.

Months later I analyzed the vocalizations spectrographically and found that the individual differences distinguished when hearing silverbacks' *wraaghs* and other calls were also apparent on the sonograms. This leaves little doubt that gorillas can identify one another by hearing vocalizations emitted by others even over great distances.

By 1969, the second year of Karisoke research, the staff and I had not yet totally succeeded in ridding the saddle terrain of cattle, so Group 5 tenaciously clung to Visoke's southeastern slopes, a region of deep ravines surrounded by steep ridges. It was often possible to track the group to the edge of a ridge and find the animals below sunbathing like so many beach bums. On such occasions obscured contacts were maintained in order to observe intragroup interactions unaffected by my presence.

On one rare sunny day, contented belch vocalizations were heard coming up from Group 5 secluded in one of their favorite bowls of rich herbaceous foliage. Quietly I crawled to the edge of the ridge and lay hidden in the brush to observe the peaceable family through binoculars. The patriarch, Beethoven, was nested in the center of the sunning circle, a great silver mound about twice the size of the females clustered around him. I could only estimate his weight as in the neighborhood of three hundred and fifty pounds and his age as probably around forty. His silvering extended along his thighs, neck, and shoulders, where it was more grizzled in color than the near-white saddle region of his dorsal surface. Other sexually dimorphic characteristics, in addition to massive size and silvering, were his pronounced sagittal crest and canines, all physical features never seen in female gorillas.

Slowly Beethoven shifted his great bulk, rolled over onto his back, gave a contented sigh, and speculatively regarded his latest offspring, six-month-old Puck. The infant was playfully tadpoling across the stomach of its mother, Effie, wearing a lopsided grin of enjoyment. Gently, Beethoven lifted Puck up by the scruff of the neck to dangle the exuberant baby over his body before casually grooming it. Puck was nearly obscured from sight by the massive hand, which eventually placed the wide-eyed infant back onto Effie's stomach.

That observation of a silverback sire with his offspring was typical of similar scenes throughout the years to be spent with the gorillas. The extraordinary gentleness of the adult male with his young dispels all the King Kong mythology.

Beethoven, as Group 5's leader, retained absolute breeding rights with Effie, Marchessa, Liza, and Idano, females he had either acquired over several years of interactions with other groups or had inherited upon the natural death of Group 5's previous leader. Beethoven was tolerant of the presence of the subordinate males Bartok and Brahms within the group. It seemed likely that they were related to him because of striking facial resemblance. However, upon reaching sexual maturity, the two younger silverbacks could not afford to remain with Group 5 because there were no breeding opportunities available to them — Effie, Marchessa, Liza, and Idano belonged to Beethoven. Subsequently Bartok and Brahms left Group 5 and became peripheral silverbacks ranging within a 300-yard radius for about nine months before becoming "lone silverbacks," when they traveled at greater distances to find suitable range areas. On such occasions both silverbacks frequently interacted with other groups as they sought to obtain females to establish their individual harems and, ultimately, their own groups.

By 1971 both Bartok and Brahms had chosen two distinct range areas adjacent to that of Group 5: Bartok on Mt. Visoke's eastern slopes high above the Elephant Tunnel, Brahms in the hills and saddle terrain between Mts. Visoke and Karisimbi.

During the past four years the Rwandese had become excellent trackers and were able to keep up with the routes of the main study groups as well as those of lone silverbacks and fringe groups we occasionally encountered in Karisoke's study area. Early one morning two trackers came to tell me excitedly that they had found a lone silverback trail immediately south of camp, and then led me to the *Mlima Moja,* or First Hill. While examining the night nest of the lone silverback, we heard an outbreak of terrified screams from the hill's base some 400 feet below. Running toward the source of the noise we briefly saw Brahms fleeing from the figure of a man — a poacher — running in the opposite direction and carrying his bow and arrows high above his head. Upon reaching the lower trail we picked up Brahms's flee route at the point where he and the poacher had met. The gorilla's trail was marked by blood-flecked leaves and diarrhetic dung deposits; the poacher's trail by increasingly lengthened stride of barefoot prints as he ran from the scene. Instinctively, both had tried to defend themselves, Brahms by charging, and the poacher by shooting his arrow into Brahms's chest.

For nearly three hours we slowly followed the injured silverback, but he seemed determined to put as much distance as possible between himself and the attack site. Brahms occasionally rested, leaving behind a circular ring of blood-soaked vegetation. I would have been convinced that he was mortally wounded except for intermittent roars, chestbeats, and foliage-breaking sounds as the frenzied animal gave vent to his shocked reactions of rage and pain.

Much later the same day, Brahms reached the lower slopes of Mt. Karisimbi, where we could not follow for fear of unnecessarily provoking him if we were seen. The next morning the trackers found that his empty night nest contained little blood but that his morning trail had climbed toward Mt. Karisimbi and headed away from Karisoke's study area.

Another year was to pass before Brahms obtained two females from Karisimbi groups. With them he eventually sired two infants as the beginning of his own group. I could not help but wonder if Brahms's experience with the unknown poacher had perhaps endowed him with extra awareness of the dangers that could threaten the safety not only of himself but of his females and offspring.

Both Brahms and Bartok had departed from Group 5, assumed

to have been their natal group, in June 1971. Six months earlier
Beethoven had acquired a nulliparous female we named Bravado,
during an interaction with Group 4. Nulliparous females — those
who have not borne offspring — usually transfer to lone silverbacks
or to small groups because the rank order of females corresponds
to the order in which they are acquired by the dominant silverback.
For this reason I was quite surprised to find Bravado in Group 5,
which already had an established female dominance hierarchy among
Beethoven's older mates — Effie, Marchessa, Liza, and Idano.

For ten months following her transfer to Group 5 Bravado never
seemed an integral part of the familial group. Then in October 1971
Bravado was able to renew her acquaintanceship with the members
of her natal group during a two-day interaction between Groups 4
and 5. The encounter occurred just behind camp within an area
known as "Contact Ridges." The two ridges, separated from one
another by a small ravine about 100 feet wide, marked the boundaries
of both groups' ranges at that time. Groups 4 and 5 were likely to
meet in the region where the ridges offered the silverbacks of each
group maximum visibility of one another, enhancing the magnitude
of their impressive displays.

Beethoven was a far more experienced group leader than Uncle
Bert, the silverback of Group 4. He seemed almost indulgent of
the younger male, who continuously strutted, chestbeat, and broke
tree branches along the top of his side of the ravine. Uncle Bert's
displays were also accompanied by prolonged hootseries usually given
before his chestbeats. The hootseries, a vocalization given by silverbacks during interactions, may carry for nearly a mile throughout
the forest.

During the first day of their interaction, Beethoven responded
to only a few of Uncle Bert's hootseries, and the adult females of
Group 5 seemed equally uninterested in the younger silverback's
displays. Bravado, however, was instantly attracted to her old group
and crossed the wide ravine, followed by young Icarus and Piper
of Group 5. Once on Group 4's side they cavorted wildly with some
of the younger animals on the slopes below Uncle Bert. Although
it had been ten months since all had met, it was obvious that Bravado
was remembered by her home group. The youngsters enthusiastically
clustered around and embraced her before beginning a prolonged
play session.

Toward the end of the day Uncle Bert unwisely chose to move

over to Beethoven's side of the ravine, accompanied by a disorderly procession of his group members, as well as Bravado, Icarus, and Piper. The tyro silverback's foolhardy action could not be ignored by Beethoven, who glared at the straggled line in the ravine below him before deliberately strutting down to meet them, leaving his own group members behind. The two group leaders approached to within four feet, halted parallel to one another, and assumed rigid stances with their gazes averted. All the animals of both groups remained silently motionless as the silverbacks' tension was conveyed.

Suddenly, unable to endure further strain, Uncle Bert stood bipedally, chestbeat, and loudly slapped down the vegetation between himself and Beethoven. This was all that was needed to trigger the older male, who had been a study in tolerance until then. Roaring indignantly, Beethoven charged Uncle Bert. The young silverback ignominiously fled downhill followed by the rest of his group, all screaming hysterically. Rather than pursue, Beethoven simply stood where he was and stared down scornfully at the confused members of Group 4. Uncle Bert halted about fifty feet below and, undoubtedly feeling more secure at the increased distance, resumed displaying with chestbeats, hootseries, and runs through the foliage. Disdainfully, Beethoven turned and strutted uphill toward the top of the ridge where his family members were waiting for him. Twice he stopped and feigned feeding on thistle leaves, all stripped with slow deliberation to allow him the opportunity to survey Uncle Bert's actions. Beethoven had been followed by his young daughter Piper, but Icarus and Bravado had remained at the bottom of the ridge, staring off in a wistful manner toward Group 4.

Once more Uncle Bert blundered by returning to the base of Group 5's ridge in an effort to herd Bravado back to his own group. Angrily, Beethoven charged downhill, causing the younger silverback to retreat into the restless cluster of his Group 4. After a prolonged stance directly facing Uncle Bert, Beethoven turned and pushed both Bravado and Icarus to the top of his ridge and out of sight. Loud belch vocalizations were exchanged between the leader and his Group 5 members as they moved off to feed. After a brief rest, Group 4 proceeded in the same direction but at a lower level and silently.

Because the interaction site was just behind camp, I had expected to hear the two silverbacks exchange hootseries or chestbeats during the night. The subsequent silence led me to believe that the groups had separated to return to the core areas of their respective home ranges. Therefore, the following morning as I returned to "Contact Ridges," I was surprised to hear Uncle Bert in the process of "warming up" with vigorous chestbeats and plaintive-sounding hootseries for the second day's interaction.

Filled with misgivings about the young silverback's lack of protocol, I climbed up the ravine between the two ridges and was amazed to see Bravado again leading Icarus, Piper, and Marchessa's juvenile daughter Pantsy toward Group 4's ridge. There they were exuberantly met by Group 4's youngsters and began another carefree play session of wrestling and tumbling.

Some forty feet above them on the top of the ridge, Uncle Bert continued energetically displaying with runs, chestbeats, and hootseries but seldom drew any overt response from Beethoven. The echoes of one display barely faded before the sounds of another one started. Nearly two hours passed before Beethoven slowly arose from his sentry position and ponderously, but silently, headed toward Group 4, his females and infants left behind. Immediately Uncle Bert ceased vocalizing. He strutted up and down the ridge with such stilted and exaggerated movements that his hind legs appeared to be attached to his body by strings as they swung in arcs before hitting the ground. The odor emanating from both silverbacks became increasingly strong even from where I sat some eighty feet away. Slowly Beethoven climbed up to meet Uncle Bert, until both stood face to face, magnifying their sizes by posing in extreme strut positions with head hair erect.

After a few seconds, like mechanical soldiers the two males turned and separated, Beethoven strutting downhill and Uncle Bert upward his silent and hushed group, which Bravado had joined. Abruptly Beethoven turned and ran up into the midst of Group 4. He was forced to retreat when the entire group surged down toward him screaming excitedly. However, Beethoven was not going to be deterred from his purpose, and charged into the midst of Group 4 again toward Bravado, who knelt submissively as he approached. He grabbed the young female's neck hair before herding her out of the group now clustered together behind her. Descending the ridge, they encountered the other three Group 5 members, and Beethoven authoritatively pig-grunted at them to accompany him. They obediently did, wearing pursed-lipped, fearful facial expressions. When all were some eighty feet below Group 4, Uncle Bert broke the silence by giving a chestbeat and hootseries. Instantly Beethoven halted and turned around to glare back defiantly at the younger male before continuing to drive his wayward brood to the base of the hill. Once at the bottom, the four youngsters broke into a chasing play session as a means of releasing their tension. Beethoven obscured himself from Uncle Bert by sitting in dense vegetation, a rear-guard tactical maneuver.

After a few minutes Uncle Bert imprudently strutted downhill and was followed by three younger members of his group who comically mimicked his bold swagger. Beethoven, hidden in foliage but aware of Uncle Bert's approach, seemed to deliberate about one final confrontation. Instead, he dutifully resumed herding his runaways back to Group 5, thus ending the interaction. The following day both groups were well within their own range areas and, as is typical after interactions, resting and feeding predominated over additional travel.

This particular interaction, one of the first I was able to observe throughout its entirety, was an impressive example of the extremes to which silverbacks will go in order to avoid physical clashes. Group 5's older, more experienced Beethoven might well have inflicted serious bodily harm to the tyro leader of Group 4, Uncle Bert, had he chosen to do so. Instead, by making use of the ritualized, parallel intimidation displays, physical damage had been avoided.

Many years later, after thousands of hours spent in the field, I found that interactions between distinct social units — lone silverbacks or groups — accounted for 62 percent of all wounds observed on male and female gorillas. From sixty-four skeletal specimens collected throughout six Virunga volcanoes, I found that 74 percent of the silverback remains revealed signs of healed head wounds, and 80 percent had either missing or broken canines.

The recuperative powers of gorillas never cease to amaze me and may well be illustrated by two skulls recovered from unknown silverbacks. Embedded in each of their supra-orbital crests, I discovered a canine cusp that had been broken off from the tooth of another silverback. The two bite victims must have received their injuries during their formative years, as evidenced by the extent of bony tissue growth surrounding the region of penetration into their skulls. These two findings vividly convey the enormous strength and power of silverbacks and compel one to reflect upon just what evolved characteristics — behavioral or physical — have enabled them to function so successfully as peaceful disciplinarians within their own family group structures.

*   *   *

In August 1972, eleven months after the two-day interaction between Groups 4 and 5, Bravado gave birth to her first offspring, a winsome male infant named Curry. This was the sixth infant to be born in Group 5 since 1967. Curry, like the other newborns, had been sired by Beethoven, the only sexually mature male within the group. I hoped that Curry's birth, tangible evidence of Bravado's link with the dominant silverback, might improve the new mother's status within the group. However, she remained apprehensive of Group 5's four higher-ranking females, Effie, Marchessa, Liza, and Idano. Bravado spent even more time at the edge of the group, and thus deprived Curry of opportunities for social interactions. Only when Curry was nine months old and had developed into an active and socially inclined baby did Bravado allow Beethoven's other offspring to groom, cuddle, and play with the group's newest addition. I felt that at last Bravado's prolonged period of social ostracism had ended.

Then the unexpected occurred. In April 1973, when Curry was nearly ten months old, a tracker found the baby's broken body left on a flee trail after an interaction between Group 5 and a silverback. Examination of the corpse revealed ten bite wounds of varying severity. One bite had fractured the infant's femur and a second had ruptured the gut, causing peritonitis and instant death. During the course of measuring and photographing the remains, I found Curry's fingernail impressions remained as pink indentations in the palms of both hands. Curry was my first introduction to infanticide among the Visoke gorillas.

The morning following the recovery of Curry's body, backtracking showed that a lone silverback had charged into Group 5 during their day-resting period. The subsequent encounter between the group and the lone male must have been extremely violent, as evidenced by numerous diarrhetic dung and blood deposits along the flee trail. Curry had been dropped about five hundred yards from the interaction site, but Group 5 continued to run for nearly a mile before stopping to build crude night nests. When the group was finally contacted, it was seen that Beethoven, Effie, Marchessa, and Idano had been badly bitten, possibly by the unknown silverback.

Soon after Curry's death, Bravado's behavior changed. She indulged in highly social play interactions with the group's younger animals. Gone was the worried look of her responsible days of motherhood as she began chasing and wrestling much like a juvenile. I found her conduct difficult to understand, particularly since I had been deeply saddened by Curry's unexpected death and subjectively expected Bravado to show some sign of distress. I was yet to learn that nearly all primiparous mothers, those giving birth for the first time, upon losing their offspring by infanticide, will react exactly as did Bravado. This type of behavior might be a method by which a female seeks to strengthen her social bonds with other group members following the trauma of having her infant killed. It might also be explained by a mother's sudden return to freedom of movement after having been hampered so many months by the constant need to support her infant while feeding and traveling.

Two months following Curry's death, Group 5 had a physical interaction with a small fringe group thought to consist of only a silverback and blackback. Bravado, along with Effie's daughter Piper, then nearly eight years old, emigrated from Group 5 to the new group, which ranged far outside the study area on Mt. Karisimbi's slopes. For this reason they were never identified positively after June 1973. I was saddened by the loss of contact with the two females whom I'd known since their infancy and also by the very real fact that their future fate was forever lost to Karisoke records.

The transfers of the two females and the departure of Bartok and Brahms reduced Group 5's membership from the fifteen individuals first met in 1967 to ten by July 1973, which included an additional four viable births during the same period. Shortly after Curry's death, another incident occurred — the death of shy, elderly Idano — depriving Group 5 of yet another adult female. Idano had shown obvious signs of weakness and deterioration shortly before her death. Beethoven regulated the travel pace of his group so that she might keep up with them during the last days of her illness. He also slept near her the night she died. Her autopsy, done at the University of Butare, revealed that she had chronic enteritis, peritonitis, and pleurisy; it was concluded that she had died of bacterial hepatitis. The autopsy also showed that she had recently miscarried, probably while trying to keep up with the group during the traumatic flee routes following Curry's death.

Among the three adult females remaining in Group 5 — Effie,

Marchessa, and Liza — the dominant Effie appeared to be the more skilled mother and one of the most even-dispositioned gorillas I have yet to meet. Her patience, stability, strong maternal instincts, and outstanding closeness with Beethoven, sire of her brood, enabled her to raise her young in a highly successful manner. Balancing consistent discipline with demonstrations of affection, Effie endowed her infants with love and security during their formative years and a keen self-confidence that carried over into their adulthood. An unusual rapport existed between Effie and her three offspring: fourteen-month-old Tuck, fifty-five-month-old Puck, and Icarus, estimated to be about eleven years old in 1973. The four seemed bound into a closely knit, mini-family unit within Group 5. There were numerous similarities in their behavioral interactions with other group members as well as in their physical characteristics. Structurally, except for size differences, all closely resembled Effie in having near-identical noseprints, patches of graying hair around their necks, and strabismus, or walleyes. This condition, typical of Effie's clan, did not seem to impair vision in any manner.

The second matrilineal clan within Group 5 was headed by Marchessa, a female estimated as about twenty-five years when first met. In 1967 Marchessa, older but not as reproductively successful as Effie, had only one offspring, her daughter Pantsy, who was about seventeen months old. By January 1971, when Pantsy was about four and a half, Marchessa had given birth to a spindly male I named Ziz. Marchessa's clan also had a physical anomaly in the form of syndactyly, a familial trait marked by webbing of two or more fingers or toes. The toes of Marchessa and her offspring were affected in varying degrees. This characteristic, probably due to inbreeding, was also observed among gorillas of other groups throughout the Virungas. Like strabismus, syndactyly did not appear to handicap the animals in any way.

Ziz was decidedly a mama's boy, but Marchessa seldom was observed applying Effie's evenhanded maternal tactics. Unlike Effie's adventurous offspring, Ziz stuck close to Marchessa and often threw screaming temper tantrums whenever she was out of sight for any length of time. Well into his third year Ziz still suckled regularly and whined vigorously if Marchessa attempted to thwart his efforts.

Liza was the third adult female remaining in Group 5 by the end of 1973 and also the lowest ranking. Her oldest daughter, Nikki, was nearly seven when she transferred out of Group 5 around the time that Bravado and Piper emigrated. Nikki's transfer had occurred during the night, and resulted in Group 5's fleeing for three and a half miles from the site of their interaction with a lone male. Nikki's sudden departure left Liza with only one other offspring, Quince, a delightful three-year-old female. She, unlike her mother Liza, was freely accepted by other individuals of the group, especially when initiating play or grooming sessions. Quince had very strong maternal inclinations and, even as a juvenile, was allowed to cuddle, groom, and carry the young of Effie and Marchessa. Though only seven months older than Ziz, Quince was always solicitous of Marchessa's son when he became separated briefly from Marchessa or was not allowed to suckle.

For most infants observed over the years of research, it was found that weaning became most traumatic to the infant around the middle to the end of the second year, a time when the mother is usually returning to regular estrous cycling or has already been impregnated. It was possible that Marchessa's prolonged nursing of her offspring was a responsible factor for the length of her intervals between births. Marchessa's spans averaged fifty-two months compared to Effie's forty-three months when only viable (those in which subsequent offspring survived) births were considered.

As usual within matrilineal clans, Marchessa, Pantsy, and Ziz were commonly found together during day-resting periods but near the perimeter of Group 5. This was in marked contrast to Effie's clan, who were normally found the closest to Beethoven or even to Liza.

Pantsy began spending less time with her mother when, toward the end of her seventh year, she began showing cyclical regularity. Among adolescent female gorillas, initial perineal swellings varied in onset from six years five months to an estimated eight years eleven months for eleven females, the average being seven and a half years.

During the two to five days of each month when Pantsy began regular cyclicity, she became highly attractive to younger group members, Icarus in particular. Pantsy frequently solicited Icarus, who in 1973 was only around eleven years old and thus sexually immature. If her mounting invitations to Icarus occurred near Beethoven, sire of all Group 5's young, the patriarch was prone to separate the two by running, pig-grunting, or whacking before then mounting Pantsy himself. For this reason Icarus discreetly tried to ignore Pantsy's advances when Beethoven was in the vicinity, yet responded freely to her at other times.

As is typical of young gorilla females, Pantsy became rather coquettish when flaunting her newly acquired sexual prowess. In Group 5, as in most groups, the presence of an estrus female, either an adolescent or a reproductively capable adult, prompts a great deal of vicarious sexual activity among other group members such as mountings between individuals of the same sex or between animals of different age groups. Unisexual mountings occur twice as often between males as among females, while age-discrepant mountings occur most often when adult males mount immatures. The only two mounting combinations not observed were immatures mounting adult males or males mounting their own mothers.

As Pantsy came into sexual maturity, her maternal inclinations, never as marked as those of the younger Quince, were given an outlet when in August 1974 Liza gave birth to Pablo, an elfin-eared infant who was the first born in Group 5 after a two-year period. Pantsy, in a manner typical of young females who have not yet had their own offspring, was particularly fascinated by lively Pablo and watched for any chance to "kidnap" him for mothering practice. Pantsy's methods of transporting the gregarious baby lacked finesse, but Pablo seldom objected to being hauled around backward on Pantsy's back or even upside down in her arms. Liza always watched such activities with calm interest before placidly intervening to retrieve her son.

Liza was a good-natured, responsive mother who seemed to enjoy the antics of the little wind-up toy she had brought into the world. For Pablo, rules were made to be broken, observers were met to be entertained, and familial group members were created for his own enjoyment. Pablo's sense of frolic was infectious as his outgoing personality freely expanded within his first year, attracting many other immatures to him.

Pablo's birth raised Liza's status in the group, as shown by the increasing amount of time she was able to spend in Beethoven's proximity. The new situation was also beneficial for Liza's daughter Quince, who was forty-nine months old when Pablo was born. Quince spent more time than any other individual grooming Beethoven, an activity that further enhanced her social position within the group and reinforced familial bonds with her half brothers and sisters.

Six months after Pablo's birth Beethoven bred with his daughter Pantsy. Several months after conception, the eight-year, nine-month-old female's personality changed considerably. She withdrew from most social interactions with other group members to cling unobtrusively at the edge of the group near her mother, Marchessa. Three months after Pantsy's impregnation Marchessa came into estrus and also conceived an infant sired by the group's leader. However, her infant lived only one day following its birth in December 1975. In spite of extensive searching of the area surrounding the birth site, no trace of the body was ever found.

Pantsy's first offspring, Banjo, Marchessa's grandson, was born in October 1975. To all outward appearances the baby seemed healthy except for whining more frequently than most newborns. Pantsy's maternal inexperience was reflected by the incompetent manner in which she handled and carried her infant. She seemed both dejected and disturbed by the newly acquired responsibility.

Banjo was three months old when Marchessa lost her one-day-old infant from unknown causes. Only then did Marchessa seek proximity with Pantsy, who benefited by the protection her mother offered during antagonistic encounters with Effie's clan. Squabbles had increased in frequency, possibly because Effie was pregnant with her fifth offspring sired by Beethoven. The dominant female became unusually intolerant whenever Marchessa and Pantsy tried to usurp her to gain Beethoven's attentions.

The increase in intragroup friction was heightened by the outcome of a violent physical contact between Group 5 and an unknown fringe group in April 1976. I found the interaction site splayed with blood, silverback hair tufts, pools of diarrhetic dung, and numerous broken saplings. Following a lengthy flee trail I found the group and was horrified to see the head of Beethoven's left humerus, surrounded by exposed ligaments and fascia, protruding through the skin of his elbow. That Icarus, then about fourteen years old, had supported his father during the fierce encounter was apparent by eight deep bite wounds received on his arms and head.

Beethoven, estimated as about forty-seven years old, had been growing more dependent upon Icarus for assistance during interactions with other groups or lone silverbacks. Because Icarus was becoming sexually mature, he sought other social units as a possible means of acquiring new young females for himself. Conversely, Beethoven had long since formed his harem and was not interested in intergroup encounters. The father-and-son team proved an opportune arrangement by offering the aging Beethoven backup support while providing valuable experience for Icarus in dealing with alien silver-backs. Beethoven retained his dominance over Icarus because of their strong kinship ties.

For several weeks following the violent physical interaction, Beethoven and Icarus lay with their heads together during lengthy day-resting periods, exchanging soft belch vocalizations as though in mutual commiseration of their injuries. The son's wounds healed more rapidly than those of his father, and Icarus soon became bored by the long resting sessions that Beethoven required. The young silver-back, accompanied by many Group 5 members, often moved several hundred feet away from the day nests for feeding purposes. Beethoven was left sitting alone with his head cocked to one side, listening to his family's vocalizations rather like an old man trying to hear a weak radio set. Heedful of his duty as the group's leader and arbitrator, Beethoven would arise after ten or fifteen minutes of near-solitude to follow ponderously the group's trail. Certainly, if Icarus had entertained any ideas about forcibly taking over the group, his father's six-month recuperation period would have been the opportune time for him to have done so.

There were moments during Beethoven's recovery period when Icarus became carried away with his new prestige by wildly running through the midst of the group's females. Pantsy was the most frequent target of the young male's provocation whenever Beethoven remained behind on a nesting site. Pantsy was aided by Marchessa in the protection of her vulnerable infant, Banjo, against the running charges and threatening behavior of Icarus, Effie, and her two daughters, Puck and Tuck. Together, Pantsy and Marchessa effectively reduced intragroup tension by maintaining a distance of some one hundred feet from other group members, particularly Icarus. Marchessa was the only individual in Group 5 who shared no blood ties with Icarus, Pantsy's half brother. It was possible that Pantsy might have been a vicarious substitute for Marchessa, whom Icarus could not directly attack because of her long-term affiliation with Beethoven.

By Banjo's sixth month Pantsy had become a capable mother, carrying the infant in a protective ventral position clutched against her chest because of the frequency of Icarus' displays. Most infants are carried in the ventral position until only about four months of age, when their mothers encourage them to ride on their backs. Because of the harassment Pantsy received, it was understandable why Banjo was carried ventrally, often hidden from view. For this reason I was not disturbed one day when unable to see Banjo clearly. Pantsy was obscured in thick vegetation while feeding apart from the group along with Marchessa. However, three days later it became obvious that Banjo was missing. Pantsy had reverted to the devil-may-care wild play behavior of her youth exactly as Bravado had done three years previously following Curry's death.

The Africans and I began an intensive search for the missing infant; it was like looking for the needle in the haystack. Night after night we returned to camp empty-handed even after having covered a half square-mile area where the group had ranged over a seven-day period before and after Banjo was noticed missing. We only found evidence of intragroup quarrels in the form of broken saplings and diarrhetic dung deposits. There were absolutely no trails leading into Group 5's territory to indicate that they had encountered a second social unit.

Determined not to leave the mystery of another infant disappearance unsolved, as in the case of Marchessa's one-day-old infant, I decided to collect all the gorilla dung left in the group's night nests from the preceding week. Banjo could not have disappeared without a trace, and the only clue I had ignored in the futile trail-searching effort was dung. I was chilled at the implications of cannibalism among gorillas, though such behavior has been recorded among free-living chimpanzees. By this time I had worked with Group 5 for nine years so there was no difficulty in identifying the occupant of a night nest simply by examination of the dung deposits, a nest's structure, and its location relative to surrounding nests.

The men and I hauled knapsacks full of dung back to camp after having first bagged, labeled, and dated each individual's nest deposit. We then began the tedious chore of straining every lobe of it at Camp Creek. Hours and days were spent monotonously sifting and searching for clues that might provide an answer to Banjo's disappearance. Only after a week of dung washing did we begin to find minute slivers of bone and teeth that were definitely known to have come from the night nests of Effie and her eight-year-old daughter, Puck. Unfortunately, the quantity of bone and teeth splinters we recovered amounted to only a fraction of an infant gorilla's skeleton. An additional complication was the finding of infant hair only in the dung of Effie and Puck, neither of whom had infants at that time.

In a further effort to solve the mystery of a missing dependent infant who could not possibly survive without its mother, the men and I next collected all the dung deposits remaining on the group's trails over the seven-day period surrounding Banjo's disappearance. More tedious sifting in Camp Creek recovered additional body fragments. Unfortunately, the trail dung could not be as positively identified as the nest dung, but the lobes yielding skeletal fragments closely resembled Effie and Puck's nest dung. When all the dung was washed, we had a total of one hundred and thirty-three fragments of bone and teeth, which constituted only about a small finger's length of an entire infant skeleton. The minute sample left no significant explanation as to where the major portion of the body had disappeared; therefore it could not be definitely concluded that Banjo had been a victim of cannibalism. I still do not discount the possibility. The quantity of bone and teeth remnants recovered from nearly a week's worth of nest and trail dung peaked for a two-day period following Banjo's disappearance. This also I could not explain.

During the years of research with free-living gorillas, dung has been extensively examined at various times. One Karisoke student spent some sixteen months microscopically analyzing several hundred samples of over a thousand dung lobes and found nothing remotely resembling bone or teeth fragments. It seems much too coincidental that Banjo could vanish and bone fragments could appear simultaneously in the dung of two members of her group. Should another infant disappear, dung washing must be undertaken immediately. Perhaps then the question of the existence of cannibalism among gorillas might be answered satisfactorily.

*   *   *

Three days following Banjo's disappearance some of my sorrow over the infant's loss was lessened when Effie, known to be nearing parturition because of her vastly increased girth and protuberant nipples, gave birth on April Fool's Day in 1976. The delightful infant, a female named Poppy, was born forty-seven months after her sister Tuck, making Effie the only female to have four offspring within her group at one time. (Effie's oldest daughter Piper had transferred out of Group 5 three years earlier.)

Unlike Liza's impish male infant Pablo, then twenty months old, Poppy could only be described as beautiful, with large, soft dark-brown eyes framed by long delicate eyelashes. The infant had a lesser degree of the strabismus characteristic of Effie and her offspring.

The individual and uninhibited personalities of Effie's offspring blossomed at an early age. Each possessed a sense of deep curiosity toward natural objects found within the terrain, as well as foreign objects such as camera lenses, thermoses, and other paraphernalia I took into the field. Their interest in such objects contributed to observations of their behavior because the animals were more inclined to remain within view of the observer rather than hide behind dense screens of vegetation. It was not my intention to provide them with playthings, for this would have strongly affected their natural behavior and interactions with one another. However, there were many times when I was outnumbered by the eager youngsters' grasping hands and not quick enough to guard all of my belongings.

Within a small section of Group 5's range grows a hard, grapefruit-sized fruit called *mtanga-tanga* by the local people. It is favored by elephants, who become besotted after extensive *mtanga-tanga* binges, but the gorillas have not been observed eating the fruit. Effie's young, though, did go out of their way to climb high into trees supporting the fruits and knock them to the ground for play purposes. Puck, when only an infant, used the fruit as a display item, gripping the stalk between his teeth and beating the fruit against his chest. This resulted in a resonant, deep chestbeating sound that, try as I might, I could not duplicate. The fruit also served as a football, soccer ball, or baseball for all of Group 5's young, according to which type of game was initiated.

The day-resting periods of Group 5 continued to be prolonged while Beethoven recuperated from his arm injury. Beethoven could not seem to get enough sleep. For several hours each day he soundly dozed off, his opened mouth emitting deep snores, his stumpy legs twitching as if he were dreaming, and his facial muscles contracting whenever he heard alien distant sounds such as men's voices. During his third month of convalescence some of the younger animals, Puck in particular, became restless with the lethargic routine.

Puck, up to a point, was content in the company of Effie and two younger sisters, Tuck and Poppy, but was often among the first to show boredom during the tedious day-resting hours. He indicated his mood by tapping a forefinger up and down on an arm or by yawningly looking around for something to entertain himself. Buzzing flies were one attraction. Intent, he would sit up, grab at the insect, and if quick enough trap it within his palms. Because of being walleyed like Effie and his siblings, his eyes almost crossed as they tried to focus on the fly held several inches from his face. He then usually pinched the victim between his forefinger and thumb before tearing it into small fragments, yet closely examined each piece before discarding it. The longer the dissection process, the more rapt Puck's facial expression, until his lower lip hung down in a manner more typical of a chimpanzee than a gorilla. Once the fly was no more, Puck's lips puckered with discontent as he looked around for other sources of entertainment.

At these times Puck's substitutes were usually the contents of my knapsack — camera, lenses, and binoculars — all of which the inventive Puck used to look through and to reflect his image as well. He looked through the binoculars from the wrong side because that was the only side that could span the distance between his eyes. I felt certain he was actually looking through them rather than simply copying human behavior because of his prolonged reactions to far-distant images of vegetation or to his fingers. He would wiggle the fingers of one hand directly in front of the binoculars before swiftly lowering the glasses to regard his fingertips inquisitively, as if verifying their attachment to his hand. His perplexity about distortion of surrounding objects was as fascinating as it was comical to watch.

Puck invented an Admiral Nelson game with my 300-millimeter lens by pivoting it around in a land-ahoy fashion when scanning distant vegetation or other group members, most of whom were curious at seeing such equipment manipulated by one of their own species. Puck also played a Madame Curie game with the 300mm lens by deliberately pointing it to the ground and concentrating his gaze on foliage seen through the lens.

Some of the equipment was quite expensive, but Puck handled every item gently and zealously guarded all from other group members. Occasionally, the group would move off to feed before Puck had finished investigating an object. I soon learned not to panic when seeing a valuable lens or binoculars carried off into thick vegetation. I always felt foolish, though, when having to crawl around on hands and knees in search of a discarded piece after the group had left.

During one lengthy day-resting period nearing the end of Beethoven's convalescence, I was provided with an excellent opportunity to take close-up pictures of the relaxed animals. I refused to surrender my camera to Puck despite his insistent pulling at the Nikon hanging from my neck. After some ten minutes he sulkily gave up and moved off several feet to build his day nest. With extremely exaggerated motions he began slapping down foliage for the nest as if the task was sheer drudgery. With a deliberated *plop* he settled into the sloppily made structure but fidgeted and scowled for nearly an hour while the rest of the group rested calmly. Hoping to appease the brooding youngster, I broke one of my rules about not offering the gorillas alien objects and handed Puck a *National Geographic* magazine. I was amazed at the dexterity with which he flipped through the pages, showing keen interest in large color photographs of faces. He gave no vocal indication of whether he was either pleased or dissatisfied with what he saw, but at least he was not bored.

After about half an hour he laid the magazine down, and the group moved off to feed. Instantly Puck arose, ran at me, and slapped down on my body with both hands as if he had been scheming this act of retribution during the entire resting period. Beethoven, who was not in sight, pig-grunted at the loud and disturbing noise caused by Puck's whacks against my plastic rain clothing. Upon hearing his father's disciplinary vocalizations, the sullen youngster paused momentarily before again standing bipedally to hit me even more vigorously with both hands. That did it! Beethoven ran to both of us, pig-grunting in annoyance but stopping by the side of my completely prone body. With furrowed brows and compressed lips, Beethoven glared directly at Puck, who had taken refuge on my opposite side. Group 5's leader silently maintained his rigid stance until Puck meekly crept downhill, wearing an offended pout expression.

With quiet restored, Beethoven wove his way through the group's spectators, gathered around to watch the proceedings, and led them off for the afternoon feeding period. I let a few minutes pass before sitting up to see, much to my amazement, Puck actively masturbating. His head was flexed backward, his eyes were closed, and he wore a semismile expression while using his right forefinger to manipulate his genital area. For about two minutes Puck appeared to be obtaining great pleasure from his actions before he stopped, self-groomed, and followed the others down the trail. Thinking Puck gone for good, I began repacking my knapsack and retrieving purloined objects.

Unexpectedly, Puck came running back. He halted at my side, stood bipedally as though wanting to give one last mighty whack, deliberated, and then ran downhill to catch up with the group.

That contact has remained outstanding in my mind. It was the only time I have ever seen a gorilla in the wild actively masturbate. That Puck had apparently enjoyed the consequences of his actions was obvious, though masturbation seemed an unusual means of self-gratification to have been prompted by Beethoven's disciplinary action. The day's contact was also meaningful in that Puck had retained his grudge against me for a two-hour period. I considered this remarkable because of the length of time involved and wondered just how long gorillas, who live in a group structure, maintain their resentment toward one another after disputes or minor differences.

* * *

In contrast to Effie's Puck, Liza's gentle six-year-old daughter, Quince, seemed very distressed by Beethoven's severe arm wound. Quince, usually Beethoven's most persistent groomer, was never observed even attempting to groom her father during the six months of his recuperation. Instead, she spent a great deal of time sitting by his side gazing anxiously into his face as if trying to console Beethoven by her presence.

Low-ranking Liza and her offspring Quince and Pablo now spent much time close to Beethoven, possibly because of Quince's strong attraction to her father as well as his tolerance of his youngest son, Pablo, who seemed to consider the good-natured group leader a convenient, sturdy, silvered leaning post. Once Beethoven recovered, the worried look disappeared from Quince's face, for she was again able to resume diligent grooming of her father's vast bulk. Often she was observed just sitting near him adoringly peering into his face rather like a puppy awaiting a pat. Whenever he returned her riveted gaze, her entire body perceptively quivered. On one occasion, following a long day's nesting period, Quince reported to Beethoven's side after playing with and grooming other group members. As the young female settled close to her father to scrutinize his face, Beethoven gave a long series of belch vocalizations to announce his intention of moving off to feed. Similar responses came from the other family members and built up into a chorus of synchronous croons led by Quince and Beethoven. The group sounded more like a pack of beagles than gorillas.

Quince was able to focus a great deal of her affectionate, considerate nature on her younger brother Pablo. Pablo's mischievous personality needed a near-constant vigilance on the part of both Quince and Liza, who shared the responsibility of retrieving the venturesome infant from the laps, heads, or backs of myself and other observers who had joined the research staff.

Following the end of an excellent contact with Group 5 on a sunny day when Pablo was about two and a half years old, the group moved off to feed. Pablo, in his typically obstinate way, decided not to follow them. Instead, he settled himself cat-cozily on my lap and showed no inclination to move even when Liza returned to pig-grunt authoritatively at us both. Hoping I looked as helpless as I felt, I leaned far back to encourage her to take her headstrong son and go. Pig-grunting more harshly, Liza began pulling at one of Pablo's arms. He instantly returned her pig-grunts and grabbed tightly to my jacket with his free hand, making the situation worse. It was then up to me to pig-grunt softly at Pablo while prying his fingers lose from my jacket and pushing him into his mother's arms. As Liza carried him away on her back, Pablo turned around to look at me with an accusing pout until he was out of sight.

Pablo, like most young gorillas, was an incorrigible glove snatcher. One day he grabbed one by my side before I could safeguard it and, as if delighted with his loot, pranced back toward Beethoven waving the glove in the air. Vigorously Pablo threw the glove onto Beethoven's lap where it landed with a sharp *plop*. The old silverback leapt up screaming in terror and scattered everyone around him. Only after the animals saw no signs of threat did they resettle themselves nearby, but they remained looking quizzically at their leader for some time. Abashedly, Beethoven also returned to his day nest, feigning total lack of interest in the discarded glove.

On the days Pablo was at his impish worst, I often felt like an octopus when trying to guard my knapsack contents or field notes. Late one afternoon, following a successful three-hour contact, I laid down my notebook, which had been filled with the day's behavioral observations. Contented, I was just getting ready to put away my camera equipment when, suddenly, Pablo ran forward and gleefully snatched the notebook. I started to crawl after him, but the rascal ran directly to Beethoven's side, where he sat down and systematically tore out page after page of my carefully recorded data. I had to sit by helplessly and watch him chew each page into a pulp, while his mother, Liza, and his father, Beethoven, regarded him somewhat skeptically. In hope that something might be retrieved the following morning, I spent an extra amount of time searching through the dung in Pablo's night nest, but alas to no avail. In the academic world he might rightfully be termed a dirty data stealer.

Pablo, nearly immune to discipline, often lured other Group 5 youngsters into roughhouse play within body contact of Beethoven. Tuck, twenty-seven months older than Pablo, was one of his favorite playmates. Because of being older, it was she who bore the brunt of Beethoven's indignant responses when he was awakened from a sound sleep by the pounding and thudding of little hands and feet against his silvered bulk. At such times Pablo, a picture of innocence, was ignored as Beethoven grabbed at whatever part of Tuck's body that was within reach to clamp his huge teeth gently around her leg or arm, pig-grunting in admonishment.

The scapegoat Tuck reacted much like a human child being treated unfairly. Her face would slowly pucker up as she whimpered pathetically. If the whimpering grew too prolonged or loud, Beethoven would turn his head toward her and snap his mouth open and shut, making sharp teeth-clacking sounds that sent Tuck back to Effie for comfort followed by a session of intense self-grooming. It was as though this were a means by which Tuck was resolving her inner conflict much as human beings scratch their head or skin in disturbing situations.

In mid 1976, when Beethoven had fully recovered from his serious injury, he behaved as mischievously as a puppy off its leash. He developed a new approach toward me by feigning disregard until very near and then chestbeating, whacking foliage down on top of me, thumping the ground by my side, and even rolling on the ground nearby and kicking his heels up in the air, all the while wearing a roguish facial expression. Albeit "undignified," such behavior was a very welcome switch from his months of listlessness. There had been times when I had questioned Beethoven's chances for survival, especially when the wound was draining copious amounts of foul-smelling exudate that attracted scores of insects to his body. Because of the wound's location at the elbow, Beethoven had been unable to cleanse it orally, which was undoubtedly one reason it had taken so long to heal. Certainly the old male's return to good health was another example of the amazing recuperative powers of gorillas.

The group's patriarch was again surrounded by his females and young, leaving Icarus, then around fourteen and a half years old, to serve as the watchdog for the group. One day when approaching Group 5 for a contact, I looked up into a tall, massive *Hagenia* and saw Icarus calmly watching me as I crawled forward on hands and knees after having left the tracker behind about seventy-five feet from Group 5. Icarus resembled a giant misplaced panda, his huge silvered body lying prone on a branch and his two minuscule legs swinging freely back and forth as though he were on a swing. As I reached the base of his tree, he slid down like a plump fireman, stared serenely into my face, and proceeded to build a complex bathtub type of day nest between myself and the group. As he trustingly dozed off, I had little choice but to settle where I was, however delighted in his confidence. While the young silverback quietly slept, I could not help noticing the vast network of scars and healing gashes that zigzagged his massive head — visual evidence of past encounters with adult males of other groups.

As I sat contemplating the old wounds, the tracker I had left behind unintentionally stepped on a branch. The result was only a faint crackling sound, but Icarus instantly awoke from his sleep, stared toward the source of the noise, rose, and, much like a cat stalking a mouse, headed toward the sound, leaving me impressed with his vigilance even when apparently asleep. I was grateful that Nemeye was equally as alert as Icarus. The young tracker, realizing that he was being approached, took off at a rapid crawl before Icarus could come upon him.

By this time Banjo had been dead for about four months. I felt that Icarus had attained sexual maturity because of the intensity of his copulations with Pantsy. Beethoven no longer showed any sexual interest in her, nor did he attempt to interfere with the mountings between his two offspring. Marchessa had returned to estrus and Beethoven was far more interested in her periodic receptivity.

Though I had not realized it at the time, both mother and daughter had become impregnated within several days of one another — Pantsy by her half brother Icarus and Marchessa by Pantsy's father Beethoven. After conception, Marchessa remained on the fringe of Group 5, whereas Pantsy spent more time adjacent to Effie's clan. Pantsy showed discretion whenever approaching Icarus and her newly acquired stepmother by soft belch vocalizing and avoiding direct body contact with Effie and her daughters Puck, Tuck, and Poppy.

On one occasion, during a long day-resting period, I observed the entirety of one of Pantsy's prolonged and deliberated moves toward Icarus, who was resting in the immediate vicinity of his mother and sisters. Pantsy lay down beside the young male and, with the dorsal surface of her right hand, stroked his back and head. Icarus responded by reaching out and patting her arm hair gently, while wearing an interested facial expression. Eventually, he sat up and stared into Pantsy's eyes, his forehead furrowed questioningly and his lips contorted into a nondescript smile. Quivering, he propelled her rump toward himself and covered her body in a close embrace. The two exchanged prolonged sighs and soft, humming belch vocalizations, apparently unconscious of the presence of myself, Effie, and her curious daughters.

As Marchessa and Pantsy neared their respective parturitions, Group 5 slackened their travel pace to adjust to the increased feeding requirements of the two females. About three months before giving birth, Marchessa was always found bringing up the rear of the group and thus was the first individual I encountered when contacting the family. If out of sight from the others, the old female felt threatened by my presence and screamed in alarm or stood bipedally to chestbeat. Marchessa's chestbeating was no easy balancing act. With each thud of her cupped hands against her upper abdominal region, I expected a series of quintuplets to pop out from her immense girth. Whenever I was fortunate enough to find her resting near the group, I could not help but think that if I tied a string to one of her legs and gave a mighty puff, she would rise and sail through the air like a huge black helium balloon!

One horrid rainy day in December 1976 I found Pantsy, rather than Marchessa, bringing up the rear of the group. I was appalled to see that the entire right side of Pantsy's face had been badly injured. Her right eye was swollen shut and draining a thick mucus, as were both nostrils. Thorough backtracking revealed no evidence of another group or lone silverback having encountered Group 5. It could only be concluded that Pantsy had once again been a victim of intragroup aggression, most likely from Effie and her two older daughters. Pantsy spent the following two and a half months alone at the edge of the group, hunched over with her arms hugging her body and her chin buried against her chest.

By the end of February 1977 my two major concerns were Pantsy's health and Marchessa's bloated body. I was as certain that Pantsy was going to die as I was that Marchessa was going to have twins.

During the evening of February 27, 1977, Marchessa gave birth to a fragile male named Shinda, an African word meaning "overcome." Three nights later Pantsy gave birth to a large tawny-haired female I named Muraha after a new volcano that had recently erupted in Zaire. For the second time, Marchessa had become a grandmother.

The contrast between the two infants, Uncle Shinda and his niece Muraha, could hardly have been more striking. When first seen clinging like a tadpole to Marchessa's ventral surface, Shinda's visible skin color was pinkish, and he had only a sparse distribution of short, shiny black body hair. The one physical feature he shared in common with Muraha was the typical embryoniclike protuberant pig snout. By the end of the infants' first month, Muraha was observed alertly staring around her and seemingly able to focus her gaze on flowers or even moving objects, abilities not shared by her infant uncle.

The near-simultaneous births made Marchessa and Pantsy a strong defensive team, particularly when backed by the respective sires of their offspring, Beethoven and Icarus. For the first time I felt secure about their safety should squabbles occur within the group. I was also pleased to note Pantsy's rapid return to radiant health following the birth of Muraha.

The two births displaced Liza from Beethoven's proximity. Since Pablo was nearing three years of age, it was possible that Liza was returning to cyclicity; however, she was ignored by Beethoven, and her obstreperous son was not well tolerated by the other mothers. To Pablo, Muraha and Shinda represented two new play toys he was determined to investigate. His unruly approaches to them created numerous pig-grunting outbreaks which compelled either Icarus or Beethoven, or both, to intervene among the females with disciplinary pig-grunts.

No longer did any of the animals try to "protect" Pablo from his play solicitations with myself or other observers. On the contrary, I think the entire group would have been delighted if I had just stuck him into my knapsack and brought him back to camp until he outgrew his roguishness.

Pablo's puppylike squiggliness, coupled with his strong grip and thrashing arms and legs, made him an uncontrollable lap dog. His teeth, though small, were razor sharp and easily penetrated my jeans and long underwear. I always tried to steel myself against his playful bites because they caused me to flinch, a reflex that alarmed any animals sitting near me. Finally I found that if I stealthily pinched

Pablo he instantly released his tooth hold on me. The offended young-ster would then back off rubbing his pinched spot and stare at me accusingly. For tranquillity's sake I felt a subtle pinch was preferable to the more obvious flinch.

Even before three years of age, Pablo was exceedingly interested in the sexual activities of other animals within his group. He often tried to examine the penises of older males but was usually shoved away by Beethoven, Icarus, or Ziz. Many play sessions with Poppy, twenty months younger, were terminated by his mounting the female infant ventrodorsally. Such sex play could result in an erection for Pablo, who, with a puzzled smile, lay back and twiddled his penis. If no other youngsters were around for Poppy to play with, she sat by to watch Pablo with interest or, occasionally, even sucked his penis.

Like Liza, Effie also spent less time next to Beethoven following the parturitions of Marchessa and Pantsy. By spending more time near the fringe of the group, Effie wisely avoided subjecting either herself or her infant Poppy to squabbles. One day Effie was observed contentedly feeding about twenty feet behind the group, while Poppy, some six feet behind her mother, was solo playing and swing-ing in a *Senecio* tree. An observer was in sight of both animals and was watching Effie feed when suddenly Effie twirled around and stared at Poppy. Following Effie's alarmed gaze, the student saw that Poppy had fallen and was hanging by her neck in a narrow fork of the tree. The infant could only feebly kick her legs and flail her arms as the stranglehold began cutting off her oxygen. In-stantly Effie ran to her baby. With considerable effort she tugged at Poppy, trying to release her from a potentially fatal position. Effie was wearing a horrified expression of fear similar to that of a human parent whose child is in mortal danger. In the midst of her struggles to free her baby, Effie glanced accusingly at the observer, although it is possible that she only wanted help from any source at a time when every second counted. The observer wisely remained immobile, a difficult choice to make under the circumstances, yet the right deci-sion since any rash movement might well have triggered a hysterical group outbreak, placing Poppy in even greater jeopardy. At last Effie succeeded in releasing her infant from the tree's stranglehold. Immediately upon regaining her breath, Poppy began to whimper, then attached herself to Effie's nipple for four minutes before her mother carried her off, in a protective ventral position, toward the group, which were unaware of the drama that had unfolded behind them.

This unique observation provided only one example of the strong maternal inclinations of female gorillas. An amazing aspect of the incident was that Effie, whose back was turned toward her infant, was aware of Poppy's silent plight even before the human onlooker, facing both animals, realized anything was amiss.

Following the births of Shinda and Muraha, Group 5 traveled far to the southwest of their normally used range area. They eventually encountered a small fringe group containing two silverbacks and a blackback. The resulting physical interaction left Beethoven, Icarus, Puck, and Effie wounded. Effie's injuries were far more serious than those inflicted on the other animals and, with the exception of a deep arm gash, were located on the back of her neck, head, and shoulders; thus she was unable to groom them effectively.

Within a week her bite wounds were draining badly and, had it not been for Effie's five-year-old daughter Tuck, the injuries would have taken far longer to heal than they did. Tuck appointed herself Effie's attentive and almost overzealous groomer, pushing away other animals who interfered with her ministrations. Tuck even pushed away the hands of Effie, who, possibly because of discomfort, wanted only to be left alone. Tuck licked and probed stubbornly at the bite injuries until all had healed six weeks after their infliction.

During that brief period of time, Tuck developed an unusual head-twirling greeting that she used only when approaching Effie for a grooming session. I could never understand what Tuck was trying to communicate. The young female would go to her mother and pivot her head around so rapidly that my own eyes could barely follow the motions. After nearly a minute of head twirling, Tuck would begin grooming the wounds intently, leaving me dizzy after having tried to follow each head movement and leaving Effie looking as puzzled as I at her daughter's strange behavior — behavior never again observed once Effie's wounds had healed.

Group 5, perhaps because they sought to avoid another physical interaction with the small fringe group, traveled even farther southwest of their own range, entering an area I had never known them to explore previously. Seemingly disoriented, Beethoven led the group up into a 13,000-foot subalpine meadow region adjacent to

Mt. Karisimbi. There the animals clung to small forested strips of stunted *Hypericum* surrounded by vast open grass and swamp areas. This was a favored poacher region, because game could be killed by spears easily in the open meadow expanses or trapped by snares in the narrow forested strips separating the meadows.

The plateau was several hours' climbing distance from camp, thus the poachers were able to reset traps faster than we could cut them down. To me it seemed only a matter of days before one of Group 5's members would get caught in a wire snare. Reluctant, I finally had to make the decision to ask the trackers and an agreeable student to herd Group 5 back toward their home range adjacent to Mt. Visoke.

The herding process went well. Beethoven, prompted by unseen "poachers," led his family toward Visoke, and Icarus — now quite experienced in interactions with both humans and other gorillas — defensively brought up the rear of the group. Indeed, by 1977, only ten years after I had first met him, Icarus had become not only a father but, more important, second in command of his father's group. His self-confidence sometimes frightened me. Icarus, unlike Brahms, who had left Group 5 nine years before and had encountered poachers as a lone silverback, seemed to consider himself invulnerable. The student who supervised the herding of Group 5 from the uppermost meadows back to their home range said that Icarus seemed unusually anxious to meet his unseen followers and that, at one time, "the bushes exploded with Icarus" when the young silverback had been hiding in wait for the pursuers.

After the herding session Group 5 rested on Visoke's slopes for several days and accepted observers readily. I was relieved to have them "home," but another danger loomed. The bamboo sprouting season was imminent. The peak of this season occurs in June and December when bamboo constitutes about 90 percent of Group 5's diet. Only a fifth of the group's range contains bamboo, which lies in a region directly abutting the cultivated farmlands excised from the original park boundaries.

No buffer zone exists between the park and the tilled fields of pyrethrum, so it is all too easy for some of the villagers to set traps for antelope within the park, only a few minutes' walk from their huts. The Karisoke staff and I always tried to patrol the narrow bamboo strip along the eastern border of the Parc des Volcans to destroy any traps set since the gorillas had last been in the area.

This time the group descended to the bamboo zone before we could check it out.

My fears for the group's safety were allayed somewhat by the knowledge that Beethoven, possibly because of numerous past experiences with snares, had once managed to release four-year-old Puck from a wire noose adjacent to the bamboo zone. I felt that Beethoven had "trap sense," a confidence confirmed on other occasions when it was observed that he had led his group around traps set in the middle of game trails heavily used by duiker, bushbuck, and gorilla.

Just as the tracker Rwelekana and I were approaching Group 5, we heard a loud outbreak of villagers' voices coming from a spot on the boundary known as "Jambo Bluff." Fearing some harm had come to the animals, we rushed to catch up with them. To my relief, the group were all perched on the bluff peering down curiously at the farmers who had been hoeing in the pyrethrum fields below. Gorillas seem not afraid of human beings near the park boundary because it is an area where people are now usually found; in the heart of the forest human sounds terrify animals. Likewise, the farmers maintain an attitude of respect for the gorillas since they know there is nothing in the fields of pyrethrum to attract the wild animals.

The seasonal return of Group 5 to the edge of the park boundary always interests the villagers. They gather together yelling *"Ngagi! Ngagi!* — "Gorilla! Gorilla!" On this day Group 5, after a brief stare, left Jambo Bluff to carry on with their feeding and the Africans returned to their hoeing. However, when I climbed onto the bluff to follow the group, a new outbreak of screams and shouts came from the people below. *"Nyiramachabelli! Nyiramachabelli!"* they cried, meaning "The old lady who lives in the forest without a man." Although my new name was pleasantly lyrical, I would have to admit that I did not like its implications.

For several weeks the trackers and I followed the routine of checking for traps ahead of the group's daily route until we felt that the 2½ mile bamboo area was safe and that observations, rather than trap cutting, could resume on a full-time basis. It was about this time, late one July afternoon in 1977, that Group 5 decided to day-rest in a small teacuplike clearing surrounded by dense bamboo thickets near the park boundary. The gorillas seemed uncomfortable with their enforced togetherness. After a restless half-hour, six-and-a-half-year-old Ziz rose and wandered into the adjacent bamboo growth. Immediately he was followed by his mother, Marchessa, his sister,

Pantsy, and the females' respective infants, now riding dorsally on their mothers. As though anxious to be on the move again, the rest of the group lost no time in following the route of Marchessa's clan. Beethoven brought up the rear.

Suddenly, after all the animals had disappeared into the darkness of the bamboo canopy, a violent screaming outbreak began. The screams rose to earsplitting intensity before Beethoven gave a prolonged, harsh series of pig-grunts followed by gruff pig-grunts from others. The outburst lasted nearly three minutes and conveyed a state of hysterical terror quite similar to that caused by an unexpected contact with poachers. Since the gorillas were making no attempt to flee, poachers definitely were not the cause of their fear. A trap had to be the only explanation.

I crawled into the dense bamboo but could only make out a circular mass of black forms ahead. Beethoven, giving gruff pig-grunts, was forcing his way through the gorillas gathered around a tautly arched bamboo pole. Within a second his silvered back was obscured by huddled black bodies. Because I was kneeling on the only trail that might offer a means of retreat for the group, I slowly backed out of the bamboo tunnel into the sunlight. A few minutes passed before Beethoven, followed by Ziz and then the rest of the group, emerged into the open. Ignoring me totally, the animals began comforting one another with belch vocalizations as they headed directly back toward Visoke's slopes, their usual refuge after times of stress. From a discreet distance I tried to follow to see who might be wearing the dreaded wire noose, but I was unable to keep up with the group.

The following morning Rwelekana and I returned to the outbreak site. We found a devastated dirt circle, about seventeen feet in diameter, surrounded by broken saplings and other vegetation. The churned-up dirt was covered with diarrhetic dung, large tufts of gorilla hair, and the barefoot impressions of a single man. Someone had beaten us there. The details of the previous day's incident were clear from the evidence. A small pit, about six inches square, had hastily been filled in with loose dirt by the trap-setter apparently only that morning. The bamboo pole, the noose, and the pegs placed to maintain the noose's position until a passing animal sprang the trap had also been removed. Further, the man had made a feeble attempt to brush away the evidence of gorilla tracks surrounding the trap site.

Crawling on our hands and knees, Rwelekana and I were able

to detect a faint human trail that led to another wire trap, as yet unsprung, just below Group 5's nesting site of the preceding day. The curved arch of the bamboo and the wire it supported would have been visible to the group from their nesting site. This explained their nervousness, the crowded nest area, and the sudden reversal of route. Rwelekana and I continued crawling along the human trail adjacent to the park boundary. We found eight more unsprung wire snares and took pleasure in confiscating the potentially lethal nooses and hacking up the bamboo poles that supported them. Satisfied that this particular area was clear, we returned to the original trap site to follow the fresh human footprints made that morning. The prints led down into the pyrethrum fields adjacent to the park boundary, where we lost them in a network of paths heavily used by the villagers. It was obviously going to be impossible to track the poacher to his doorstep, so we began the long uphill climb to locate Group 5.

The trap's victim had been Ziz. The young male bore the evidence: a narrow, deep flesh wound encompassing his right wrist like a bracelet, fresh pink scraps on the palmar pads of his right hand, and long gash lines running from his biceps down to his wrist. Recollecting the events of the day before, especially Beethoven's actions, I was convinced that the old silverback had removed the wire by getting his teeth between it and Ziz's arm to pull downward until the tight noose worked free over his son's hand.

Ziz never could have released himself; his trapped hand was tautly suspended high beyond the reach of his own teeth. Beethoven's big pudgy fingers would have had difficulty working their way between the tightfitting wire and Ziz's flesh in the same way that I could never work a wire noose off my wrist when wearing gloves. Also, adult gorillas have a certain aversion to touching alien objects with their hands. Beethoven probably stabilized Ziz's arm with one hand, slid his teeth down the youngster's arm (causing Ziz's long gash wound), and then secured his teeth under the wire gripping Ziz's wrist. Finally he must have tugged the wire off Ziz's hand with his teeth; this would have resulted in the scraping marks over Ziz's palms.

Within a week Ziz no longer favored his right arm when walking or playing. The remainder of the summer spent in the bamboo zone passed without further incident. Apparently the trap-setters realized that Group 5 was being contacted by Karisoke observers on a daily basis. Because of our antipoacher patrols, the encroachers would lose more than they might gain by setting new traps.

*   *   *

Muraha and Shinda were now nearly six months of age. The contrast noted between the two infants at birth had significantly increased. Shinda remained a wizened, squeaking tadpolelike form seeking only to cling to his mother's undersurface, though Marchessa had been trying to get him to ride on her back since he was three months old. His niece Muraha, on the other hand, was not encouraged to ride dorsally by Pantsy until the end of her fourth month.

From the day of her birth Muraha was a fluffy, chuckling, playful ball of vitality. Pantsy, unlike Marchessa, gave every evidence of totally enjoying her offspring. During day-resting periods Pantsy, with a broad smile, often dangled the baby over her head until both mother and daughter were chuckling, a sound much like human giggling. Pantsy's dangling of Muraha had a functional purpose. The action stimulated the release of the infant's liquid yellow-colored feces, a normal excretion for very young infants whose primary food intake consists of their mothers' milk. Marchessa was never observed dangling Shinda in the same manner, which might have explained why her abdominal hair was thoroughly stained reddish-yellow by the time Shinda was about four months of age. Pantsy also spent a great deal of time allowing Muraha to solo-play on her vast body, using her mother's abdomen as a slide and her extremities as "wrestle-partners." The baby's delight in such play activities was often indicated by her "Bugs Bunny" type of toothy grins.

Muraha at the age of three months was teething and constantly trying to gnaw on foliage stalks. Her uncoordinated efforts to obtain the stems resembled a drunk person seeing in multiples while reaching for a martini glass that refuses to remain in one place. At the same age her uncle Shinda seemed content to gaze blankly at the vegetation around him and, with considerable effort, pick discarded foliage bits from his mother's lap.

By her fourth month Muraha was able to totter away from her mother for distances up to ten feet, though most gorilla infants remain within arms' reach — some six feet — of their mothers until about six months old. Muraha's forearms supported her fairly well but her hind legs often collapsed like two overcooked noodles spread out behind her. The locomotion of all young gorillas develops similarly.

From the day of their birth, gorilla infants must rely upon their hands and arms to grasp their mothers' ventral surface, particularly in stressful situations such as rapid travel, when their mothers cannot support them. Their shorter legs and stumpy toes are used only to maintain a less effectual grip around the broader girth near their mothers' stomach region.

When four months old Muraha gave me an unforgettable memory. Group 5 were day-resting peacefully on Visoke's lower slopes when I contacted them. I sat about eight feet away from Pantsy, who was lying within arms' reach of Icarus. Muraha, snuggled in between her mother and father, looked at me with interest if not with some degree of calculation before getting up to wobble toward me. Both Pantsy and Icarus sat up with quizzical facial expressions when the baby left them. Muraha crawled closer and closer, her hind legs splayed out behind her as she clumsily traversed the tangled ground foliage. The baby's lopsided smile broadened as she neared my legs, which might have resembled a bluejeaned mountainous challenge to her diminutive height of twelve inches.

Once next to me Muraha delicately touched my jeans with the extended fingers of her right hand, then put her fingers to her nose to sniff. Icarus and Pantsy seemed curious about their offspring's exploratory efforts, and I found myself holding my breath, not daring to move. Just as Muraha was ready to begin her ascent onto my lap, Pantsy nonchalantly stood up, yawned, and looked at me with an almost apologetic facial expression. She strolled casually over to my side, feigned an interest in Muraha's rear by smelling and licking it, and gently picked up the baby to carry back to her nest site. I thought Pantsy's behavior most diplomatic.

Pantsy and Icarus resettled themselves, but the "wanderlust" glint returned to Muraha's eyes. She left Pantsy's side once more, came wobbling over, and crawled onto my legs. Pantsy, again seeming rather abashed, came forward to retrieve the baby, averting her gaze from mine. After another pretext of examining the infant's rump, she tucked Muraha under one arm and slowly went out of sight into surrounding dense vegetation. My thrill of that remarkable display of trust has never diminished.

Although willing to allow me a few brief seconds with Muraha, Pantsy remained wary when the younger animals of Group 5 tried to get near her infant. When only fourteen months old, little Poppy sought to "mother" the two four-month-old babies just as Quince and Pablo had done with her a year previously. Though only an infant herself, Poppy was extremely capable of shamming an "oh-hum" approach by mincing forward a few steps, then sitting to self-groom or yawn studiously before advancing again, trying to get close enough to tackle one of the infants if they were out of their mothers' arms. Shinda, still barnacled to Marchessa's bulk, seemed less attractive to other Group 5 youngsters. This might have been attributed to gender differences between the infants, or to Marchessa's unpredictability.

Quince, whose maternal inclinations were the strongest of any young gorilla female I have yet observed, appeared deeply dejected when Marchessa or Pantsy denied her their babies to cuddle, carry, or groom. At seven years Quince had only a year to go before becoming an adult herself. She and her three-year-old brother Pablo shared chimpanzeelike sulk faces. When deprived of access to the two infants, both stalked off with their lower lips hanging down, an unusual expression among gorillas.

For Pablo the early months of 1977 were traumatic times. Not only was he being weaned, but he was also on the receiving end of a growing tide of discipline from other group members. It was as though he wondered what had happened to the world he had ruled so well up to his third year. Had he been a human child, I imagine he might have wrapped up his favorite belongings — field notes and film — in a bag and set out to find a new family, one that would appreciate him.

Pablo's best days were the two to three each month when his mother came into estrus. At these times Liza withdrew from her sideline position at the group's edge and coquettishly sought playful interactions with other group members. When she returned to regular cycles during Pablo's third year, her breasts were extremely uneven. Her left, the one favored by Pablo, became vastly distended and lumpy but her right one hung empty and flat. Pablo was able to suckle more freely on Liza's estrus days because she was distracted by the increased amount of attention from other members of the group. Beethoven, however, still ignored her invitations to copulate.

Although Tuck frequently went out of her way to placate Pablo by play or grooming when he fretted, it was becoming clear that Effie was nearing the end of her patience with him because of his persistence in trying to roughhouse with Poppy. One day Pablo impudently bit Effie lightly on her arm when she pig-grunted at him for pulling at one of Poppy's legs. Had he nipped any other adult group member, he would have been severely rebuked, but Effie simply resumed cuddling Poppy. Boldly Pablo approached Poppy again, and sat at a safe distance to direct a moue at Effie. Minutes later Poppy, of her own accord, came over to him and began whining when the play became rough. Immediately Effie pig-grunted and Pablo briefly retreated. After being comforted by her mother, Poppy returned and the entire process was repeated. This time, however, Effie grabbed Poppy away, mock-bit Pablo, and returned to her nest with Poppy securely tucked under her arm. Seemingly frustrated at life's injustice, Pablo sat down and beat his head for nearly a minute, his face contorted into a grimace and his eyes squinted shut. Finally he stood, glared at Effie and myself, and went off whimpering to find his mother, Liza, at the edge of the group.

Poppy was Group 5's "little darling." There was something winsome and appealing about her. She could do no wrong. Unlike Pablo, she was not interested in alien objects such as cameras or film but was content with objects within her environment. Discarded bird nests held a special fascination for her. They could be beaten against her body or on the ground until nothing but shreds remained. She also enjoyed laboriously plucking nests apart strand by strand for the same result.

Poppy occasionally liked to perch daintily on the laps of observers, as if wanting to be cuddled. Usually whenever she "honored" myself or students with her attention, we received pig-grunts or threatening stares from Beethoven, Effie, and other group members. More often than not, Beethoven would leave his nesting site to come to Poppy and gently butt her away from us with his massive head. The younger group members, Puck, Tuck, Quince, and Pablo, were equally concerned when Poppy had settled with observers and would often retrieve her to carry back to their midst. Such group supervision of Poppy was in marked contrast to the animals' lack of interest whenever the adventurous Pablo was interacting with humans.

Effie's third daughter, Tuck, had had to forego her mother's attention after Poppy's birth, but I seldom detected any traces of jealousy in the young female. Only on a few occasions when Puck was engrossed in grooming Effie, who in turn was intently grooming Poppy, was Tuck seen wearing a morose facial expression. This I called her "Woe is me, the middle sibling" look. For the most part Tuck's affectionate good nature overcame tendencies to brood. When she wanted to be in physical contact with her mother, she simply squiggled up against Effie's body, even when her mother's arms were monopolized by Poppy. If Poppy was off playing with other animals, Tuck often took advantage of the situation by making a beeline for Effie's arms, wearing an intensely puckered face as though thinking, "She's all mine at last."

Effie's dependable temperament contributed not only to the secure development of her offspring but also the maintenance of her top rank among Beethoven's harem of three breeding females. On one rainy day-resting period both Effie and Pantsy built large bathtub nests out of bushy *Hypericum* branches and settled down as comfortably as possible during the downpour. Beethoven had made himself only a slipshod nest, into which he settled resignedly as the rain increased. About a half-hour passed before he began appraising the more comfortable positions of Effie and Pantsy. Abruptly he stood up, strutted to Effie's side, and stared down accusingly at her. Effie shifted positions and pretended to ignore him. With somewhat of a miffed expression Beethoven then strutted some twenty feet over to Pantsy and again stood in a stilted, intimidating position nearly on top of the young mother, leaving little doubt as to the nature of his intent.

I fully expected Pantsy to ooze submissively out of the far side of her nest in obedience to Beethoven's postural command. Instead, she made it quite clear that he wasn't going to pull rank on her, looked directly at him, and harshly pig-grunted. With as much dignity as he could muster, Beethoven promptly withdrew and plodded back to Effie's nest to stand by her side looking somewhat chagrined. Gently he placed his hand of her shoulder and nudged her, but Effie only tightened her embrace around Poppy. Tuck, also in the nest, squeezed even closer to her mother.

Beethoven, thoroughly drenched, somehow managed to squeeze his massive bulk into the jampacked nest against Effie's backside. A large portion of the old male's body was left hanging over the rim, and he looked anything but comfortable. It seemed he was willing to accept the compromise Effie's tolerance had allowed.

Since both Effie and Marchessa had dependent infants well under the age of two years, Liza remained the only member of Beethoven's harem of three who came into regular cyclicity, though she continued to be ignored by him after nearly a year of periodic receptivity. Her growing demands for attention caused frequent squabbles as she wildly ran through the group whacking nearby individuals who didn't manage to get out of her way. The only animal Liza was observed to mount regularly was Puck, who remained passively indulgent of the older female's sexual attentions. As Quince neared her eighth year, when she would be classified as an adult, Liza's estrous periods tended to coincide with those of her daughter. Unlike her mother, however, Quince was willingly mounted by both Ziz and Icarus whenever she wasn't involved with her obsessive "mothering" of the infants Shinda and Muraha.

In December 1977, nine months after Pablo had been fully weaned, Liza resumed intent grooming as well as nursing of Pablo and appeared to have a plentiful milk supply. Her behavior was typical of females who were nearing parturition or those who have had nonviable births. Recently Liza had been spending much of her time feeding on the sidelines of the group, the typical behavior of impregnated females. I felt that she probably had had an abortion that the trackers and I had failed to discover in the group's nesting sites. Liza's reentry into the bulk of the group and her solicitous attention toward Pablo lasted for six months. Then the unexpected happened.

Following an unobserved physical interaction with the fringe Group 6, who ranged around Visoke's eastern slopes, Liza was missing. I thought it unbelievable that Liza would transfer to another group and leave eight-year-old Quince and nearly four-year-old Pablo behind in Group 5. Therefore, I was amazed upon finding her comfortably ensconced in Group 6 as though she had always belonged.

Liza was the second adult female known to have emigrated from the group in which she had first been contacted and in which she had conceived offspring with the silverback leader. Both transfer females had in common the fact that they were wasting breeding time within their respective groups. This was apparent because both had returned to regular cycles following the transitions of their offspring from infant to juvenile stages; yet their mounting solicitations to their mates had been ignored for nearly a year. Additionally, both of their juveniles, even at forty-six months of age, had been observed occasionally suckling before being left behind by their mothers in their natal groups. It may be suggested that prolonged lactation is responsible for inhibiting conception since the females gave birth within fourteen months following their transfers to the new groups. Not only were the emigrations reproductively advantageous for the two females but their rankings were also enhanced. Each was among the first-acquired females of her group, and female dominance order depends upon acquisition order.

To my surprise Pablo showed few signs of depression following Liza's departure in July 1978. He spent much more time in Beethoven's proximity during the day and nested with him at night. Pablo also had his sister Quince to groom or cuddle him, and he went to her frequently for quasi-maternal attention.

Quince became an adult female in the same month Liza emigrated. Her gentleness and concerned regard for others, notably her father, brother, half brothers and sisters, continued as it had been when she was only an infant herself. It seemed to me that Quince was made to be a mother and that Icarus was destined to sire her first offspring.

Much to my consternation, Quince became despondent following her mother's transfer out of Group 5. Although Quince was observed incessantly grooming Beethoven and young group members, much of her special vitality and spontaneity slowly and inexplicably faded.

One morning, three months after Liza's departure, several bloody tissue deposits were found in Quince's night nest. Since she was only eight years old, I thought she was much too young to have conceived or to have aborted an infant, though the possibility could not be discounted. As the days passed it became obvious that Quince's physical condition was rapidly deteriorating, but she appeared determined to carry on with her usual grooming and maternal attentions toward the members of her family.

Two weeks following Quince's abnormal nest deposits, her weakened body was forcing her to marshall nearly all of her energy toward keeping up with the group. This she did by shakingly crawling on her knees, wrists, or elbows, obviously in pain.

Beethoven was the only member of Group 5 who showed obvious concern over the young female. He slowed the group's travel and feeding pace to allow her to keep up with them and defended her against increasing abuse from group members. The weaker Quince grew, the more frequently she became the recipient of runs, kicks, and pig-grunts, particularly from Effie's three daughters. Quince then made weak and pathetic attempts to defend herself by pig-grunting, feebly kicking, or biting the family members she had befriended throughout her life.

It is possible that there is a likely explanation for what to a human

appears as unreasonable and heartbreaking abuse. Because of her
weakness, Quince was not able to react as might a healthy animal
to the hostile behavior directed toward her. She was physically incapa-
ble of responding with either submissive or forceful defensive actions
that those around her sought to elicit. The weaker Quince became,
the more persistent the attempts of the others who were unable to
evoke customary conventional reactions from her. In my opinion
the concept of "cruelty" could not be ascribed to the treatment
Quince received because both her "attackers" and she were acting
anomalously during the terminal stages of her illness.

In October 1978, twenty-five days after the blood had first been
found in her night nest, gentle Quince died. I found her still warm,
emaciated body lying under a log in the midst of Group 5's night-
nest site. The other animals were circling about 150 feet beyond.
As quickly and quietly as possible, porters removed Quince from
the cool darkness of her canopied forest death site. The Ruhengeri
hospital later identified the cause of death as malaria. I buried the
body of the young female several hundred feet in front of my Karisoke cabin in the soil of her mountain home.

* * *

The death of Quince left Pablo as the only visible evidence that
Liza had ever existed in Group 5. Pablo, a happy-go-lucky clown,
remained the group's jester even following his sister's demise. He
slept with Beethoven during the night and frolicked away the days
in play, particularly with his current favorite, two-and-a-half-year-old
Poppy. Effie expanded her tolerance of Pablo's presence to the point
where she allowed him to huddle next to her along with Tuck and
Poppy on rainy, cold days. Both she and six-and-a-half-year-old Tuck
groomed him regularly. The four-year-old Pablo seemed perfectly
content without a mother or full siblings.

Perhaps because of his close association with Effie's clan, Pablo
developed the same degree of inquisitive nature as they had. Once
he and Tuck found a baby duiker nesting alone, almost concealed
in thick foliage. Pablo gave full vent to his curiosity by pulling at
the fawn's legs and hair, poking its body or lifting its head up to
his nose to smell while the duiker repeatedly whined. Tuck, equally
as interested in the animal as Pablo, contented herself with staring,
smelling, or stroking the trembling body. Tuck unsuccessfully tried
to prevent Pablo from molesting the baby antelope by pig-grunting

at him and pulling his hands away. After nearly an hour of inquisitive behavior, both Tuck and Pablo lost interest in their find and, having pulled away all of the screening vegetation from the infant, left the baby lying in the open, where I hoped it would be found by its mother before poachers chanced along.

On other occasions young gorillas were observed playfully stalking duiker in a mischievous manner rather than if seriously intent upon capturing them. Nearly anything that moved — from duiker to frogs — seemed to serve as an enticement to a brief chase. Oddly enough, caterpillars and chameleons were not pursued but either were struck at or cautiously pushed away.

Four months before the death of Quince, in June 1978, Puck began spending increasingly longer periods away from his mother Effie and his sisters Tuck and Poppy, for reasons I did not understand. I was more likely to encounter him either feeding or traveling in the rear of Group 5 rather than near others.

On one such occasion I met the young male dawdling about twenty yards behind the group in an area predominated by *Vernonia* saplings and where the animals had not traveled for at least fourteen months. The sparse understory contained little in the way of gorilla food vegetation, thus I was surprised to see Puck suddenly stop to stare with concentration toward the top of a dying *Hagenia* tree. The first thirty feet of the tree trunk were enveloped by dense tangles of vine and shrub growth. The midsection of the tree showed some semblance of life, but beyond that point it was topped by dead barren branches.

For some three minutes Puck continued staring up in a calculating manner, then purposefully tackled the thick growth of vines at the tree base. About fifteen feet above the ground Puck paused to plot his upward route before reaching the uppermost dead branches. There he dislodged a large bark slab with both hands and revealed a vast honeycomb alive with bees. The moment the bark was dislodged, Puck descended like a four-legged fireman. Once on the ground, he literally galloped off after his group. I ran equally fast in the opposite direction as the *Hagenia* tree instantly oozed wild, angry, buzzing bees that darkened the sky overhead.

After making a large detour I caught up with Group 5, only to find Puck calmly feeding in their midst as though nothing had happened. The day's observation was of particular interest because of the memory retention Puck had shown of a beehive situated out of view from other group members, who, like himself, had not been in the area for about fourteen months. Puck was a never-ending source of amazement to me.

On November 14, 1978, Puck gave birth! Upon hearing the news from an incredulous student who had made what he thought would be a routine contact with Group 5, I exclaimed, "It can't be!" Thus Puck's first infant gained the name Cantsbee.

The young "male" Puck became an exemplary mother and made Effie the second grandmother in Group 5. Puck endowed her infant with the same type of security as her mother had given her. By the end of 1978, Effie's clan consistently were found in protective proximity of one another: Effie, who had been impregnated by Beethoven shortly after Cantsbee was born, eleven-year-old Puck with her male infant Cantsbee, six-and-a-half-year-old Tuck, and thirty-two-month-old Poppy.

After a three-month period of calm within Group 5, Effie aborted a fetus estimated at between two and three months post-conception. The miscarriage followed harassment from a French film team that had pursued Group 5 relentlessly for a two-week period. Until the abortion, Effie had averaged forty-three months between births. Following the unnatural interruption to her breeding regularity, Effie did not give birth again until June 1980. The newest baby, Effie's sixth, was named Maggie and, like Effie's earlier offspring, had been sired by Beethoven.

At the time Maggie was added to Effie's brood, Marchessa's clan also appeared to be doing well under the stable leadership of old Beethoven and his son Icarus. By August 1980 Marchessa and Beethoven's fourteen-and-a-half-year-old daughter Pantsy was successfully raising their forty-one-month-old granddaughter Muraha and was due to give birth in December 1980 to another grandchild, who would be named Jozi, both grandchildren having been sired by Icarus. Marchessa's other two offspring, nine-and-a-half-year-old Ziz and forty-two-month-old Shinda, both sired by Beethoven, completed Marchessa's matrilineal line.

However, even after thirteen years Marchessa's clan remained subordinate to Effie's in spite of the fact that Pantsy had merged bloodlines with Effie's son Icarus. For the first time, the subordinate matriline shared the gene pool with the dominant matriline because of Icarus' matings with Pantsy. Marchessa remained the only member of Group 5 sharing no blood ties with Icarus.

On August 5, 1980, a student contacted Group 5 and found most of the animals feeding on Visoke's slopes. After about half an hour, hootseries and chestbeats were heard from Icarus in the saddle area some one hundred feet below. Group 5 headed toward the source of the noise, followed by the student. Icarus was found running, chestbeating, and whacking at the foliage that surrounded the still form of Marchessa, who lay under a *Vernonia* tree. The old female was either dead or in a comatose state and, mercifully, unaware of what was occurring.

The group members gathered around to watch the actions of Icarus. Each animal, with the exception of Effie, briefly examined Marchessa's body. After nearly two hours of displaying, Icarus dragged the limp form from under the *Vernonia* tree and began pounding on it with both hands. This abuse continued for three more hours and was only occasionally interfered with by Beethoven whenever Marchessa's body was dragged. Marchessa, much like young Quince, twenty-two months before, was far beyond responding to the abuse her form was receiving. Icarus' "attacks" became more frenzied. In addition to beating on her, he began jumping up in the air to land with all of his weight, either feet first or rump first, on Marchessa's still body.

The following morning the members of Group 5 were still clustered around Marchessa. Icarus had apparently dragged her several more yards during the night. Examination of the night nests showed that little Shinda had nested with his father, Beethoven. Icarus was found continuing his bodily abuse of Marchessa. Only when he took periodic rests was Shinda pathetically able to try to suckle or crawl under his mother's cold, unresponsive arm. Other young animals of Group 5 carefully investigated Marchessa's corpse by sticking their fingers or tongues into her mouth and anus. Effie's fifty-two-month-old daughter Poppy mounted and thrust against Marchessa and also beat and pounced upon the defenseless corpse. Muraha groomed her grandmother's body whenever Icarus was resting from his almost ritualistic attacks. All had in common the possibly instinctive compulsion to draw forth a response from the deceased form, so long a vital member of Group 5.

Marchessa's autopsy, done in the Ruhengeri hospital, showed that she had numerous hydratic cysts in her spleen which might possibly have proved fatal. More important, the autopsy revealed that she was neither in estrus nor pregnant.

Above: A silverback yawns contentedly on his day nest made from *Lobelia* leaves. Resting periods may occupy about 40 percent of a gorilla's day and are prolonged in rare sunny periods or during heavy rain and hail storms, when animals seek the protection of trees and thick vegetation. (*Dian Fossey*)

Overleaf: During day-resting periods, group members seek closeness to their silverback leader — in this case Beethoven of Group 5. The clustering of group members results in more frequent social interactions than might occur during travel or feeding periods. The young adult in the foreground basks in the sun. (*Dian Fossey*)

Above: Fruits, especially the wild blackberry (*Rubus runssorensis*), form about 2 percent of the gorillas' diet. A juvenile is using her incisors to strip leaves cautiously from a prickly blackberry stem. (*Dian Fossey*)

Opposite: One of the more common herbaceous plants is celery (*Peucedanum linderi*) shown here. It, like many other favorite gorilla foods, is very succulent; this is why gorillas are seldom observed drinking water in the wild. (*Elisabeth Escher*)

Above: An adult female stands bipedally and uses her incisors to gnaw at tree bark, which when decayed harbors insect larvae favored by gorillas. (*David Watts*)

OPPOSITE

Top: Unhabituated gorillas flee silently or resort to actions meant to intimidate, such as chestbeating, on the discovery of an observer. This excitable Kabara female, two weeks from parturition, precariously trees to chestbeat when discovering my presence. (*Dian Fossey*)

Bottom: Rafiki, the majestic silverback leader of Group 8, about fifty years old when first met in 1967, was well silvered along his neck and shoulders, his back, and down to the sides of both thighs. (*Dian Fossey*)

The young silverback Peanuts, after a ten-year period, still bears the evidence of a severe head-bite wound incurred during a violent interaction with another group. (*Stuart Perlmeter*)

Just as every human being has distinct fingerprints, every gorilla has a distinct noseprint — linear indentations above the nostrils and the shape of the nasal wings. Compare this noseprint sketch of Peanuts with the photograph above. Because close-up photography is ordinarily impossible with unhabituated gorillas, observers first draw rough noseprints, usually with the aid of binoculars, of unidentified individuals in the field. Gradually the drawing is improved as habituation progresses to allow final identification photographs. (*David Minnard*)

A sexually immature female, seen here, usually reacts passively when mounted by a sexually immature male within her group. However, sexually mature females nearly always are the solicitors of copulations from the one or two mature males of their groups. (*Peter G. Veit*)

Above: Syndactyly, webbing of digits, is an anomaly suggestive of inbreeding. This physical characteristic was seen in Marchessa and her offspring in Group 5. (*Peter G. Veit*)

Opposite: Strabismus, or walleyes, is characteristic of another matrilineal line in Group 5, that of Effie and her offspring. Like syndactyly, strabismus does not appear to hamper the activities of the affected animals. (*Peter G. Veit*)

During a newborn's first few weeks of life, the skin of its face, palms, and soles is usually pink or tan, though pigmentation spots may remain on the soles of an infant's feet for two years. (*David Watts*)

Cantsbee, at the age of three months, suckles her mother's breast. Weaning periods become the most traumatic for gorilla infants around the age of two and one half years, when their mothers are usually returning to regular cyclicity. (*David Watts*)

Titus, two weeks old, is being carried in a protected position against his mother's (Flossie's) chest. From birth infants may travel ventrally until about four months old, when they begin riding dorsally. (*Dian Fossey*)

Above: There is some degree of competition between matrilineal clans for the privilege of grooming a group's dominant silverback. Silverbacks themselves, however, seldom groom others, for they have no need to reinforce their social status. (*Dian Fossey*)

Opposite: For nearly a year following her transfer to Group 4, Macho was the lowest-ranking female in the group and visually conveyed her apprehension of the other animals to her son Kweli, aged eight months. (*Dian Fossey*)

Below: In a classic submissive posture, Samson, bending on his forearms, rump in the air, and gaze averted from his rigidly stanced father (Rafiki), conveys apologies for having been rowdy. (*Dian Fossey*)

ABOVE

Top: Augustus has been the only wild gorilla thus far observed hand-clapping, an activity she retained through her fourth year. (*Dian Fossey*)

Bottom: During an infant's first months of life most play activities are in body contact or within arm's reach of the mother. Six-month-old Cleo pulls on her mother's head hair; her facial expression indicates her exuberant mood. (*Dian Fossey*)

Humanizing the sequence of events surrounding Marchessa's death, albeit tempting, is a serious error. Her death probably has closed the reproductive doors for Beethoven, whose remaining female, Effie, had a two-month-old infant, Maggie, at the time of Marchessa's death. Effie will not be ready to conceive for at least another two and a half to three years. It might be considered likely that Marchessa's unrelatedness to Icarus fomented his agitated actions, but this remains pure speculation. Among other gorilla groups, when members have died of natural causes unrelated silverbacks or near-related group members have never been observed attacking or abusing the bodies of the deceased individuals in the way that Icarus abused that of Marchessa.

* * *

Recounting the history of Group 5, I find myself somewhat over-whelmed by the mosaic of memories — humorous, perplexing, sad, tender, loving. Throughout these years, thirty-one individuals have contributed toward Group 5's gene pool and the long-term stability of their family unit. Of the fifteen members first met, only four, Beethoven, Effie, Icarus, and Pantsy, remain.

I am greatly privileged to have been able to observe the growth and development of gorillas like Icarus, Pantsy, and Puck throughout the learning periods of their infancy into adulthood as they gained the experiences, sometimes traumatically, needed to become success-ful parents themselves.

More than any of the five Karisoke study groups, the members of Group 5 have taught me how the strong bonds of kinship contrib-ute toward the cohesiveness of a gorilla family unit over time. The success of Group 5 remains a behavioral example for human society, a legacy bequeathed to us by Beethoven.

5 | Wild Orphans
Bound for Captivity:
Coco and Pucker

OME FOURTEEN MONTHS after the founding of the Karisoke Research Centre, the approach of the holiday season toward the end of 1968 made me quite apprehensive of the impending increase of poacher activities in the Parc des Volcans. During the holidays the demand for poached meat and trophies rose along with the prices paid for illegal black market killings. A successful poacher could gain a substantial income during this period. At that time the only antipoacher patrols conducted in the Parc des Volcans were those organized by myself and performed by a staff of park guards on condition that I pay their wages and supply them with food and uniforms — requests I willingly fulfilled in hope of replacing their lethargy with self-motivation toward their job.

For this reason I was delighted one day shortly before Christmas to see the park's Rwandese Conservator arrive at camp unannounced, accompanied by several guards. Their arrival indicated to me that the park service was ready to accept responsibility for antipoaching efforts. It therefore came as no little shock when the Conservator asked me to help him capture an infant gorilla from one of the study groups. I was stunned. The Conservator, who told me he had never entered the Parc des Volcans until my arrival in September 1967, said that city officials visiting Rwanda from Cologne, Germany, had requested a mountain gorilla for the Cologne Zoo. In turn, they promised to donate a Land-Rover and an unspecified amount of money for conservation work in the Parc des Volcans. The Conservator's request was made as casually as though he were asking for the time of day.

At great length I explained why Rwandese officials should not even consider such a proposition. I emphasized the strength of gorilla family bonds and stressed that the majority of any group's members would have to be killed before they would allow one of their offspring to be captured by poachers. The likelihood of mass slaughter did not seem to impress the young park official charged with obtaining the young captive gorilla as much as the possibility of international repercussions that might result should he attempt to go through with the capture. He appeared to reflect seriously on these matters, and this was, I thought, the end of the proposal. My naïveté blinded me to the basic fact that most officials considered the gorillas as goods for barter to be used whenever they saw fit for political or material goals. Conservation positions were filled by persons not trained or sensitive to the complexities of protecting the wildlife or the parkland from local poachers or international delegations. In the three countries sharing the Virungas the gorillas were never the first order of conservation business.

The holidays were accompanied by the expected high mortality toll of slaughtered antelope, buffalo, and elephant. To my knowledge, however, no gorillas had been directly affected by the vast poacher intrusions. By February 1969 the research and the habituation of Karisoke's study groups were progressing smoothly. Even the cattle and human encroachment within the Parc des Volcans were substantially reduced. Then, on March 4, a friend from the nearest town of Ruhengeri came to camp to tell me that a young gorilla had been captured by poachers about six weeks earlier and was now confined in a small wire cage in the Conservator's office.

Immediately I went down the mountain and drove to the rambling old barrack buildings that served as offices for numerous officials connected with the park in those days. In the small open square behind the ramshackle building were a series of sheds, the largest of which now served as a garage for the Conservator's new Land-Rover. Nestled between the Land-Rover and a stack of wood was a coffinlike box surrounded by swarms of laughing people, mainly children. A discarded wire cage lay nearby. Pushing away jeering children, I slowly released the door bolt of the wooden box in an attempt to see the captive, who had retreated as far in the back of the dark interior as it could get. Instantly, the little black furry form hurled itself toward me, shrieking in fear and rage. Quickly I slammed the door shut as the people crowded around laughing loudly at an injustice they did not understand.

I had the box carried into the Conservator's room, where relative quiet prevailed. Against his wishes I opened the coffin door, this time to let the baby out. Once more the small ball of fluff came hurtling forward. Before the Conservator could move she sank her teeth into his leg. She next ran to the windows where the people had gathered to cheer the action noisily. The frightened gorilla baby beat on the panes with such force that I was convinced the glass was going to shatter. She shed pools of diarrhetic dung as she ran back and forth between the windows, and, because of her state of dehydration, stopped to lick it up. With an ashtray filled with water I was able to lure her back into the box.

I only briefly questioned the Conservator about the manner in which he had acquired the young gorilla, being desperately anxious to get the baby to camp as soon as possible. Every minute spent talking seemed to be a minute less of her life, if she could survive at all. Without any abashment whatsoever, the Conservator admitted to having asked the leading poacher of the park, Munyaru-kiko, to organize a group of poachers to make the capture. What money passed through whose hands at this point I do not now know, nor did I then care. The men had climbed Mt. Karisimbi and selected a random group containing an infant. Later I learned that ten members of the gorilla group were killed in the capture.

The young animal was wired on to bamboo poles by her arms and legs and carried to a small village near the park boundary. Kept for two weeks in the specially made wire cage the park guards had ready for her, allowed neither standing nor turning space, she was fed corn, bananas, and bread until brought to Ruhengeri. There she was transferred to the coffinlike wooden box. Soup was added to her menu because her captors were worried about her condition and did not know what else to do for her.

I shall never understand how the orphaned infant managed to survive the confines of the cage, her meager diet, or the infected wounds caused by the wire bindings. Somehow she had found the will to live an additional two weeks in Ruhengeri before I heard about her. Not wanting to waste more time in the Conservator's office, I informed him that I would be taking the baby back to camp with me. He showed no remorse in letting her go. He seemed more than happy to let me bear the responsibility of the captive's fate — probable death.

The infant's most immediate need was liquid, vitamins, and glucose, though I intended to reintroduce her to natural vegetation as soon as possible. Unfortunately, to procure the necessary medicines I had to go to Kisoro, Uganda, thus delaying the gorilla's journey to the mountains and camp another day.

Lily the Land-Rover managed to carry the box from the noise of Ruhengeri to the relatively quiet home of a European couple who lived fairly near the park boundary below Visoke. Once there, I transferred the baby gorilla into a child's playpen and nailed it shut in preparation for the following day's trek up to camp. The stressful transfer from one container to the other was accompanied by the infant's bellows of combined rage and fear as she was subjected to yet another ordeal. Once she was secured in the playpen, I gave her bits of *Galium* vine and thistle, which she immediately ate before dozing off in an exhausted slumber. I slept next to the pen in order to comfort her whenever she cried in her sleep.

During the long night I firmly made up my mind that, provided she lived, I would release her to the wild, most likely to Group 8, rather than allow her to be put into another wire cage at the Cologne Zoo. I estimated her age as between three and a half and four years — old enough to be able to survive in the wilderness under the protection and care of adult gorillas. I named her Coco in memory of an old female in Group 8 who had recently died of natural causes.

The following morning the second stage of little Coco's journey began. The forty-minute drive on the rough lava-rock road from the Europeans' home to Mt. Visoke's base was agonizing. The infant screamed in pain and fright almost every jolting minute of the way. At the base of the mountain I hired eight porters, who took turns carrying the pen up to camp. Once we had climbed the first steep slope away from the noise of the *shambas* and had reached the other side of the rock tunnel, Coco showed a keen interest in the familiar forest surroundings. Occasionally she emitted plaintive whines such as those heard from wild gorillas when separated from their mothers. I wondered if the infant was recalling memories of life as she had known it before her capture. There was no way I could comfort her except to make reassuring gorilla vocalizations and periodically halt our trek to put some vegetation into her pen during the long five-hour climb to camp along the slippery elephant track of mud that serves as a trail.

The Karisoke staff had prepared the second room of my cabin for the captive's arrival. In a note sent to camp by a porter, I had asked that wire mesh be nailed over the windows for the protection of the glass as well as of Coco, and that a wire door be installed between her room and mine. I had also asked that the floorboards of Coco's room be covered with gorilla feeding and nesting vegetation, and that *Vernonia* saplings be wedged between the floor and the ceiling for climbing purposes. By the time Coco and I arrived at camp, the room had been transformed into a miniature facsimile of a gorilla habitat.

"Chumba tayari!" The men yelled to me excitedly that the room was ready as the porter line merged from the thick forest into camp's open meadow. I was indeed impressed by the transformation their efforts had wrought to the cabin. Two pans of water weighted down with large rocks so that the dehydrated captive could not drink too much at one time were placed inside. Then, with many screams and orders in Kinyarwanda, the porters managed to squeeze the playpen through the doors of the cabin and deposit it amid the trees that now sprouted between the floorboards. Suddenly, the baby and I were left alone together in blissful quiet.

Cautiously, I pried off the cover of the pen, unsure of the reaction to expect. Would the infant be timid, aggressive, or lethargic? I was thrilled when Coco straightaway left the pen and dazedly walked over the vegetation, patting the leaves and stalks as if to reassure herself they were real. Because of her weakened condition she made only one feeble attempt to strut by my side to indicate that she intended to be in charge of this new situation. She then stood and stared at me intently for nearly a minute before very hesitantly crawling onto my lap. I badly wanted to hug her but did not do so for fear of jeopardizing the first trust she had been able to place in a human being.

Coco sat on my lap calmly for a few minutes before walking to a long bench below the windows that overlooked nearby slopes of Visoke. With great difficulty she climbed onto the bench and gazed out at the mountain. Suddenly she began to sob and shed actual tears, something I have never seen a gorilla do before or since. When it finally grew dark she curled up in a nest of vegetation I had made for her and softly whimpered herself to sleep.

I had to leave the cabin for about an hour but expected to find her still sleeping upon my return. Opening the door, I found instead complete chaos. The "gorilla-proof" matting the men had nailed over the camp's stock of food supplies, stored in shelves along one wall of Coco's room, had been torn away from the storage cupboards. In the midst of an array of tin cans and opened boxes Coco sat contentedly sampling sugar, flour, jam, rice, and spaghetti. My momentary dismay at the havoc she had created was instantly replaced with delight upon realizing that she somehow had had the curiosity and energy to create such a mess.

During the next two days Coco ate increasing amounts of natural vegetation, *Galium,* thistles, and nettles, and — after a battle of wills — she accepted a milk mixture containing all the medications I considered essential for her health. Yet she continued to cry frequently, particularly when looking out of the window of her room. One day vocalizations from Group 5 were heard from the slopes behind camp, causing Coco to keen more than ever. Quickly I turned on the radio as loud as possible to drown out the sounds of the group, but Coco continued to gaze up at the slopes for most of the day, whimpering softly, with the knowledge that her own kind were near.

On the third day the youngster's partial satisfaction with her new environment diminished, and she took a sudden turn for the worse. The same type of transition has occurred with every newly captured gorilla I have ever known. Each seems to have just so much courage and will to survive, but all too often the brutal trauma of the capture coupled with the physical neglect of the captors are too much to overcome. Help usually comes too late. Coco completely stopped eating, her dung seeped out in liquid bloody pools. She lay huddled in a pile of nest vegetation shaking uncontrollably. Nothing I could do, including playing tape recordings of other gorillas, could shake her out of this semicomatose lethargy. I started her on antibiotics, but there was no evidence of response to any of the medications, and she continued to fail at an alarming rate.

On the sixth night of Coco's stay at camp, I carried her to bed with me for what I assumed would be her last night alive. Warmth and security was all that I could now provide. At five o'clock the following morning, instead of finding a corpse in my arms I found Coco still alive and both of us lying in a bed soaked with diarrhetic dung. She seemed somewhat more lively, and I hoped that her crisis might have passed during the night. After giving her the medicine, I carried her outside into a large wire enclosure adjoining her room so that she might have access to the sun. The door was closed between the room and the cage so the men and I could sanitize every inch of her room before putting in fresh foliage and saplings.

While we were working I suddenly heard porters' voices approaching camp. I ran outside and saw about six men carrying what looked like an oversized beer barrel suspended between long poles supported on their shoulders. The head porter handed me a note from a friend in Ruhengeri who had come up to see Coco when she had been so terribly ill. "They've captured another gorilla. They want you to take care of it but didn't know how to send it up to you, so I improvised this thing. Hope all is going well with the first. Doubt this one will survive either."

Incredulous, I paid the porters and had the barrel brought into Coco's room, leaving Coco outside in the sun-filled pen. It seemed highly probable to me that a week ago the Conservator had been anxious to get rid of Coco because he had thought her nearly dead. He may well have had the second captive as a backup for the Cologne Zoo. Much later I learned that the second gorilla similarly had been captured on Mt. Karisimbi, but from another group than Coco's. It had come from a group of about eight animals and, like those of Coco's group, all the family members had died trying to defend the youngster.

I pried off the nails of the barrel. Unlike Coco, the newcomer refused to leave her container and huddled farther in the rear. Although I moved Coco back into the room to serve as an enticement, the new animal did not respond; so I left the two alone and watched their interactions from my room. Coco showed great curiosity in the new arrival and peeked into the barrel giving soft grunts whenever the newcomer moved slightly. Occasionally they would reach out to one another, only to withdraw swiftly whenever their hands nearly touched.

Still the captive showed no inclination of emerging from her barrel. Finally I went into the room, laid the barrel on its side, and gently spilled her onto the foliage-covered floor. I was horrified to see that the young female, estimated as between four and a half and five years old, was far more emaciated than Coco had been. Pus was oozing from what appeared to be *panga* wounds on her head, wrists, and ankles and from deep cuts where she obviously had been bound by wire. From the extent of her infections and deterioration, I judged she had been caught about the same time as Coco but had had to spend nearly a week longer with her captors.

Because of my presence the new captive retreated as far away as possible to huddle in a dark corner under a table. Coco began strutting back and forth in front of her, a form of locomotion typically used by gorillas during introductory periods. I was pleased to see Coco, only hours after her extreme illness, show some renewed interest in life, but I doubted that Pucker — named for her morose and dejected facial expression — would ever show similar vitality.

I brought in fresh celery, thistle, *Galium,* and a tray of wild blackberries that the camp staff had collected. Pucker showed a spark of interest in the familiar foods, but, possibly because they evoked memories of the past, she began giving plaintive, whining cries exactly as Coco had done in the same situation. As though sympathizing with the newcomer's reaction, Coco whined and pursed her lips in response before going over to choose a few ripe blackberries from the pile. I left their room to watch them through the wire door. Pucker slowly approached the favored fruit to reach hesitantly for a berry. A small pig-grunting squabble broke out between them as they had their first experience in "sharing." Then both youngsters grabbed as many berries as their hands could hold and took their loot into opposite corners of the room to feed upon while continuing to pig-grunt mildly at each other. I knew right then that competition might well be the secret to their survival.

Their first day of getting acquainted was a combination of mildly antagonistic behavior coupled with a pathetic sense of awareness of their need for one another. By day's end, however, they cuddled together in a single nest of vegetation, cried softly, and fell asleep in a close embrace.

Three days passed before I succeeded in persuading Pucker to accept the same fortified milk formula I was giving Coco. I only managed then because Pucker was falling into the same dangerous state of lethargy as Coco had following her arrival at camp. Two other similarities existed between the captives: like Coco, Pucker willingly accepted the natural forest vegetation and she had a liking for bananas. Their mutual fondness for bananas, a fruit that does not grow in the wild state within the Virunga Volcanoes, indicated that they had been held by poachers long enough to develop a taste for the alien fruit given to them by their captors.

During Pucker's period of severe illness, Coco greatly increased her dependency on me for cuddling, holding, and play. If her demands for attention weren't met immediately, her whines often grew to loud, shrieking temper tantrums, which caused Pucker, from her secluded nesting perch high on an empty storage shelf, to give weak belch vocalizations as if trying to comfort the younger captive. The constant stimulation of Coco's presence and her absolute acceptance of me provided Pucker with the incentive needed to overcome much of her despondency. She slowly resumed her interest in food and even showed enough gumption to keep her share of the favorite food items when Coco tried to take them away.

Those were indeed trying days. To make matters worse, the new cook quit when I asked him to help out with formula preparation and bottle sterilization. He haughtily informed me in Swahili, "I am a cook for Europeans, not animals." Some of the other men were also on the verge of leaving because of the constant demands for fresh foods from the forest and the removal of even fresher dung from the room. Mutarutkwa, still illegally grazing cattle in the park quite some distance from camp, was of tremendous help at this time.

One day a young tracker, Nemeye, and I were going about our daily, time-consuming chore in search of the biggest, juiciest, ripest blackberries — a scarce commodity even when in season — when we encountered Mutarutkwa. After watching our unskilled efforts for a while, the Mututsi bounded off like a gazelle and disappeared into dense brush. Within about half an hour, during which Nemeye and I had gathered scarcely a dozen berries, the tall herdsman returned as silently as he had left. With a shy smile, he extended his hands to us. Each contained large *Lobelia* leaves full of the most magnificent lush blackberries I had ever seen. Wondering and grateful, I accepted them. They were only the first of many that Mutarutkwa took pleasure in providing for the captive gorillas, saving camp residents hours of searching in the forest.

Happily, Coco had developed a craving for the medicated milk pan given three times a day, not only for her own but for Pucker's portion as well. Pucker was puzzled at Coco's eagerness to drink a concoction that she obviously found distasteful. Once again, competition settled the problem. The more insistently Coco demanded Pucker's medicated milk pan by pushing and pig-grunting, the more assertive became Pucker's efforts to defend it. Eventually, with the wriest of faces, Pucker drank the strengthening liquid down before Coco had a chance to get at it.

I was deeply grateful for Coco's unintentional assistance because, even after eight days, Pucker still did not like me to touch her, nor had she approached me, in total contrast to Coco's initial behavior.

Pucker, estimated to be about a year older than Coco, was far more introverted and depressed by the changes brought about in her capture than the younger female. Despite my constant reassurance that she wasn't going to be further harmed, Pucker remained extremely apprehensive, particularly when people were heard or seen near the infants' room or outdoor cage.

Pucker's first approaches toward me were made under the guise of "protecting" Coco, who had to be increasingly distracted by means of rough and tumble play from tearing up the wall and ceiling matting in her favorite game, "house demolition." Coco had discovered that the matting was both shredable and chewable and that her persistence in such activities allowed access to the whole cabin ceiling. It took several weeks to reinforce the ceiling sufficiently to restrain her exploratory efforts and restore my half of the cabin to its original pee-free state.

On the occasions when I had to entice Coco teasingly away from her intended mischief, Pucker usually tried to drag the infant away from me by running, pig-grunting, and either hitting or lightly biting at my legs. It seemed apparent that Pucker only wanted to be included in the play, or even be cuddled, although she absolutely refused to trust me to that extent. The jealous and somewhat "neurotic" behavior seemed pathetic to me. Undoubtedly, it was warranted by the terrifying experiences of her capture and subsequent confinement with her captors. With a great sense of guilt, I continued focusing my attention on Coco, not simply because she demanded it but also in an effort to coax Pucker out of her withdrawn state. I felt that eventually the element of competition for attention would work as well as had the competition for food.

One night, nearly two weeks after Pucker's arrival, she sidled up to Coco and myself as we were playing on a bench in their room. Pucker grabbed one of Coco's arms, attempting to pull the chuckling youngster off my lap. When Coco refused to budge, Pucker, smacking her lips nervously, tried to pry one of my hands from Coco's body. Gently I began stroking Pucker. Immediately she flinched, apprehensively stiffened her body, and turned her head away. By touching me, Pucker had made at least a very small beginning with a human being. Within two days she was able to follow up this approach by briefly clawing my hands whenever I was holding on to Coco. I tried to apply medication in the form of salve to her badly infected hand wounds whenever she remained near, but more often than not, she outmaneuvered me and withheld her trust for the remainder of the day.

* * *

The day finally came when Coco was healthy enough to be allowed the freedom of the tree stands and meadow glades around camp. Pucker had not yet sufficiently recovered from her wounds, and I doubted my control over her. On our first venture outside the cabin I had to carry Coco on my back, since she seemed overawed by the vast amount of open space around her. Even when we perched on a huge *Hagenia* tree trunk laden with *Galium* and other gorilla foods, she wouldn't leave my lap.

The thirty-minute reintroduction to the "wild" took us no farther than some 165 feet from the cabin. Pucker, who had gone into the outdoor cage to watch Coco and me leave, began whimpering softly. The cries soon built up into loud sobs terminated by screams and shrieks as Coco and I moved farther away. I was obliged to return much sooner than intended, although Pucker's cries seemed to be ignored by Coco. The howling subsided upon my obedient return, but in typical Pucker style our reentry was totally ignored as she feigned a sudden interest in feeding. Her behavior might have been considered comical because of its similarity to that of spoiled human children; nevertheless, I felt it indicated a deep sense of deprivation.

I took Coco out alone for several more days but always to the accompaniment of Pucker's cries and screams from the cabin. Coco was rapidly becoming used to the environment around camp, a very un-gorillalike area consisting mainly of open meadows, scolding chickens, and my ever patient and playful young dog Cindy. Coco took delight in running after the chickens and grabbing for their tail feathers whenever they were not fast enough. She also delighted in riding Cindy piggyback style and chasing her in endless circles until they both collapsed dizzily in an exhausted heap of tawny and black-haired bodies.

On the morning I was ready to take both babies out, the park guards came up to camp unexpectedly. The two youngsters were in their outside run when the guards noisily arrived carrying spears and guns. Terrified, the infants fled into their room, climbed onto the highest shelf, and clung together for the remainder of the day. The guards demanded that I immediately turn the gorillas over to them for the trip to the Cologne Zoo. After an hour I was able to get rid of the intruders by convincing them that the gorillas were not well enough to leave — the truth, at least, in Pucker's case. Two days passed before the babies could be coaxed out of their room.

At long last both gorillas were well enough to cavort freely in the relatively unconfined freedom of their natural habitat with Cindy as their willing and dependable mascot. Cindy's eternally good nature and sense of playfulness had gained the complete trust of both Coco and Pucker. I was considerably impressed by this, because dogs had undoubtedly been used by the poachers during the gorillas' captures. The ability of the infants to accept either human or dog after their brutal experiences seemed extraordinary. Perhaps one reason the captives took to her as readily as they did was that Cindy, who had not seen another dog for more than two years since her arrival at camp, had forgotten how to bark. This, coupled with her gentleness, gave my dog even less resemblance to the dogs the infants had probably already encountered.

When the two youngsters were really wound up to play, Cindy had a rough time coping with their high spirits. She was pinched, bitten gently, whacked, ridden upon, poked, de-whiskered, smelled, suckled, chased, and all but torn apart during their wild outdoor play sessions. There were times I wondered whether Cindy knew if she was dog or gorilla. To her, the most obvious difference must have been made painfully clear when Coco and Pucker took to the trees, leaving her grounded yet determinedly tackling the tree trunk with all the wrong body equipment.

The time spent outdoors with the gorillas and Cindy formed the highlights of most days whenever weather permitted our excursions away from the cabin. The captives never quite got over their apprehension of the wide meadows around camp, but we had to cross the open areas to get to the tree-clump mounds that offered so much in the way of climbing and play facilities as well as a diversity of foods. Dozens of such clumps were fairly accessible to camp, but to go back and forth to them I nearly always had to carry Coco in my arms, while Pucker would either insist on a piggyback ride or cling to one of my legs. The combined weight of the two, once they were fully recovered, totaled one hundred and eight pounds, wiggly burdens that made the meadow crossings rather an agonizing trek, from my point of view.

I tried to encourage Coco to use her own legs, but that was like trying to teach an elephant how to fly. If left behind, she usually began crying, giving a plaintive series of muted *hoo-hoo-hoos* that slowly built up into shrieking temper tantrums, compelling me to return to pick her up. Pucker was far more independent and, as I feared, I did not have absolute control over her.

From her very first day out of the cabin Pucker often tended to gaze up at the mountain slopes with expressions of longing that seemed to be mixed with slyness. It was as though she fully realized that gorillas did not frequent meadows but belonged in thick vegetation on the slopes or in the forest. Pucker's inner conflict was triggered one day when a loud outbreak was heard from Group 5 high on the slopes behind camp. Both youngsters were playing in the meadow below. Of course I did not have a radio with me to drown out the vocalizations.

Without hesitation Pucker began running toward the mountain in the direction of the sound, Coco willingly tagging along behind. By the time I caught up with them, they had started climbing up the slopes; rather like tourists, they had chosen the most visible path. In this case it was a heavily used elephant trail. Coco, who somehow had managed to get in the lead, halted before the first huge water-filled elephant footprint and was about to give up when Pucker, seeing me creep up from behind, gave her a mighty shove, sending Coco chin-deep in the muddy contents of the bog. That was all Coco needed to send her fleeing back to me. Pucker then had little alternative but to follow.

Even without hearing outbreaks from Group 5, Pucker continued to try to lure Coco onto the mountain slopes. Fortunately, Coco could usually be bribed with bananas to come back to me. A vast supply of these were always secreted in my pockets.

One day while walking in a new area, Pucker suddenly ran toward a large cluster of *Hagenia* trees on the edge of the forest leading to the mountain. Coco leapt from my arms in rapid pursuit — which was unusual. I thought they were making a dash for the mountain and was hastily taking out the bananas when both infants halted below one of the larger trees. They peered up at the tree like children looking up a chimney on Christmas eve. I had never seen them so fascinated by a tree, nor could I determine what it was that so strongly attracted them. Suddenly the two began frenziedly climbing the huge trunk, leaving me even more puzzled. About thirty feet above the ground they stopped, pig-grunted at one another, and avidly started biting into a large bracket fungus. Previously I had noted these shelf-like growths, which protrude from *Hagenia* tree trunks and rather resemble overgrown solidified mushrooms. They are rare throughout the forest, and before acquiring Coco and Pucker I had never observed wild gorillas being interested in them. Try as they might, neither Coco nor Pucker could pry the fungus from its anchorage on the trunk, so they had to content themselves with gnawing chunks out of it. A half-hour later only a remnant remained. Reluctantly they descended, but as we walked on they gazed longingly back at the tree with the fungus elixir. Needless to say, the next day everyone in camp was asked to search the forest for bracket fungus!

Another rare food item that evoked squabbles between Coco and Pucker was the parasitic flowering shrub *Loranthus luteo-aurantiacus* belonging in the mistletoe family. Fortunately for the gorillas, the staff knew exactly where to find abundant supplies of the delicacy.

My studies with the gorillas showed that larvae and grub matter were often obtained from the inner hollow dead stalk material, but I was amazed to see the two captives ignore such treats as blackberries to search for worms and grubs. They often appeared to know exactly where to peel the slabs from live and dead tree trunks to find abundant deposits of larvae. Even while licking one slab clean, purring with pleasure over their feast, they were ripping off another slab for more burrowed protein sources. Worms, when discovered, were immediately torn in half — a rather revolting sight to watch — and each half was chewed with gusto, though not always ingested. After realizing that Coco and Pucker craved such food I included boiled hamburger in their diet, which they ate before any of their cherished foliage or fruit.

Coco and Pucker's outdoor freedom carried over into the security of their room, where eventually nearly every conceivable item of natural gorilla food was introduced three times a day, along with medicine given on a routine schedule.

Usually the pair awoke voluntarily around 7:00 A.M. They weren't the least bit reticent about informing me they were awake by boisterously banging on the wire door between our rooms. After the three of us exchanged good morning hugs, I gave them their milk formulas in two separate pans bolted down to the top of the playpen. Then, food such as bananas and wild blackberries was tossed into the outside run to get rid of the babies during the time it took us to scrub the floor and shelves of the room and discard every bit of vegetation and other debris left over from the previous day. During that time other members of the staff were collecting fresh vegetation for feeding and nesting purposes, so that when the runway door into the room was finally opened the gorillas could return to a "fresh forest," albeit one that smelled slightly of disinfectant.

If the weather was overcast or cold, they spent about an hour feeding contentedly before making their nests in the new vegetation. If it was sunny, they demanded to be taken outdoors, where they could unleash their pent-up energy in roughhouse wrestling, chasing, and tree-climbing.

Between 12:30 P.M. and 1:00 P.M. I would bring them back to the cabin to repeat the early morning routine of medication, favorite food, and fresh foliage. Afternoon activities were again dictated by weather, though the two ruffians were more content to rest during this time of day. At 4:00 P.M. old foliage was discarded for new along with piles of leafy *Vernonia* saplings for me, and later for them, to use as their night-nesting materials. The 5:00 P.M. schedule was much the same except that the youngsters were left alone for an hour to feed. During this time their croons of pleasure and belch vocalizations nearly drowned out the noise of my typewriter in the adjoining room, lending an air of serenity and contentment to the near end of each day.

Once Coco and Pucker had eaten their fill, the four of us, including Cindy, set the cabin frame shaking as we chased, tumbled, and wrestled within the miniforest of their room. I recall those hours as some of the most joyful I have ever known at camp, because Pucker, somewhat inhibited during the day when other people were around the cabin, became ebullient and outgoing when just the four of us were alone together.

During these relaxed sessions I learned a great deal about gorilla behavior that I had not gained previously from the free-ranging animals who had yet to become totally habituated to my presence. Tickling between Coco and Pucker provoked many loud play chuckles and also lengthened their play sessions. Tentatively, I first tried out tickling on Coco, and after receiving a very receptive response tried it later with Pucker. After a few weeks I changed approaches from mild "tickle-tickles" to drawn-out "oouchy-gouchy-goo-zoooom" tickles, much like those given by parents or grandparents when zeroing in with a teasing finger for a child's belly button. The term "oouchy-gouchy-goo-zoooom" is not in any dictionary, yet it seems to be an international and interspecific term that can evoke laughter and smiles from both human and nonhuman primates. Later I had occasion to tickle free-living gorilla youngsters in the same manner and was able to elicit the same delighted responses that Coco and Pucker had given. This was done very rarely for it was always necessary to keep in mind that the observer should not interfere with the behavior of the wild subjects.

When I felt that they were tiring from our strenuous sessions, I broke off the leafy tops of the *Vernonia* branches to place on fresh beds of moss on the highest storage shelf. The final positioning of the foliage signaled night-nest time to the infants. After about seven weeks, Coco and Pucker were able to construct their own night nests and showed selectivity in their choice of the fullest branches for their nests. That was exactly the type of independent behavior I had been hoping for, a necessity if the two were going to be reintroduced to the wild. During the night stillness I often was saddened by the thought of the inevitable separation between myself and the two captives, yet thrilled to imagine them as members of Group 8, free to spend the remainder of their lives in the forests of their birth.

Suddenly we again had an unexpected visitor in the form of the Conservator whom Coco had bitten so justifiably in Ruhengeri about seven weeks earlier. The behavior of the two youngsters upon his arrival epitomized my own feelings exactly: Coco hid and Pucker went to the door separating my room from theirs and slammed it shut, an act I thought rather amusing.

The Conservator had made the long climb to camp to demand the infants' prompt release to the Cologne Zoo. For the second time I insisted that they were not well enough to travel. While I was desperately insisting for more time, the sound of play chuckles, chasing, and wrestling unfortunately were heard from the next room. I silently cursed them both for choosing this inopportune time to play, even though pleased that they would play in spite of the Conservator's presence. The harder I pleaded to keep them, the more insistent he was about taking them. He claimed that the Cologne Zoo was exerting pressure on him for the animals, sick or not. What he did not tell me was that the zoo was giving him a trip to Germany, ostensibly to act as a "companion" for the gorillas. He was to be honored by zoo and city officials upon his arrival. For a man who had never been out of his own country, this was indeed an exciting prospect.

After a circuitous and argumentative conversation, the Conservator stated flatly that if I did not immediately relinquish Coco and Pucker, he would send Munyarukiko and other poachers to capture two more young gorillas. He had called my hand. That same day I sent a cable to the Cologne Zoo officials telling them that they could have the captives once I felt the infants were well enough to make the long journey.

Sending that cable in order to avoid further slaughter was one of the biggest compromises I had to make during the years of my gorilla research. At that time there were few regulations concerning exportation or importation of endangered species. The Conservator's intentions to capture more young gorillas left me no alternative but to relinquish Coco and Pucker. Once the man had left, I went into the infants' room and received enthusiastic welcomes from both. Hugging them to me, I felt like a traitor.

The days of feeding and playing went on as usual for Coco and Pucker. Their lively behavior resembled that of two rowdy little girls at summer camp when time seemed endless. For myself, most of the joy of watching them develop into near-normal, playful juvenile gorillas was gone. Knowing their time in the forest was so limited and their future so bleak was constantly depressing, particularly since there was nothing I could do to prevent their fate. I wrote to Cologne imploring the zoo director to allow me to reintroduce the gorillas to the wild in a "foster group," but received a flat negative in reply.

Several weeks after the Conservator's visit the park guards returned to camp to demand the gorillas, but this time in a very aggressive manner by waving rusty rifles at me, the youngsters, and the Rwandese camp staff. By now Coco and Pucker freely accepted the men on the staff but retained a deep fear and timidity around unknown Africans. For this reason I was quite surprised when the gorillas tried to "attack" the guards by screaming and beating violently at the wire between themselves and their would-be captors. Their actions were just the cue I needed to tell the guards they were more than welcome to enter the room to collect the gorillas but that they couldn't expect my help. Nothing, including the wrath of the Conservator, could possibly have induced the men to enter the pen of the displaying gorillas, and in a few minutes they left. Afterward I learned that they told the Conservator the juveniles were still much too sick to travel.

A few days later the Conservator arrived at camp accompanied by guards carrying a small, newly built, coffinlike box intended for the captives' air-cargo flight from Kigali, Rwanda, to Brussels International Airport. From there they would fly to Cologne. The only opening to the box was a small twelve-inch door. Ventilation had not even been considered. The Conservator moreover had the effrontery to ask that I pay for the container. Eventually he left, content with the equivalent of thirty dollars in his pocket. I was slightly pleased only because I had gained several more weeks until the terrible departure day by explaining that the box would have to be entirely rebuilt.

Robert Campbell, a *National Geographic* photographer, came to camp about this time to undertake extensive photographic documentation of the wild gorillas as well as of Coco and Pucker. With Bob's help, the staff and I were able to reconstruct the crate. After enlarging the box and its entry door and drilling dozens of large ventilation holes along the sides and top, we put the structure into the gorillas' room so that they might grow accustomed to it. The crate served as a dispenser for their special foods and for the milk formula. Within a few days the youngsters invented a chasing game around it. Coco proved herself the cleverer of the two. She discovered that a quick reversal of her chasing direction inevitably caught Pucker offguard, resulting in a delightful head-on collision. Coco also enjoyed hiding in the back of the box in the middle of a chasing game, leaving Pucker to run wildly around the exterior several times before realizing where Coco had gone. As much as they obviously enjoyed their new oversized play toy, I found it a constant reminder of our pending separation as well as a preview of all the trauma that lay before them.

When the dreaded departure day came, Bob Campbell was willing to accompany the babies to the small Ruhengeri airport, from where they would fly to Kigali to leave Africa for ever. All the preparations for their journey had been completed. Numerous pages of instructions for their care between Ruhengeri and Cologne were provided. Tins containing their milk formula were strapped on to the crate's sides, and a fresh selection of forest vegetation, the last they would ever eat, was packed up. I also placed two large bracket fungi inside the crate. The instant the unsuspecting youngsters ran in to grab them, the door was latched shut. The porters who would carry the box down arrived a few seconds later. That was all I could endure. I ran out of the cabin, ran through the meadows of our countless walks, and ran deep into the forest until I could run no more. There is no way to describe the pain of their loss, even now, more than a decade later.

*   *   *

For a number of years a member of the Cologne Zoo staff periodically informed me of Coco and Pucker's welfare through bulletins and photographs. The photographs revealed all too clearly that the captives were just barely tolerating their caged environment. While writing this book I learned that Coco and Pucker, within a month of each other, died in 1978 in the Cologne Zoo.

6 | Animal Visitors to the Karisoke Research Centre

THE FIRST YEARS OF RESEARCH at Karisoke were much like the tremendously rewarding first six months at Kabara because I was able to concentrate primarily on daily observations of the gorillas with few interruptions from the outside world. Countless days were spent tracking and observing — usually through binoculars — the shy and as yet unhabituated gorillas. Evenings were spent sitting on the cot in the tent typing up the day's notes on an improvised table made from a packing crate. Usually I was surrounded by dripping clothes hung from lines along the top of the tent as near to the warmth of the hissing kerosene lamp as I could safely arrange them.

I thought of the lamp as a friendly genie, particularly when stepping outside the tent into a bitterly cold, ink-black night. It was awesome to think of this as the only speck of light, other than perhaps occasional poacher fires, within the entire Virunga mountain range. When contemplating the vast expanse of uninhabited, rugged, mountainous land surrounding me and such a wealth of wilderness for my backyard, I considered myself one of the world's most fortunate people.

It was impossible to feel lonely. The night sounds of the elephants and buffalo, who had come to drink at nearby Camp Creek, combined with the squeaky door-screech choruses of the tree hyrax (*Dendrohyrax arboreus*), encompassed me as part of the tranquillity of the nights. Those were magical times.

Some six hundred feet from mine was the tent for the three Rwandese assistants, who collected water and firewood and whom I eventually trained as gorilla trackers. Following the natural deaths of Lucy and Dezi, the chickens from Kabara, the crew presented a replacement couple, Walter and Wilma, who shared my tent for several months. Walter was no ordinary rooster. Every morning, much like a dog, he followed me into the field several hundred feet from camp. Every afternoon he would come running to meet me with greeting clucks. At night he roosted on the carriage of my typewriter, never fluttering a feather while being shuttled back and forth across the keyboard.

After a year and a half, the tent was beginning to bulge at the seams. Some European friends from the Rwandan towns of Ruhengeri and Gisenyi decided to build a small one-room cabin for me, complete with an oil-drum fireplace and windows. The thought of a cabin disturbed me at first, for it represented permanency at a time when I still bore the scars of my exodus from Kabara in Zaire. In spite of my lack of faith, the first Karisoke cabin became a reality in only three weeks of cooperative work on the part of all of us. A near-constant line of porters carried up ruler-straight eucalyptus saplings from the villages below to form the cabin's supporting framework. Tin sheeting (*mbati*) was brought from Ruhengeri for the exterior, and handwoven Rwandese grass mats served as insulation for the interior wall, roof, and floor surfaces. From Camp Creek, the four-foot-deep stream running through the meadow, rocks, gravel, and sand were collected to form a stout setting for the very functional oil-drum fireplace. The original crew members and I spent hours sanding and polishing broad planks for work tables and bookcases. We then made curtains from brilliantly colored local cloth for the finishing touches of the first real home I had had since leaving America. Eight additional cabins, each more elaborate than the ones preceding them, were to be constructed during the years that followed. None, however, came to mean as much to me as the first simple structure.

The cabin conveyed a new sense of security. I was finally convinced that I could welcome the company of a puppy needing a home. I named the two-and-a-half-month-old part-boxer female Cindy because of her habit of lying with her nose immersed in the cinders of the fireplace. In no time at all she became an integral part of camp life. (Indeed, two years later she would befriend Coco and Pucker, the gorilla orphans.) A great affection quickly developed between the staff and Cindy. She never lacked for human attention when I was out with the gorillas each day. The puppy spent hours playing with Walter, the rooster who thought he was a dog, a teasing pair of ravens, Charles and Yvonne, and even the elephants and buffalo that came to Camp Creek after dusk. Brilliant moonlit nights brought out the gamin in Cindy whenever she heard elephants trumpeting and bellowing around the waterhole. If let out of the cabin, she always ran directly toward the nearest elephant herd, some fifteen to twenty animals, to run playfully between their legs. I'll never forget the sight of the tiny puppy yapping and nipping at elephants' heels like a wearisome fly, yet somehow avoiding being flattened out into an elephant-sized pancake.

Late one afternoon, when Cindy was about nine months old, I returned to camp to be welcomed only by the clucking Walter. Neither Cindy nor any of the staff were to be found. Several hours passed before the men returned dejectedly with the news that Cindy had been "dog-naped" not too far from camp by either cattle grazers or poachers. We followed her pawprints along a muddy trail until they merged with footprints of some six to ten barefooted men and then disappeared entirely. It did not take skilled trackers to see that the puppy had been picked up and carried off.

Though uncertain as to whether poachers or cattle grazers were responsible for taking Cindy, I decided I could retaliate by rustling some cattle grazing illegally in the meadows near camp and hold them as ransom for Cindy's possible return. With no small difficulty, the men and I herded several-dozen head of long-horned Watutsi cattle back to camp, where we began building a corral around the trunks of five huge *Hagenia* trees. While the men cut saplings to fill in the spaces between the trees, I drew on former occupational-therapist skills to weave a knotty fishnet stockade, using every available piece of string in camp. About midnight the ludicrous structure was completed and judged strong enough to hold the seven cows and one bull, all that remained from our initial captive herd. We pushed eight uncooperative rumps into the flimsy corral, nailed tin sheeting over the entrance, rekindled the campfires surrounding the stockade, and wearily began a night-long vigil against rustlers, the cattle's legal owners.

Under a brilliant starry night, the entire scene resembled a bizarre Hollywood Western. The cattle were bawling from the confines of their fire-lit corral, their indignant bellows joined by the curious snorts of passing buffalo and elephants on their way to Camp Creek. The usual tranquillity of the night was also broken by the shouts of my staff, whom I had asked to yell out into the surrounding forest, in Kinyarwanda, that I would kill a cow a day for every day that Cindy was missing. In between messages, I dozed fitfully, dreaming on and off of lassoing elephants from the backs of buffalo or trying to confine elephants within a string stockade. It was nearly dawn before Mutarutkwa, whose cattle I had taken, came out from the shadows of the forest surrounding camp to call out a message to us concerning Cindy's whereabouts. During the night Mutarutkwa had taken it upon himself to track the real culprits and found that poachers, led by Munyarukiko, had taken Cindy to an *ikibooga* high on the slopes of Mt. Karisimbi.

That morning I "armed" my camp staff and Mutarutkwa with fire-crackers and Halloween masks for Operation Rescue Cindy. Marine-style, the four men charged the poachers' *ikibooga*, threw firecrackers into the main campfire, and, during the confusion, retrieved Cindy while the poachers fled from the attack site. Cindy's rescuers later told me that she was anything but a dejected captive. They had found her happily ensconced in the midst of all Munyarukiko's dogs cheerfully chewing away on buffalo bones remaining from the poach-ers' kills. With Cindy restored to camp, I gratefully returned the cattle to Mutarutkwa, thankful that the turn of events had not called my bluff.

Nine months later Cindy was again stolen by poachers led by Munyarukiko. This time they took her directly to their Batwa village near the park boundary below Mt. Karisimbi, where she was tied up alongside their hunting dogs. Mutarutkwa's dignified father, Rutshema, rescued Cindy and returned her to me, commenting bit-terly on the thieving ways of Batwa poachers. Over the years I came to owe a great deal to this Watutsi family. Indeed, Mutarutkwa was later to join my staff and lead antipoacher patrols throughout the parklands of the Virungas.

A year after Cindy's arrival, and shortly after Coco and Pucker's departure for the Cologne Zoo, I acquired a new companion at camp. At a gas station in the lakeside town of Gisenyi, a shifty-eyed man carrying a small basket sidled up to my car door. He asked the equivalent of thirty dollars for the contents. For a while I feigned lack of interest; then, casually taking the basket, I found inside a small blue monkey (*Ceropithecus mitis stuhlmanni*) about two years old, fearfully huddled at the bottom more dead than alive. Instantly I grabbed the basket, started the car, and threatened the poacher with imprisonment should he ever capture another animal from the park, where such animals are legally protected. As the man fled, I gazed down into a pair of huge brown eyes looking shyly up at me. Thus began a love affair that was to last for eleven years.

All I needed was a monkey living in an environment unsuited for it. There are golden monkeys (*Ceropithecus mitis kandti*) and blue monkeys living in the lower bamboo zones of the Virungas to as high as nine thousand feet, but not to the ten-thousand-foot altitude of Karisoke.

The captive, named Kima, an African word meaning "monkey," was carried to camp the next day, and thereafter camp life was never the same, a fact to which anyone who has ever visited or worked at Karisoke would readily attest.

Kima soon learned to thrive on fruits and vegetables, brought up by porters from Ruhengeri's open market, as well as bamboo shoots carefully selected from lower park altitudes where others of her species live. Within a month Kima had also developed a decided liking for human food such as baked beans, meat, potato chips, and cheese. Her hors d'oeuvres menu expanded to include glue, pills, film, paint, and kerosene.

Kima's natural simian destructive inclinations were mitigated somewhat by her eventual acceptance of "babies" that I first made from old socks. Later I supplied her with stuffed toys from America. The koala bear, with its big shiny button-nose and dark eyes so much like her own, was her favorite. Deprived of others of her own kind, Kima spent hours each day grooming and carrying her stuffed "babies" around camp. Because I do not believe in confining animals, Kima was given the freedom of both house and forest, though she never strayed far from camp. My cabin became extremely neat, for anything left unguarded was bound to end up on top of a *Hagenia* tree or be shredded to pieces.

Kima, Cindy, Walter, and Wilma formed an unusual welcoming committee upon my return from the gorillas to camp each afternoon. During the chilly nights Kima stayed inside my cabin, usually in a wire cage with a two-way door permitting her access to the outside. It was always a most delightful, cozy feeling to type up field notes near the crackling fireplace at night with the two pets dozing nearby, the sounds of the owls, hyrax, antelope, buffalo, and elephant outside.

Two years following Kima's arrival, I had to spend seven months at Cambridge University. During my absence, Kima lost an eye in an accident. She survived that trauma only to succumb nine years later while I was teaching at Cornell University in 1980. Neither camp nor my life will ever be the same without her loving, albeit somewhat impish, personality.

In August 1980 I returned to Karisoke after an absence of five months and found Cindy, then almost twelve and a half years old, near death. Cindy instantly recognized me but could only hobble weakly and feebly wag her tail in greeting. Together we went to the mound near my cabin where Kima had been buried, and Cindy laid her head down on the wooden plaque marked KIMA. It was then that I made the decision to bring Cindy back to America, where she has become "habituated" to civilization and regained her health. Now accustomed to the noise of planes and cars, Cindy remains puzzled only by cats, unknown to her previously, and the raucous barking of neighborhood dogs. During her life in Africa she had only heard barking occasionally from poachers' dogs near camp and, consequently, never acquired the habit of barking. Even now, socializing with other canines, Cindy does not bark.

* * *

Over the years camp harbored other animals, visitors from the surrounding forest. One clear night in 1977 I looked outside the cabin window and thought my vision badly impaired. There was a giant rat (*Cricetomys gambianus*) feeding on chicken corn. Rufus, as I named him, had a body and tail each about twenty inches long. I wondered where the animal could have come from, for, although such rats were common in villages below, it was a long hike to make just for grains of leftover corn. Several weeks later Rufus was joined by Rebecca, then Rhoda, Batrat, and Robin. Soon every cabin had its own family of rats, each reproducing at an alarming rate until I was compelled to deprive them of the sources of food that had attracted them in the first place.

By the end of 1979 camp had grown into a scraggly town of nine cabins separated by small meadow glades and nearly concealed from one another by natural thickets of herbaceous foliage that grew profusely under the shelter of large *Hagenia* and *Hypericum* trees. Walking between the cabins was a sure guarantee of seeing a growing number of duiker (*Cephalophus nigrifrons*), bushbuck (*Tragelaphus scriptus*), and buffalo (*Syncerus caffer*). The antelope and buffalo had begun to seek camp proximity as a refuge against poachers. I had never envisioned the shy ungulates becoming habituated to the presence of humans; they are far more subjected to hunting pressures than are gorillas, yet Karisoke seemed to be their last refuge, an unprovisioned one at that.

I named the first duiker resident Primus, after the sparkly and tasty local beer. When she first came into camp with a flicking white tail, huge brown jeweled eyes, and a moist black quivering nose Primus was about eight months of age. The buds of her horns were submerged in a clownish black headtuft of hair, later growing into two delicate needle-sharp spikes. In the first months of her stay Primus never associated with any other duiker, leading me to believe that she probably had been orphaned. In time it seemed that Primus also had identity problems in not knowing if she was a duiker, chicken, or dog. She frequently followed the chickens around camp because they provided a feather alarm system by clucking whenever supposed danger threatened.

Primus often spent cold overcast days curled up around camp's central outdoor fireplace, something no other duiker has ever done. On bright sunny days she participated in the usual duiker games of head butting, playing hide-and-go-seek in the foliage thickets, and wild chasing pursuits that often ended up in mountings with other duikers. She also playfully chased the chickens or Cindy and was chased by the mischievous Kima. My dog had long since been trained to understand that duiker are not to be pursued, so she was totally perplexed when Primus took after her in a teasing manner. On many occasions Walter, Wilma, and other chickens would be scratching idly on the main path between the cabins when Cindy would come running along with Primus on her heels, and often even Kima thereafter. What a flurry of feathers, fur, and screeching these incidents caused!

In time Primus began chasing the humans at camp, taking special enjoyment in seeming to tease the housemen whenever they were balancing a load of dishes or laundry on their heads. Since Primus had never been pursued by humans, she exhibited caution but little fear of unknown people. Of course, this is what a game park should be.

Primus gave much joy, and occasional surprises, to many guests at camp, especially some Africans. One day I was showing the grave-yard of gorilla-poacher victims to a group of important Rwandese that included armed soldiers. Our hum of conversation attracted Primus, who emerged from dense vegetation to stroll casually through the crowd on her way to the meadows. Immediately everyone stopped talking. As the men watched the duiker delicately browse, I allowed myself to hope that someday poaching would be a thing of the past and that animals of the park might be able to put their trust in all humans.

On another occasion an extremely surly poacher, temporarily held at camp, was being escorted along the main path between the cabins and saw Primus dozing under a tree at the trail edge. The poacher's astonishment at seeing a duiker quietly lying only a few feet from him was, in a sense, comical. It was also poignant, since the man showed a deep personal delight in having been trusted by the antelope, an animal he had only recognized as prey before.

Bushbuck were far more reclusive than duiker in that they were seen usually only in the early morning or dusk when grazing around camp. The largest resident family of bushbuck grew to seven individuals headed by a huge male of advanced age. His hair was grizzled black and from a distance, in dim light, he resembled a buffalo because of his enormous size. It was remarkable that he or his aged mate had managed to elude the traps, hunters, and dogs throughout their environment.

I discovered aged male bushbuck frequently lead solitary lives except for occasional dependence upon duiker. This relationship was seen at camp, as well as in many other sections of the forest. Duiker serve as sentries by moving ahead of the bushbuck and giving penetrating, whistle-like alarm calls when spotting potential danger. The signaling system might imply that duikers' senses are keener than those of bushbuck. I think it more likely that this satisfactory arrangement evolved to allow more browsing time for the far larger bushbuck than constant solo vigilance would have permitted.

One of the most memorable bushbuck incidents around camp instantly brought to my mind Jody's words from *The Yearling:* "Pa, I done seen me something today!" As usual, upon awakening I looked out of the cabin windows and observed a sight more credible to a Walt Disney movie than to real life. All the hens, led by Walter, were tiptoeing awkwardly toward a young male bushbuck. The chickens' heads bobbed like yo-yos from their stringy necks. The curious antelope minced toward them with a metronomically twitching tail and quivering nose. Each chicken then made a beak-to-nose contact with the bushbuck as both species satisfied their undisguised interest in the other. Just about this time Cindy innocently came trotting up the path and froze, one foreleg suspended in the air, as she gazed at the weird spectacle before her. Her presence was too much for the young buck, who bounded off with a bark, his white-flag tail held high.

Much like the antelope, buffalo around camp were distinctive because of personality traits or physical variations. One lone male that appeared to be in his prime was noticeable because of his pink mottled muzzle and gregarious inclination toward humans. He thus gained the name Ferdinand, and first introduced himself one evening about dusk. Two of the staff and I were completing some carpentry work in front of the cabin, abusing the twilight calm with the noise of hammering and sawing. Just as I felt a faint trembling sensation beneath my feet I turned to behold the incredible sight of the massive bull trotting toward us. One of the men ran into the house immediately, but the second remained outside with me to watch Ferdinand. Stopping about eighteen feet away, the bull stared with unabashed curiosity and not the slightest bit of antipathy or fright. It was as though he wanted to be entertained. My helper and I resumed our hammering and sawing, and Ferdinand watched contentedly for another five minutes before turning away to feed slowly, without as much as a backward glance. Since that time I have met Ferdinand a number of times around Karisoke, most frequently in the early morning, and he continues to react mildly when seeing humans pass him on the camp trail. Like the duiker and the bushbuck, this buffalo is another gift of trust from the forest.

A second bull, an ancient animal initially accompanied by an elderly female, had as startling an appearance as Ferdinand had a personality. From rump to withers his body was scarred like a road map, with countless healed wounds, possibly the results of encounters with poachers, traps, or other buffalo. The heavy boss on his head must have at one time been twice its size, but had been relentlessly chipped away over the years of his life. The remnants of the horns themselves gave evidence of decades of battles and remained only worn and shattered nubbins.

I named the old bull Mzee, meaning "old one" in Swahili. It was an impressive sight to see Mzee following his aged female, who, until her disappearance, seemed to serve as a watchdog when his eyesight began to fail. During his second year around Karisoke the old bull, upon hearing my voice in the early evening hours, would slowly feed toward me as if wanting company and, eventually, allow me to scratch his withered rump. Early one morning the woodman found the old buffalo's body lying in a small grassy hollow next to Camp Creek under the towering silhouettes of Mts. Karisimbi and Mikeno. I could not imagine a more fitting spot for Mzee's final resting place. The serenity of the surroundings matched the dignity of the bull's character. Although he had lived in the shadow of poachers, he had managed to defy them in death.

Ten years before Mzee's natural death, when there were no regular patrols conducted from Karisoke, several buffalo in their prime met horrid endings at the hands of poachers very near camp. The first killing occurred during the second holiday season I had been in Rwanda and before I realized the devastation holidays brought to the park animals. I made the mistake of leaving camp for several days around Christmas 1968 and returned to find that my staff had locked themselves up in my cabin for safety's sake. Nearby, I found the remains of two poachers' dogs smashed up against the side of Camp Creek. The entrails of a buffalo led to a nearby hill where the carcass had been skinned and carried off by poachers. According to the staff, the poachers' dogs had chased the buffalo out of the forest, onto the meadows, and into the creek in front of my cabin. The bull, fighting for its life, managed to gore two dogs to pieces before losing his battle to spears flung from the poachers led by Munyarukiko. This was the last time I ever left camp unprotected during the holiday season.

The second buffalo slaying occurred several months later when the camp staff reported hearing the bellows of a "cow" in pain not too far below Karisoke. Taking a small pistol with me, I followed the men toward the source of the noise and found an adult buffalo wedged into the forked trunk of an old *Hagenia* tree. I regret that poachers also had heard the agonized wails of the trapped animal and had hacked off both its rear legs with their *pangas*. We found the poor beast desperately trying to stand on its two hind stumps amid a wallow of blood and dung. Still, the bull was able to toss his head boldly and snort at our approach. To have to kill such a valiant example of courage, an animal that could fight so bravely even to the last second of its life, was difficult. I returned to camp with the knowledge that one more magnificent creature of the Virunga Volcanoes was dead.

* * *

By early 1978 I had organized effective weekly poacher patrols capable of walking for miles and camping in bivouac tents, or even under trees when necessary, in their untiring efforts to clear the parklands of poachers. These excellent Rwandese, under the guidance of Mutarutkwa, brought many additional animal guests to Karisoke for recuperation periods — animals such as duiker, bushbuck, and hyrax that had been left to die in poachers' traps.

In 1978 the Zairoise park authorities allowed me to bring a juvenile gorilla, estimated at from four and a half to five years of age, to camp for possible recuperation. The young male had been caught in a wire antelope snare some four months previously, and his emaciated, dehydrated body already had been invaded by gangrene from the remains of his deformed, pus-filled stump leg. The youngster was doomed when I received him, yet I could not help admiring the Zairoise Conservator for having done all that he knew how to do for the terminally ill victim, and for his hope that it might be reintroduced to the wild rather than sold to a European zoo.

Everything done years earlier for Coco and Pucker was repeated for the new captive, who was immediately installed in a cabin stocked with fresh gorilla foliage and blackberries. The young gorilla responded miraculously to the familiar foods by trying to eat and walk. He was even able to give contented belch vocalizations upon recognizing the sounds, odors, vegetation, and surroundings of the forest. For six days the youngster valiantly fought advanced pneumonia, dehydration, shock, and septicemia before dying. It would be ludicrous to say that he died peacefully, but at least he had had a brief chance to return to the mountains of his birth rather than die alone on a cement floor in a wire cage, unloved, unwanted, and uncared for. Had he lived, I would have named him Hodari, a Swahili word meaning "brave." An autopsy revealed that the gorilla's lungs were only gray-white nonporous lobes. The medical team of the Ruhengeri hospital wondered how he had survived as long as he had.

Because the young gorilla's captors had been found near the park boundary below the southern-facing slopes of Mt. Mikeno, I asked the patrols to concentrate their efforts in that general area and to seek remnants of the youngster's group. Nothing was ever found. Nevertheless, many poachers and traplines were encountered in the saddle area between Mts. Karisimbi and Mikeno, and patrols rarely returned to camp without scores of wire snares or some of their victims.

Late one afternoon, several months after the death of the young gorilla, the Africans came back to camp carrying a black animal. I rushed to meet them, thinking they had found yet another gorilla-trap victim. Only when the men neared the cabin could I see that Karisoke's newest poacher victim had a long, feebly wagging tail, two sharply pointed, alert ears, and unusual emerald-colored eyes. The middle-aged female dog had been caught that same morning in a wire antelope trap and was found struggling hopelessly in the snare when the patrol chanced upon it. The wire had ripped the dog's leg to the bone and was working its way into the bone when the men released her and gently carried her back to camp. With their assistance, I dressed the horrible wound. I noticed that her other leg bore two narrow bands of white hair several inches above the paw, indicating that she had previously recovered from other trap injuries.

For three months she patiently endured daily leg soakings and bandage changes. For a while I feared her lower leg was going to have to be amputated. Although I had encountered many poacher dogs over the years, this was the first one that quickly accepted the strangeness of a white person. Her friendliness, trust, and complete composure at being inside a cabin with hissing kerosene lamps, type-writer, and radio sounds convinced me that she, like Cindy, might have been stolen from Europeans by poachers. She adapted easily to camp life, but I could not allow her freedom outside because of the numerous antelope — particularly Primus — who also considered camp home. Hunting was in the dog's blood and there was nothing I could do to curb her impulses to chase antelope, Kima, or the chickens. Kima soon learned that the newest camp addition was con-trolled by a leash and took delight in jumping up and down on my cabin's tin roof to tease the dog whenever it was out.

After the leg healed, I remained in a quandary as to what to do with the dog. While I was debating her future, an ABC television crew came to Karisoke to film gorillas in mid 1979. The crew's presence was a great treat. I had a touch of bush fever and welcomed the addition of nine faces from the outside world. Among them was Earl Holliman, an actor who had long been active in American organizations concerned with humane care of domestic animals, in particular Actors and Others for Animals. After hearing the dog's story, Earl immediately named her Poacher. One night he asked me, "How do you think Poacher would like to live in Studio City,

California?" From that moment on I almost believed in miracles. Within a few weeks Poacher was on a jet bound for Hollywood, where she was picked up by a veterinarian for a complete medical checkup. She continues to live there with Earl as a television star in her own right, earning sizable sums of money for animal causes by her appearances. The Karisoke staff remain justifiably proud of their part in having shaped Poacher's Cinderella fate.

7 | The Natural Demise of Two Gorilla Families: Groups 8 and 9

URING THE FIRST TWO MONTHS of study at Karisoke, my daily contacts with the gorillas were fairly evenly distributed between Group 4 — which ranged on the southwestern and western Visoke slopes under the leadership of a silverback I had named Whinny — and Group 5 — led by Beethoven on the southeastern mountain slopes. The composition of the two groups totaled 29 animals, but half of them had not been fully identified, and I could only speculate about the degrees of relatedness between the older individuals. My speculations were based on the frequencies of close affiliative associations between group members compared with aggressive, antagonistic reactions. Physical similarities such as noseprints, hair coloring, and evidence of syndactyly or strabismus were also extremely important in determining kinship ties within any group. The cohesive nature of gorilla groups fortunately provides one with a high degree of reliability regarding each offspring's sire. The early days of the research were spent trying to clarify the composition of the two main groups and seeking clues that would reveal the genetic connections between the individuals of the study groups available to me.

In this period, a third group entered the study area for the first time since my arrival and was subsequently named Group 8. (Group 6 was a fringe group; "Group 7" was a mistake — a failure to recognize Group 5 members on an occasion when they were feeding apart from one another.) I first saw Group 8 through binoculars from some five hundred feet up on Visoke's slopes. Even at that distance it was possible to distinguish an extensively silvered old male, a young silverback, a handsome blackback in his prime, two young males, and, bringing up the rear, a doddering old female. Unaware of my presence, they slowly ambled and fed throughout the nettle zone adjacent to Visoke's slopes before crossing a wide cattle trail that led into the forest. While watching the group I could not help being impressed by the manner in which all of the animals periodically paused in their feeding to allow the elderly female to catch up with them.

The following day I tracked Group 8 into the saddle area west of Visoke and contacted them from a distance of about sixty feet. They gave me the calmest reception I had ever received from an unhabituated group. The first individual to acknowledge my presence was the young silverback, who strutted onto a rock and stared with compressed lips before going off to feed. I named him Pugnacious, Pug for short. He was followed by the extremely attractive blackback, who nipped off a leaf to hold between his lips for a few seconds before spitting it out, a common displacement activity known as symbolic feeding and indicative of mild unease. After whacking at some vegetation, the magnificent male swaggered out of sight into dense foliage seemingly quite pleased with himself. I named him Samson. Next, the two young adults scampered into view and impishly flipped over on to their backs to stare at me from upside-down positions, giving the impression they were wearing lopsided grins. In time they were named Geezer and Peanuts. When the elderly female came into view, she gazed briefly at me in a totally uninterested manner before sitting down next to Peanuts and maneuvering her patchy rump almost into his face for grooming. I named her Coco because of her somewhat light chocolate-colored hair, and it was in her memory that the first Karisoke reclaimed captive was named sixteen months later.

Lastly, the old silverback came forward. In all my years of research I never met a silverback so dignified and commanding of respect. His silvering extended from the sides of his cheekbones, along neck and shoulders, enveloped his back and barrel, and continued down the sides of both thighs. Having little to go by in comparison, except for zoo gorillas, I estimated his age as approximately fifty years, possibly more. The nobility of his character compelled me to seek a name for him immediately. In Swahili, *rafiki* means "friend." Because friendship implies mutual respect and trust, the regal silverback became known as Rafiki.

Geezer and Pug closely resembled one another in having slightly pig-snouted profiles unlike the facial characteristics of the other three males or Coco. Physical traits, coupled with the affinity of the two males, suggested common parents. I thought it likely that their mother, who would have had to be an elderly Group 8 female because of their ages, had died before my arrival into the study area. For the same reasons of striking physical resemblances and rapport, Coco was thought to be the mother of Samson and Peanuts, both males sired beyond any doubt by Rafiki.

Coco and Rafiki often shared the same nest, resembling a gracefully aging old married couple who needed no words to strengthen their respect of one another. Coco's serene presence among the males of Group 8 frequently prompted mutual grooming, a social and functional activity involving meticulous hair parting done either orally with the lips or manually with the fingers in search of ectoparasites, dry-skin flakes, and vegetation debris such as burrs. Usually, after Coco's initiation, most of the other Group 8 members would follow suit and, within minutes, there might be a chain of intently grooming gorillas.

Response behavior — actions prompted by the presence of a human being — were only occasionally given by Group 8 members and seemed to convey elements of braggadocio, daring, and curiosity rather than aggression or fear. This unusual group with no young to protect seemed to accept or trust my presence from the start and to "enjoy" the break in their daily routine that I offered. Samson, in particular, reacted more than others but seemingly with a sense of self-enjoyment. Peanuts often tried to mimic Samson's actions. The two resembled a chorus line when standing upright for almost simultaneous chestbeats closely followed by several kicks of their right legs. When finished with their repertoire, they stood and stared at me as if gauging the effect of their performance. Samson also relished the noise made when he broke branches, and his massive weight guaranteed him many satisfying crashes. Once he climbed a tall dead sapling directly over my head. Like a logger, he deliberately preplanned the direction in which the tree would be felled. After several energetic bounces and swings, he managed to bring the tree down right by the side of my body before running off with a smug grin.

* * *

Often I am asked about the most rewarding experience I have ever had with gorillas. The question is extremely difficult to answer because each hour with the gorillas provides its own return and satisfaction. The first occasion when I felt I might have crossed an intangible barrier between human and ape occurred about ten months after beginning the research at Karisoke. Peanuts, Group 8's youngest male, was feeding about fifteen feet away when he suddenly stopped and turned to stare directly at me. The expression in his eyes was unfathomable. Spellbound, I returned his gaze — a gaze that seemed to combine elements of inquiry and of acceptance. Peanuts ended this unforgettable moment by sighing deeply, and slowly resumed feeding. Jubilant, I returned to camp and cabled Dr. Leakey I'VE FINALLY BEEN ACCEPTED BY A GORILLA.*

Two years after our exchange of glances, Peanuts became the first gorilla ever to touch me. The day had started out as an ordinary one, if any day working from Karisoke might be considered ordinary. I felt unusually compelled to make this particular day outstanding because the following morning I had to leave for England for a seven-month period to work on my doctorate. Bob Campbell and I had gone out to contact Group 8 on the western-facing Visoke slopes. We found them feeding in the middle of a shallow ravine of densely growing herbaceous vegetation. Along the ridge leading into the ravine grew large *Hagenia* trees that had always served as good lookout spots for scanning the surrounding terrain. Bob and I had just settled down on a comfortable moss-cushioned *Hagenia* tree trunk when Peanuts, wearing his "I want to be entertained" expression, left his feeding group to meander inquisitively toward us. Slowly I left the tree and pretended to munch on vegetation to reassure Peanuts that I meant him no harm.

Peanuts' bright eyes peered at me through a latticework of vegetation as he began his strutting, swaggering approach. Suddenly he was at my side and sat down to watch my "feeding" techniques as if it were my turn to entertain him. When Peanuts seemed bored with the "feeding" routine, I scratched my head, and almost immediately, he began scratching his own. Since he appeared totally relaxed, I lay back in the foliage, slowly extended my hand, palm upward, then rested it on the leaves. After looking intently at my hand, Peanuts stood up and extended his hand to touch his fingers against my own

---

* Nine years after Dr. Leakey's death in 1972 I learned that he had carried the cable in his pocket for months, even taking it on a lecture tour to America. I was told that he read it proudly, much as he once spoke to me of Jane Goodall's outstanding success with chimpanzees.

for a brief instant. Thrilled at his own daring, he gave vent to his excitement by a quick chestbeat before going off to rejoin his group. Since that day, the spot has been called *Fasi Ya Mkoni,* "the Place of the Hands." The contact was among the most memorable of my life among the gorillas.

Habituation of Group 8 progressed far more rapidly than with other groups because of the consistency of Rafiki's tolerant nature and the important fact that the group had no infants to protect; thus they did not need to resort to highly defensive behavior. Their "youngster" was old Coco, who received solicitous attention from the others. Coco seemed to be even older than Rafiki and had a deeply wrinkled face, balding head and rump, graying muzzle, and flabby, hairless upper arms. She was also missing a number of teeth, causing her to gum her food rather than chew it. She often sat hunched over with one arm crossed over her chest while the other hand rapidly patted the top of her head in a seemingly involuntary motion. Sitting in this manner, with mucus draining from her eyes, her lower lip hanging down, Coco presented a pathetic picture. I suspected that her senses of hearing and seeing were considerably dulled by age.

The remarkable displays of affection between Coco, Rafiki, Samson, and Peanuts could be described as poignant, though this was not surprising when one considered the number of years the family had probably shared together. One day I was able to hide myself from the group feeding on a wide open slope 130 feet away from me. They were widely spread out with Rafiki at the top, moving uphill, and Coco far at the bottom, wandering erratically on a feeding course that led away from the rest of the group. Rafiki suddenly stopped eating, paused as if listening for something, and gave a sharp questioning type of vocalization. Coco obviously heard it, for she paused in her wanderings and turned in the general direction of the sound. Rafiki, out of sight from her, sat and gazed downhill. The other group members followed his example as though they were waiting for her to catch up. Coco began climbing slowly, stopping occasionally to determine their whereabouts before again meandering in the general direction of the patient males. Once within sight of Rafiki, the elderly female moved directly to him, exchanged a greeting series of soft belch vocalizations until reaching his side. They looked directly into each other's face and embraced. She placed her arm over his back and he did likewise over hers. Both walked uphill in this fashion, murmuring together like contented conspirators. The three young males followed the couple, feeding along the way, while the young silverback, Pugnacious, watched them intently from a farther, more discreet distance. He too then disappeared out of sight over the top of the hill. I did not let Group 8 know of my presence that day since I felt that to intrude upon them for an open contact would have been improper.

Working on Visoke's western slopes usually gave me the opportunity to contact Groups 4 and 8 on the same day within an area of nearly two square miles. Alternating contacts with Groups 4 and 8 provided almost daily knowledge of their respective range routes and locations. Thus, in December 1967, I was puzzled to hear a series of screams, *wraaghs,* and chestbeats coming from an unknown group located about halfway out in the five-mile-wide saddle area between Mts. Visoke and Mikeno, a region that only Group 8 had been known to frequent.

The search was started for the "ghost group," which, when finally found, was named Group 9. The dominant silverback, one in his prime, perhaps twenty-five to thirty years old, was named Geronimo. He was a most distinctive male, with a triangular red blaze of hair in the middle of his massive brow ridge and luxuriant blue-black body hair that framed bulging pectoral muscles resembling steel cables. Geronimo's supportive male, about eleven years old, was a blackback named Gabriel, because he was usually the first to spot my presence and inform the group with chestbeats or vocalizations. The degree of physical resemblance between the two adult males suggested they probably had a common sire. One young adult female was all too easy to identify because of a recent trap injury which had rendered her right hand useless. The hand, with its swollen pink fingers, hung limply from the wrist and was frequently cradled by the young female. Within two weeks the young female became adept at preparing food by using her right arm or foot to stabilize vegetation stalks and her mouth or left hand for the more intricate tasks, such as peeling or discarding unwanted parts of a plant. She was able to climb or descend trees by hooking her right arm around branches and tree trunks instead of using her injured hand. Within two months after first being observed, she was no longer seen with the group and was assumed dead. The dominant female among Geronimo's harem of four was named Maidenform because of her long pendulous breasts. Each of Group 9's four adult females had at least

one dependent offspring, which indicated Geronimo's degree of reproductive success.

The addition of Group 9 to the study area provided a total of
48 individuals in four distinct groups, a population with both an
adult male-to-female ratio and an adult-to-immature ratio of 1 : 1.1
at the start of 1968.

By this time Coco, the aged female of Group 8, could no longer
be considered capable of reproduction. Peanuts, estimated as nearly
six years of age, had probably been her last offspring. Group 8,
therefore, had no breeding females, and Rafiki, the old but still potent
silverback leader of the group, sought physical interactions with
Group 4, which contained four females who were either approaching
or had recently reached sexual maturity.

Encounters between distinct social units increase in frequency when
range areas overlap, or if there is a disproportionate ratio of males
to females, as was the case on Visoke's western slopes during the
early years of the study. It was not long before Groups 4 and 8
met for a physical interaction instigated by Rafiki after following
Group 4 for several days.

The two groups first met in a section of ridges separated by deep
ravines at the edge of Group 8's range on the southwestern-facing
slopes of Visoke. Climbing toward the loudly vocalizing animals, I
looked ahead and saw what appeared to be an aerial act of five
flying silverbacks: three from Group 4 and Rafiki and Pug of Group
8 leapt from tree to tree, charged parallel to one another, chestbeat,
and broke branches along the ridge with crashing, splintering sounds.
Their powerful muscular bodies varied in shades from white to tones
of dull gray, and formed a vivid contrast to the green forest background. So engrossed were the displaying silverbacks that they
seemed unconscious of my presence.

Hoping to remain unnoticed, I crept to a nearby *Hagenia* tree
and found old Coco resignedly huddled against the tree trunk —
one hand tapping the top of her head and the other arm crossed
against her chest. She glanced at me calmly and heaved a big sigh
as if expressing patient tolerance of the commotion going on around
her. Occasionally Peanuts rushed down to her side to reassure himself
that she was there. After brief embraces he would rejoin Group
8's second young adult male, Geezer, with chestbeats directed toward
the three silverbacks of Group 4.

Excitement, rather than aggression, dominated this first observed physical interaction between Groups 4 and 8. While watching the discretion of the parallel displays between the two dominant silverbacks — Rafiki of Group 8 and Whinny of Group 4 — I received the impression that both were equally experienced and were thus capable of avoiding overt combat because of mutual respect based on numerous previous interactions. Late that afternoon the two groups separated, though they continued to exchange hootseries and chestbeats for several hours, communications that seemed to become more taunting as the distance between the two familial units increased.

Two months later, in February 1968, Rafiki had ceased trying to interact with either Group 4 or Group 9, which were then also ranging on Visoke's western slopes. Old Coco had weakened, and because of her difficulty in keeping up with the group, Rafiki adjusted their travel and feeding pace to meet hers. On February 23 I found no sign of either Coco or Rafiki after contacting Group 8. Only the four males — Pug, Geezer, Samson, and Peanuts — were to be seen wrestling playfully together as carefree as boys at a summer camp. Backtracking the group's trail, I found that Coco and Rafiki had night-nested together in connecting nests for the past two nights, but I completely lost all trail sign after that. Two days later Rafiki returned to Group 8 alone. Coco's body was never found.

The old female's disappearance and assumed death resulted in a lack of cohesion among the five males. Their intragroup squabbles became more frequent and they resumed interactions with Groups 4 and 9, whose ranges overlapped their own.

Group 8's first encounter with Group 9 was held only days after Coco's disappearance and several ridges away from where she last had been seen. The tracker and I came upon Group 9 at unexpectedly close range, giving my assistant just time enough to dive under a large *Hagenia* tree before the gorillas became aware of us. Because of tall vegetation, I climbed into the same tree to gain a better view of Group 9. Within moments loud brush-breaking sounds were heard coming from below. Hiding myself in the tree's heavy vine growth, I was surprised to see Rafiki leading his bachelor band directly toward Group 9 without the chestbeats or hootseries that usually precede an intergroup encounter. The only obvious evidence of excitement at the initiation of the contact was the overpowering silverback odor, most of which was coming from Rafiki. Almost immediately Samson and Peanuts began mingling with three young adults of Group 9. Rafiki calmly made a day nest directly below me in the hollow bole of the *Hagenia,* unaware of the presence of myself or the tracker. Previously I had considered a gorilla's sense of smell superior to that of a human, but this observation did not support the supposition.

After nearly thirty minutes of quiet, my accidental breaking of a tree branch sounded like a pistol shot in the stillness of the resting period. Rafiki jumped from his nest and glared upward through the heavily vined skirts of the tree. Then the majestic silverback strutted deliberately around the trunk before posing stiffly some four feet below me. In an accusing manner he stared into my face nervously chewing his lips, one indication of his stress. Trying to act as innocent as possible and with anxiety only for the cramps in my legs, I gazed at the sky, yawned, and scratched myself while the old male indignantly displayed around the base of the tree unaware of the tracker huddled out of sight only several feet away from him.

Although curious about Rafiki's tolerance of a human's presence, the members of Group 9 eventually moved off to feed after contributing their own chestbeats and alarm vocalizations to this unexpected encounter. Rafiki instantly followed them, though I couldn't help but feel he had enjoyed being the intermediary between a gorilla-habituated human and a nonhuman-habituated group of gorillas.

* * *

The northwestern slopes of Visoke offered several ridges of *Pygeum africanum* trees shared by both Groups 8 and 9. The fruits of this tree are highly favored by gorillas, though such site-specific food prompts competition and increases opportunities for interactions between distinct social units. Groups 8 and 9 often met along the ridges for prolonged interactions because of their interest in obtaining the fruits.

Rafiki, more dominant and experienced than Geronimo, usually established Group 8's claim to the most prolifically fruiting trees higher on the slopes and Geronimo's Group 9 raided the lower-ridge trees. It was an amazing sight to watch the 350-pound silver-backs climbing onto thin tree limbs about 60 feet above the ground and harvesting with mouth and hands as many fruits as they could collect before climbing down to sit close to the tree trunks to gorge on their yield.

On one occasion Peanuts and Geezer, bored with the long feeding period, playfully galloped downhill toward several of Group 9's immature youngsters. The two Group 8 males failed to see Geronimo bringing up the rear of his group. Giving harsh pig-grunts, Geronimo immediately charged uphill. This caused the two young males to brake to a stop and momentarily stand bipedally, their arms around one another, their expressions panicstricken. Then both rapidly turned and ran back toward their group, all the while screaming fearfully. Geronimo pursued them to the top of the ridge, where he encountered Rafiki, who was running down to the defense of Peanuts and Geezer. Discretion prevailed when Geronimo turned heel and herded his group away from the bachelors.

The absence of Coco, coupled by frequent interactions with other groups, increased the unrest among the all-male Group 8. Pug and Geezer finally left their natal group to travel together on Visoke's northern slopes in a range area not too far removed from that of Group 8. Their departure left Rafiki only with his and Coco's presumed progeny, Samson and Peanuts. For nearly a year, however, squabbles continued between Rafiki and his oldest son. The friction occurred most often when the three males interacted with other groups and Samson's excitement grew beyond Rafiki's toleration. The old male had little difficulty in subduing Samson by either running or strutting directly toward his sexually maturing son, who would immediately assume a typical submissive posture by bowing down on his forearms, his gaze averted from his father and his rump upward. Rafiki needed only to maintain his stilted pose for a few seconds, his head hair erect, his gaze directed toward Samson, before temporary harmony was restored within the group.

Three and a half years after Coco's death Rafiki acquired two females, Macho and Maisie, from Group 4 during a violent physical interaction in June 1971. During the encounter Peanuts' right eye was permanently injured from a bite wound inflicted by Uncle Bert, the young silverback who had inherited the leadership of Group 4 three years previously following the death of his father, Whinny.

With the acquisition of the two new females Rafiki seemed invigorated. He staunchly defended his harem against Samson, thereby causing more friction between father and son. It was obvious that Samson was wasting breeding years by remaining in his natal group. He was prompted to leave just as Pugnacious and Geezer had done nearly a year before. Samson became a peripheral silverback, one who travels three hundred to six hundred feet from his natal group before setting out to establish his own range area and gain experience from interactions with other gorilla groups in order to

acquire and retain his own females. Both peripheral and lone travel
are usually necessary stages for any maturing male unless breeding
opportunities are available within his natal group. Samson's departure
left Rafiki with Maisie and Macho, the two young females taken
from Group 4, and with young Peanuts.

Unexpectedly, Samson returned from his distant ranging area and
managed to take Maisie away from Rafiki in September 1971. Fourteen months later Maisie and Samson were observed with a newly
born infant. In June 1973 Rafiki proved his own virility when his
only female, Macho, gave birth to a female infant named Thor.

Group 8 remained an oddly composed group, consisting of Rafiki,
his young mate Macho, his eleven-year-old son Peanuts, and his newborn daughter Thor. Seemingly content with his little family, Rafiki
no longer sought other groups. When Thor was about six months
old Rafiki was observed in one last interaction with Group 4. I noticed
that the regal old silverback's chestbeats and hootseries lacked reso-
nance and strength, though his physical appearance seemed as impressive as ever. Possibly he had been avoiding other groups because
he realized his physical limitations brought on by age.

*   *   *

In November 1971, five months after Rafiki had taken Macho from
Group 4, the trackers and I began an intensive search for Group
9, whom we had not seen for seven months. They were finally found
in the saddle area between Visoke and Mikeno, in almost exactly
the same spot where they first had been contacted four years earlier.
Instead of the thirteen robust individuals I had expected to find,
only five remained in Group 9. The once powerful body of Geronimo
had become gaunt, his muscular chest concave, his blue-black body
hair dull and patchy. His right hand was deformed and contracted,
perhaps as the result of a trap injury, and more wounds were visible
along back and thighs. I might never have recognized Geronimo
had it not been for the faded vestige of red hair in the center of
his forehead and the presence of Maidenform, one of the four females
he previously had in Group 9. I tried to conceal myself from their
view, but after an hour the ailing male knew that a human being
was nearby. Wearing a troubled facial expression, Geronimo, with
tremendous physical effort, kept trying to stand bipedally to scout
the surrounding area. His fear odor was strong, alarming his two
females and their young, who clustered near him ready to flee. I

had to reveal my presence but was satisfied when Geronimo seemed to recognize me and his group slowly resumed feeding farther west in the saddle area toward Mt. Mikeno.

I never saw Geronimo again, though Maidenform and several other females of Group 9 were later observed in two different groups ranging on Visoke's northwestern slopes and the saddle area west of Visoke. Whether poachers or natural causes were responsible for Geronimo's ultimate disappearance, I shall never know, but I feel that he died of natural causes. For several years his dung had become increasingly mucoid, often crawling with *Anoplocephala cestoda,* and he certainly had not appeared well when last seen. His death, of course, meant the end of Group 9 as a distinct social unit, because no gorilla family group can endure without a silverback leader.

With Group 9 no longer using the northwestern slopes of Visoke, chances for interactions between Groups 4 and 8 decreased markedly as additional range area reduced the amount of overlap between the two groups. Rafiki, however, remained content to pass the days slowly with his oddly assorted small Group 8, although Peanuts sometimes wandered alone as far as eight tenths of a mile away as if in quest of other groups for social interactions.

The social environment of little Thor, now eleven months old, was in definite contrast to that of Group 5's gregarious young with their numerous peer affiliations. The absence of playmates deprived Thor of rich learning opportunities. Her motor skills lagged about three months behind most eleven-month-old gorilla infants raised with the stimulus of interacting with others of their own age. Thor had to rely upon her mother, Macho, for tactile play and surrounding vegetation for solo play. Thor weighed about seven pounds less than the average eleven-month-old and was seldom seen more than ten feet from Macho at an age when other infants often played out of sight of their mothers. In addition to the lack of social incentives, Thor might have been less venturesome because she was Macho's firstborn and her mother therefore lacked previous experience in handling offspring.

My beloved Rafiki, a friend for seven years, was never able to witness the development of his last offspring beyond her eleventh month. In April 1974 the regal monarch of the mountain died of pneumonia and pleurisy, leaving Macho, Thor, and Peanuts as the sole remnants of Group 8. For about six days before his death Rafiki moved and fed very little, but during this period Macho and Peanuts circled around the weakening old silverback within distances of one hundred to two hundred feet for feeding purposes.

I received the news of Rafiki's death in Kigali, Rwanda, upon return from Cambridge, England. A student, on his way back to England, knocked at my hotel door carrying a large plastic bag that seeped liquid and an odor of putrefaction. Without preamble the student stated, "This is Rafiki's skin and I want to take it home with me." The ghoulish statement hit me with shattering force. This gruesome violation of the majesty, strength, and dignity of Rafiki seemed an intolerable sacrilege. I promptly confiscated the trophy, revolted by the request.

Rafiki's young silverback son Peanuts, about twelve years old at this time, was found traveling with Macho and Thor. Four weeks later, the inevitable happened. With the regal old leader dead and only the inexperienced Peanuts in "command" of Macho — an adult female without strong group affiliations — Uncle Bert led Group 4 into what had been Group 8's range. Peanuts was certainly no match for Uncle Bert. Twenty-seven days after Rafiki's death, eleven-month-old Thor was killed during a violent physical interaction between the two groups. Uncle Bert bit the infant fatally in the skull and groin, both typical infanticide wounds causing almost instant death. Macho carried Thor's body the remainder of the day before leaving it about thirty feet from her night nest. Eleven days following the infanticide, Macho was observed copulating with the sexually immature Peanuts. Five months later Uncle Bert took Macho away from the young male in yet another violent physical confrontation.

There followed a nineteen-month trial-and-error period for Peanuts, as he sought unsuccessfully to obtain females from other groups. Like all young silverbacks without breeding opportunities in their natal groups, he needed this travel period alone in order to gain experience in interactions for the acquisition of other individuals to begin his own group and to develop the necessary leadership skills required to hold his new group together against intrusion by other, more adept silverbacks. I found it distressing to encounter Peanuts wandering through the forest alone since I could easily recall him as a frolicking youngster living within his small family group.

Finally, in November 1975, Peanuts was found traveling with a younger animal I named Beetsme because of uncertainty as to the new gorilla's sex or background. Beetsme showed an unusual tolerance of observers and, since it had been acquired from the northwestern slopes where Group 9 had previously ranged, I felt that the animal was possibly one of Geronimo's offspring who had now matured to an estimated age of about ten years. For two months Beetsme and Peanuts wandered together until Uncle Bert again intervened and took Beetsme into Group 4.

Possibly to avoid further encounters with Uncle Bert, Peanuts shifted his range to the northern slopes of Visoke and out of Karisoke's study area. An entire year passed with only an occasional sighting or trail sign to confirm that Peanuts was still traveling alone. Then, in March 1977, Peanuts was found with five other adults, three of whom strongly resembled Geronimo's females. About fifteen years old now, Peanuts was considered sexually mature, but his vitality had dwindled considerably. The young male had never fully recovered from the bite wound received during the interaction in June 1971 when his father, Rafiki, had succeeded in acquiring Macho and Maisie from Group 4. The right side of Peanuts' face remained swollen, and his right eye was draining profusely. I thought it unlikely that Peanuts would be able to hold the females he had gained and that, indeed, Group 8 had ended with the death of noble Rafiki just as Group 9 had come to a close with Geronimo's disappearance and assumed death.

8 | Human Visitors to the Karisoke Research Centre

URING THE YEARS when there were four main study groups around the Karisoke study area, I was more than content to live for months without seeing anyone other than the camp staff, Kima, Cindy, and gorillas. After several years and some publicity about the research center, our peace became threatened by unannounced and uninvited strangers from the outside world. One day I had returned home early from Group 5 and was typing the day's observations when loud pounding at the door made the cabin shudder on its foundations. Opening the door, I saw a fairly attractive American man leaning against the door frame. Adorned with long beard and hair, wearing skintight bluejeans — not ideal apparel for trekking through the mountains — he faced me.

The stranger said, "I've come to see the gorillas."

His demanding tone triggered my own sense of hostility and I waved my hands in the air toward the saddle area south of camp and replied, "Go find them."

He retorted, "I'm going to stay right here and follow you the next time you go out for the gorillas, no matter when that is."

I answered, "You're going to have a long wait," and quietly shut the door. The stranger stalked off to rejoin his porter some sixty feet from my cabin, where both settled down to eat sardines and bread.

Quickly I convened the camp staff to plan the first of many tracking games designed to get rid of intruders. Twenty minutes later two of the men and I pretended to be sneaking out of camp to follow the gorillas. As we had hoped, the American hastily repacked his knapsack, tossed it to his porter, and followed stealthily behind us. Making sure that we had left ample tracks, the men and I walked for about thirty minutes before I hid in vegetation near the main trail. Several minutes later the American crept past, followed by his burdened porter, both trying to keep up with the camp staff. My men succeeded in leading the intruders on a three-hour route in and out of some of the most difficult ravines within the working area. I returned to camp feeling only slightly guilty for having misled the demanding stranger.

The onslaught of uninvited tourists, reporters, and photographers to camp was totally unexpected. Because Karisoke Research Centre is located in a public park, many intruders consider camp cabins to be public property. The doors and windows of the cabins are sometimes forced open and the Rwandese staff asked to perform chores for total strangers who attempt to turn the study center into a miniature tourist site at high season. One student, sitting in her outhouse, looked up to find a sightseer making the moment memorable with a long-lens camera! There were people who deserved hospitality, but whenever it was extended notice seemed to spread like wildfire that Karisoke was open to all comers, and even more strangers arrived unannounced.

Late one afternoon a large group of tourists appeared to demand cabin space and my services as their personal guide to the gorillas. The houseman pronounced me to be in Zaire while the woodman simultaneously insisted that I was in Uganda. Sensing some trickery, the strangers stubbornly pitched their tents about two hundred feet from my cabin. For three days and nights I was trapped in my cabin, emerging only for hygienic reasons and daily gorilla contacts. To sneak away, I had to borrow the woodman's clothing, wear a black knitted cap, and carry a light load of firewood until out of sight of camp.

One of the most unforgettable Karisoke intruders showed up in the summer of 1971 and was well on his way to my cabin before the staff could deter him. I was engrossed in map work when I heard a distinct British voice shouting, "Hello there. I say, anyone home?" Totally disbelieving my ears, I went outside and could only gawk at the incongruous figure approaching the door. He wore a dark wool suit, white shirt, loosened tie, city shoes, carried a briefcase, and looked every bit like a commuter who had got off at the wrong tube station. After a stilted conversation I learned the man was a freelance reporter from one of London's biggest scandal sheets and that he was determined to interview me. Instead, I pacified the journalist with tea, cookies, and two *National Geographic* articles I had written about gorillas and returned to my cabin to work. While he was "interviewing" the articles outside, an outbreak of hootseries and chestbeats was heard from Group 4, who were having an interaction with a lone silverback on Visoke's slopes just behind camp.

Denied his interview, the reporter left and I thought no more about him until some six weeks later when I received a copy of the newspaper he represented. Splashed across the front page was a photograph of myself complete with an unbelievable account concerning the gorilla research and the personal peril the reporter had undergone to obtain his on-the-spot story. The highly exaggerated article described his courageous solo climb through a jungle teeming with lions, tigers, and hyenas — a combination of animals that would be extraordinary anywhere outside of a zoo. He recounted how, upon his arrival, he had found my cabin surrounded by gorillas and how I had called them out of the forest to camp. The bogus article concluded ". . . and the locals call her Nyiramachabelli" — "the old lady who lives in the forest without a man."

Television crews constituted only a few of the interruptions by outsiders to the research at Karisoke. For the most part TV teams gave as much as they received, and their presence was usually sorely missed upon departure. This was especially true with the ABC team of nine, including Earl Holliman, and the "Wild Kingdom" crew, Warren and Genny Garst. The generosity of these people provided Karisoke with a generator, a refrigerator, and numerous other treasured gifts in the form of food, clothing, and equipment. Each of these teams also gave companionship and expressed concern for the fate of the mountain gorillas. Other television crews came to camp with schedules uppermost in their minds, unable to think beyond the lenses of their cameras or the needs of their own comfort. These people left camp residents with feelings of bitterness. In addition to uncaring professional photographers were uninvited tourists who insisted upon seeing the habituated gorillas of Karisoke Research Centre's study groups. Most of them tried to avoid both camp and Nyiramachabelli. Usually these tourists came in large, unruly aggregations and bribed Rwandese park guides to lead them to gorillas even though I had an agreement with the Parc des Volcans administration that the research animals were not to be disturbed by tourists.

Because of the distribution of Group 5's range near the eastern boundary of the park and the main porter trail to Karisoke, it was largely Group 5 that bore the brunt of tourist invasions, particularly during summer vacations and on weekends throughout the year. This held true even after other gorilla groups were semihabituated specifically for the sake of tourists.

On many occasions students or I went out to contact Group 5, only to find the animals' flee route filled with diarrhetic dung left as they had tried to get away from the mobs of people following them. The park guides soon learned to hide from me, but they were not reluctant to push Karisoke students aside. On several occasions guards threatened to shoot their guns in the air to frighten the gorillas if the students did not allow tourists to see the study animals.

Just as Groups 4, 8, 9, and a newly formed group (Nunkie's family) often had to keep diligent watch against poachers within more remote sections of the Virungas, Group 5 was forced to maintain vigilance against tourists. Icarus and Beethoven soon learned that they could scatter tourists by bluff-charging even though the guides' guns pointed directly at them — two silverbacks who sought only to defend their family against a growing and demanding public.

In their greed to obtain photographs, tourists and uninvited film crews came to pose almost as much of a threat to the gorillas as poachers did. One French film team, noted earlier, relentlessly pursued Group 5 daily for six weeks. The trauma caused Effie to abort. Group 5 then moved from their normal range within a poacher-free area to flee into the interior of the park, where tourists seldom ventured but where poacher traps were abundant. The French team triumphantly returned to Paris to air an acclaimed television documentary, leaving Group 5 to recover slowly from the Gallic invasion and the Karisoke staff to herd the group out of the trap zone.

* * *

Two years after establishing the Karisoke Research Centre, and while taking care of Coco and Pucker, I grudgingly had to admit that I was only one person. If I wanted the research and the conservation aims of Karisoke Research Centre expanded, I was going to have to enlist the aid of student assistants. Trying to be helpful as always, Dr. Louis Leakey sent a twenty-year-old American who thought he wanted to pursue field work in Africa. He collapsed at my feet following his three-hour climb to camp. Between heavy panting breaths he whispered, "I'm not going to be able to take this." In the heartland of the forest he had immediately realized he could not cope with the solitude combined with the physical exertion required by the rugged terrain. My heart sank as I listened to him, but I did not then realize just how exceptional this young man was in recognizing straight off that gorilla research was not meant for him.

I was yet to learn that the symptoms manifested by persons who were to arrive at Karisoke and find themselves unable to adapt to the work at camp or census studies were strikingly similar to those of some astronauts undergoing isolation-training for outerspace missions. The malaise may include sweating, uncontrollable shaking, short-term fevers, loss of appetite, and severe depression combined with prolonged crying spells. I termed the condition "astronaut blues," a very real sickness. Once I realized just how seriously it affected some individuals, I never tried to encourage them to remain at camp and pursue field work.

The second person to arrive at Karisoke was Bob Campbell, the *National Geographic* photographer who documented Coco and Pucker's last days at Karisoke so thoroughly. On and off for a period of nearly three years, Bob was of considerable help in keeping up with the four main study groups, antipoacher patrols, cabin building, training more Rwandese staff assistants, and in repairing the kerosene cooking stoves and lamps. It was always so frustrating to return to camp at dusk after a grueling rainy day of climbing and taking pages of notes to find that one's cabin lamp was not functioning because of a broken mantle, needle, or some mysterious malaise that required a total dismantle of various parts like springs, coils, washers. Bob Campbell was one of the few persons ever to come to Karisoke who had the patience to teach the camp staff — unfamiliar with kerosene lamps — how to keep the "genie of the forest" operable at a time when such lamps or parts were not available in Rwanda. Since one of my strongest beliefs was, and remains, that field notes should be typed up and analyzed the evening of the day in which they are taken, perfectly functioning kerosene lamps became a near obsession of mine. Kerosene cooking stoves were equally temperamental but took second place on my list of priorities simply because typing notes was more important than filling an empty stomach. A stomach can wait to be filled but daily impressions of contacts with the gorilla grow stale if not immediately recorded.

As research among the Karisoke study groups expanded I became more curious about the fringe groups ranging outside the study groups as well as the number of gorillas in the Virungas. When George Schaller ended his excellent field study in September 1960, he had estimated the total population of mountain gorillas as being between 400 and 500 animals. Regrettably the political situation at that time made it impossible for Schaller to conduct an accurate count of the gorilla population in the Rwandese sector of the volcanoes. Because of the six months spent at Kabara in 1967, I had been able to correlate three Kabara study groups with three that Schaller had studied six and a half years previously. The comparison was made on the basis of some similarities in the groups' composition, photographs of certain outstanding individuals, and, in particular, group range boundaries.

The most obvious change that had occurred in the three groups during the interval between Schaller's and my study periods at Kabara was the reduction in number of gorillas from 20 to 12 individuals. This meant a loss of at least 12 animals, since four were known to have been born after Schaller's period of study. Another outstanding difference was the change in the ratio of adults to subadults from 1.2:1 to 2:1. There had also been a marked reduction of four, three, and one square miles respectively in the ranges of each of the groups.

These realizations made it essential to obtain an updated count of the remaining Virunga gorillas. Land encroachment was expanding, so I felt it necessary to know precisely where gorilla populations were concentrated in order that long-term conservation efforts might be enforced in those areas.

In 1969, with the help of Alyette DeMunck and Bob Campbell, I began census work — actually counting gorillas and evaluating gorilla group ranges within the Virungas. Census-takers lived in bivouac camps with virtually everything in the way of food and extra clothing carried in knapsacks on our backs. The small tents, a portable typewriter, a small lamp and stove, a few pots, water containers, and sleeping bags were carried by two porters. The duration of each temporary camp depended as much on access to water as on the frequency of gorilla sign discovered within a reasonable four-hour walking distance from camp. Eventually, with the assistance of students, recruited mainly through the mail, the arduous work was expanded and continued on a near yearly basis. The physical challenges involved covering each of the six Virunga mountains from saddle to summit, exploring every gully, ravine, and slope. If it had been an easier task it might have been attempted after Schaller's initial groundwork. Personally, I found the explorations throughout the volcanoes some of my most memorable forest experiences — the challenge of the search, the thrill of encountering a new gorilla group, the awesome beauty of the mountains revealed by virtually every turn in a trail, and the pleasure of making a "home" with only a tent and the benevolence of nature.

Long before attempting to recruit European and American students for the work, I trained a number of Rwandese to track gorillas and handle the less complicated chores, like securing water and firewood, needed to maintain bivouac camps. On daily treks into the forests it was necessary to record both old and new signs of gorilla use and include feeding or nesting remnants and dung deposits. All these signs were entered on a contour map to show frequency with which gorillas use an area over time.

At Karisoke later census-takers were made familiar with the relationship between dung size and age and sex classification of an individual, though there was some unavoidable degree of inaccuracy when distinguishing between the ages and sexes of immature animals. Whenever fresh gorilla trail (one less than four days old) was encountered, it was to be followed by a contact with the group, or at least by concise night-nest counts. I preferred that the age and sex of these shy gorillas be confirmed by census workers with no less than five consecutive night-nest counts of each group. Night-nest counting, although tedious, was absolutely necessary to determine the presence of infants, so often obscured during contacts with unhabituated groups, as well as to determine the presence of peripheral males — gorillas who might build their nests several hundred feet away from the main cluster of group members' nests.

Once a group was contacted, observations with binoculars allowed noseprint sketches to be made of the more forward individuals. These simple line drawings of nostril shapes and wrinkle patterns indented along the bridge of a gorilla's nose distinguished the individuals of one group from those of another, especially when the groups were similar in size. Such graphic records were supplemented by written descriptions of variations in behavior and vocalizations, traits that helped to identify members of distinct groups.

In the summer of 1970 Dr. Leakey sent a second student to Karisoke as a census-taker to continue the count that Mrs. DeMunck,

Bob Campbell, and I had begun a year earlier. For two weeks the young man was introduced to the main gorilla study groups, instructed in basic Swahili, and versed in daily camp routines. Once he seemed at ease with the work, Bob and I, along with Rwandese porters, trekked with him to the northern slopes of Visoke to set up the first student-operated census camp. I had chosen a site called Ngezi, Kinyarwanda for "a place where the herds water," where gorillas were known to abound.

We chose a lovely campsite near a small lake that was visited nightly by vast herds of elephant and buffalo. Bob and I spent three days exploring adjacent terrain with the student. We found no fresh gorilla spoor but were able to account for numerous nesting signs about a week old. Satisfied that the work would prove valuable, Bob and I returned to Karisoke to concentrate on the four main study groups. In the following weeks, a steady stream of porters from Ngezi brought me distressing reports of the young man's activities, many of which were not related to census work. I was left with no alternative but to send him back to America.

The next eleven years saw some twenty-one census workers come to Karisoke to train for research work in various outposts within the Virungas. The majority of these students did not succeed in coping with the rugged demands of the work. Many returned home after only brief stays. Unwittingly I had expected anyone who applied for census work to be as ecstatic as I over the wonder of the mountains and privilege of meeting gorillas. It never dawned on me that exhausting climbs along ribbons of muddy trail, bedding down in damp sleeping bags, awakening to don wet jeans and soggy boots, and filling up on stale crackers would not be everyone's idea of heaven.

Most census workers, as well as Karisoke Research Centre's eventual research assistants, were chosen by mail application accompanied by impressive academic references. In-person selection was rare because I seldom left camp during the initial days of the study and spent only brief periods in America or England. However, I interviewed applicants whenever possible.

It soon became obvious that it was absolutely impossible to determine how even the most promising individual was going to function in the remoteness of the forest. Certainly all candidates felt that previous camping and backpacking experience in America or Europe, plus a sincere interest in studying the gorilla, qualified them for field work in the Virungas. Although I stressed the discomforts and social isolation students would undergo, the combination of their enthusiasm and my optimism that newcomers would succeed led to varying degrees of unpleasantness between myself and some of the individuals who ended up at Karisoke. Another cause of difficulty between us was my inclination to think of Karisoke as a functioning whole, while, understandably, many students had only their own interests to consider. Conflicts between student researchers and myself arose most frequently over the need to carry on with antipoacher patrols, notably when traps were found in the study area; or over the building of additional cabins and the maintenance of the present cabins and equipment such as lamps, stoves, and typewriters; or over the ever-demanding task of keeping up with daily field notes for Karisoke's long-term records. Indeed, in most field stations, differences between residents may grow out of all proportion when individuals from totally different backgrounds are obligated to live together in a secluded environment. The conflicts may have been more marked at Karisoke than at other research facilities because of the inclement weather, the high altitude, the boring diet, and — for many students — the social isolation.

The selection of study topics posed little problem — there was so much to be learned about gorillas' behavior and the ecology of the Virungas. Other than research in general, a number of students had no specific interests and were free to choose from a list of projects that included dominance behavior, infant development, maintenance activities (feeding, grooming, nesting, maternal care), vocalizations, interactions between groups, ranging strategies, parasitology, and botany. The National Geographic Society continued its very generous maintenance of Karisoke Research Centre, myself, and the African staff; doctoral students were usually supported by their universities or agencies with whom they made private funding arrangements before arrival at Karisoke. Whenever special needs arose by way of equipment or funding of short-term research projects, the L. S. B. Leakey Foundation ever so kindly assisted. Both the National Geographic Society and the Leakey Foundation supported general research assistants, though I did not like to request money for researchers' air fares or salaries. I had never been salaried and felt that the research should be its own reward.

*   *   *

In early 1975 an absent-minded professor came to Karisoke for a concentrated three-month botanical study project. All his travel expenses, equipment, and supplies had been paid for by a grant I had obtained. Two stipulations of the grant were that the new equipment was to remain at Karisoke when the botanist departed and that a full report of his findings was to be prepared for publication within a reasonable period following his return to the United States. It is to be regretted that neither condition was met. Within eight days of the botanist's arrival, he unintentionally burned his cabin to the ground by hanging flimsy plant-drying racks over the wood-burning stove. All of the new equipment, the furnishings, my irreplaceable botanical library, other rare books, and Karisoke's new shortwave radio were lost in the fire. The camp staff and I fought the blaze for hours, hauling buckets of water from Camp Creek about eighty feet away. By the end of the day, little remained of the cabin and its valuable contents except for a charred, soggy, smelling mess. The Africans and I had been subjected to severe smoke inhalation and other injuries. We had just collapsed near the smoldering ruins when the botanist returned from his day in the field. In so many brief expletives, he expressed annoyance at the temporary interruption of his study. For myself and the Rwandese staff, this was the first of several disasters that were to befall the camp we had built together in the wilderness.

A second cabin was destroyed when a student left clothes drying on top of the fireplace. For some weeks following this fire the person responsible conscientiously set about rebuilding the cabin. Her efforts helped me regain some degree of faith in the caliber of the people who came to work with the gorillas.

Another student had no sense of direction even with a compass and marked trails. The trackers and I came to accept his idiosyncrasy, though we spent many hours organizing and conducting search parties for the wanderer. In spite of Karisoke's regulation that everyone was expected back at camp by five-thirty (except under extenuating circumstances when accompanied by a tracker), the staff and I soon learned to look for the student in the most unexpected places, often in the totally opposite direction from where the tracker had left him, and always at night. Yet this shy and somewhat solitary person developed an excellent rapport not only with gorillas but also with the mischievous Kima and Cindy. Over the ten months of his stay we made a number of field contacts together, and I was pleased to find that he put the needs of the gorillas above his own. He never pushed the animals beyond the limits of their tolerance, a practice that many students failed to follow.

Conflicts in respect of the rights of the gorillas caused some problems between myself and several students who were primarily concerned with obtaining observational data for their doctoral degrees. One person made it a habit to defecate when in the midst of Group 5, apparently not realizing the possibly dire consequences such unsanitary actions might inflict on the group's members. When I criticized this behavior, I was angrily told that observations could not be interrupted even by the calls of nature.

One twenty-year-old proved himself a highly valuable census student and finally worked from camp as a doctoral candidate. Because of his initial dedication, I left him in charge of Karisoke during a brief stay at Cambridge University. In my absence he proved to be an asset in the maintenance of both field research and camp records. For about a year and a half, the young man alternated periods at Karisoke with visits to his university in England. Then one day, possibly because of his increased self-confidence and experience in the field, he made the near-fatal mistake of trying to outbluff a female buffalo standing on a trail some five feet above him. He snorted at her in an attempt to frighten her away, but instead received the full impact of her justified anger. The buffalo charged, rolled on him, and gored his body numerous times. Although near death, the student somehow made it back to my cabin, where he passed out from loss of blood. I had good reason to be thankful for my hospital training while treating shock, severe deep puncture wounds, and multiple lacerations. In the midst of his delirium the student muttered, "I was bloody damn stupid." I nursed him around the clock for four days before he was able to travel to England for necessary surgery.

There were some students who came to feel as at home in the Virungas as I and who also cherished the animals of the forest above their personal interests. In the summer of 1976 I was driving back to the base of Visoke after depositing four census workers near the base of Mt. Mikeno in Zaire. No other students remained at Karisoke, and I was worried about how the African staff and I were going to be able to keep up with the main study groups, the fringe groups, and maintain antipoacher patrols. I stopped to offer a heavily laden hitchhiker a lift. Tim White, an American traveling around the world on his own, turned out to be just about everything anyone would wish for in a field-station assistant. Tim expected to pass one day in the mountains; he stayed ten months at Karisoke. He repaired cabins and equipment, tracked study groups, fringe groups, census groups, and, after some frantic lessons, typed up his field notes each night. His even temperament was a boon, not only to myself but to the Rwandese staff and some students who eventually learned from him.

Although at first a dedicated pacifist, Tim soon realized that the illegal slaughter of animals and the presence of poachers in the park could not be tolerated at Karisoke. I quickly came to rely upon his help with antipoacher patrols. When new students arrived at camp, Tim decided to continue his travels. He ended up spending nearly six years in Africa, and for eighteen months volunteered his services at a mission hospital in Liberia. I feel that everyone who meets Tim White must be greatly affected by his basic goodness and selflessness. For giving so totally of himself when we needed it, Karisoke will always remember him.

When Ric Elliot applied from England to work at camp, his sparing use of the words *I* and *me* in his letters conveyed the impression that he was a person who wanted to contribute toward the aims of Karisoke rather than someone who needed the experience for his own purposes. The ten months Ric spent at camp proved this to be so. Although their backgrounds differed, both Ric's and Tim's grandfathers had been carpenters and each of the young men truly enjoyed building and maintaining cabins and equipment. Ric was especially interested in veterinary medicine and was of great help in expanding gorilla autopsy work and parasitology studies. His departure left a void for the Rwandese camp staff and myself.

A year and a half after Ric departed, Ian Redmond, also from England, eagerly took up the parasitology project where Ric Elliot had left off. Ian relished the time spent hunched over a microscope in pursuit of new species of gorilla nematodes and cestodes. He was fanatical about the work, and his enthusiasm managed to dispel even the initial perplexity of the Africans, who like myself were somewhat overawed by Ian's dedication and the hundreds of bottles, trays, and plastic specimen sacks that accumulated from his work. Ian's curiosity extended to all animals of the forest, from elephants to frogs. Soon Karisoke began to resemble a natural history museum as he painstakingly collected and categorized bones and other bits and pieces of deceased animals, birds, and insects. I made it a point to enter Ian's cabin as infrequently as possible, never sure of what I might find added to his odoriferous collection!

The Africans truly loved Ian Redmond. His favorite way of ending a day was to sit around their fireplace to share an evening meal of corn, beans, and potatoes or whatever vegetables were on hand. No European was ever more at home in the forest than Ian. He thought nothing of going out on poacher patrols or of making census counts. Ian would easily cover as much as ten miles a day when tracking either gorillas or poachers, and, when far from camp, he often spent the night comfortably huddled up under a huge *Hagenia* tree with moss for a mattress and a poncho for a blanket. The Africans accompanying him on such excursions never complained, because his enthusiasm was contagious. He usually wore shorts even in nettle fields. I found this rather puzzling, since Ian was not a pretentious person. One bitterly cold day as he was cheerfully setting out in shorts and sweater, I asked Ian just what he was trying to prove. He answered slowly, as though embarrassed.

"Dian, when you wear shorts in the field, you are more aware of your surroundings. You can sense the difference between the saddle area with its soft vegetation, the marsh plants of the meadows, and the bleakness of the alpine zone." His speech faltered as he tried to expand on his feelings, suddenly shy as though he had revealed too much of the depth of his sensitivity for all that nature had to offer.

For Ian, no gorilla group was too far away from camp to investigate, nor were any traplines too distant to destroy. Shortly before Ian had to return to his family in England, a tracker told us of a fringe group encountered on the opposite side of Visoke. With no qualms whatsoever, Ian and the tracker left camp early the next day to attempt to identify the individuals and to make several night-nest counts. The terrain on that side of the mountain consists of numerous ridges, favored by poachers for trap-setting. The fringe group sought by Ian had avoided the ridge section by moving off the mountain into the adjacent saddle.

The day's trek was an unusually long one. Ian and the tracker were patiently following the gorillas' trail when they encountered three newly set wire duiker traps. While routinely breaking the bamboo poles and confiscating the wire snares, they heard saplings being cut only about 150 feet away from them. Both Ian and the tracker dropped out of sight behind a small knoll to wait quietly for the trap-setters to leave so that they could cut down the freshest traps. When all became quiet, Ian was on the verge of standing up to see where the poachers had gone when three spear tops suddenly came bobbing into view only a few feet away. Until that moment, neither party had been aware of the location of the other. It was ironic that the poachers had chosen to climb onto the very knoll concealing Ian and the tracker.

Slowly Ian stood up to show the poachers that he was completely unarmed. Nonetheless, the unexpected closeness of a *Bazungu* (European) startled the encroachers. In the shock of the encounter two men fled. The third, facing an eye-to-eye contact with Ian, dropped his *panga* and with both hands plunged his spear directly toward Ian's heart. Instinctively, Ian put his left arm over his chest and crouched down. His left wrist took the full force of the spear and undoubtedly saved his life. When he realized what he had done, the poacher, as Ian put it, "showed a clean pair of heels."

The wrist wound was serious, but after binding it up, the student and tracker went on to cut down the additional traps. Only then was Ian willing to return to camp and go down to the Ruhengeri hospital for treatment. His wrist eventually healed, although it was never the same as before.

Tim White, Ric Elliot, and Ian Redmond remain outstanding among the students who worked at camp as the three who contributed to Karisoke not for personal accolades but for what they could do for the gorillas and active conservation within the Virunga Volcanoes. Not only do I remember them as very special people, but they are also well remembered by the African staff as among the best friends we have known.

It goes without saying that camp never could have existed without the loyalty of the African assistants who came to believe that this remote outpost, the Karisoke Research Centre, could further the dual aims of conservation and research. We came to share a common goal by working together for the future of the wildlife of the Virungas. The nature of our work was as simple and direct as our start with only two tents in 1967 had been. We saw the expansion of our aims with the support of Africans from all walks of life, people such as Paulin Nkubili, Mutarutkwa, and many other Zairoise and Rwandese who joined the antipoacher patrols. These men, much like Tim, Ric, and Ian, believed that their individual conservation efforts need not be heralded nor widely acclaimed: the satisfaction came simply from doing. As the forest will always be my real home, these men will always be my truest friends. They learned from me as I learned from them. Together we made the dream of the Karisoke Research Centre an actuality.

Our biggest camp sprees occurred around Christmas time when the entire camp was decorated with candle-lit trees, garlands made from tinfoil, popcorn, and other homemade ornaments. Beneath the "big" tree in my cabin would be mounds of wrapped gifts I had brought from various trips overseas for the staff and their families. At least fifty Rwandese and Zairoise showed up at Karisoke for our Christmas celebration, bringing their wives and children all dressed in their very best clothing. We spent the day eating, drinking, and singing carols in Kinyarwanda, French, and English — accompanied sometimes by the wailing of the many babies unaccustomed to so much commotion.

On the third Karisoke Christmas, the staff unexpectedly asked me to take a seat while I was in the process of passing out soft drinks to the children. Mukera, the head woodman, also a talented drummer and dancer, hauled out a big African drum kept in a corner of the cabin's living room. He initiated the first of the song-and-dance routines that were to become a large part of our Christmas parties for the following years. Each of the men on the staff had composed his own song and original dance pattern describing events that had occurred during the previous year. As each man sang, danced, and drummed his individual composition for me, I was overwhelmed by their creations. In the years following, I taped their singing and dancing — dancing that literally stamped the sawdust out of the floorboards. These tapes remain among my most treasured memories of Rwanda and Karisoke.

9 | Adjustment to a New Silverback Leader: Group 4

A S JOYOUS AS CAMP LIFE OFTEN COULD BE, the gifts of the forest were by far the most meaningful, particularly once the trust of the gorillas had been gained. I felt that Karisoke research was off to a propitious start when, on its first day of existence, two Batwa poachers introduced me to a gorilla group they had heard vocalizing from Visoke's slopes while hunting duiker in the meadows west of camp. The two men led me to the group, which I named Group 4.

For nearly forty-five minutes Group 4 was unaware of my obscured presence on the opposite side of a ravine about ninety feet away from them. With binoculars I was able to distinguish three very distinctive animals. The aged, dominant silverback was the first to see me peeking from behind a tree. Whinnying in alarm, he fled back to his group, at times even somersaulting on the steep slope to get there faster. The old leader was named after his strained and raspy vocalizations, which sounded rather like a horse's whinnies. I had never heard this vocalization previously from a gorilla, though it was heard once by George Schaller. Later I realized that the sound was atypical and, in Whinny's case, caused by advanced lung deterioration.

Whinny was followed by a bright-eyed, inquisitive ball of fluff who came to be known as Digit because of a twisted middle finger that appeared once to have been broken. The close facial resemblances and the youngster's strong dependency on Group 4's dominant silverback led me to believe that Digit was Whinny's son. Digit associated with none of the four adult females within the group and it seemed likely that his mother had died before I met the group in September 1967.

On our first meeting, Digit, estimated at about five years old, gave the impression of wanting to take a second look at the cause of the alarm, but obediently heeded Whinny's warning call and scurried to catch up with him. Then all the animals fled from sight except an adult female who maintained a rear-guard position. Looking as though she had just swallowed a mouthful of vinegar, she glared at me with compressed lips as she stood in the rigid strut stance usually adopted only by adult males. Because of her facial expression, I could not help thinking of her as an old goat. Soon Old Goat became her name.

Besides Whinny there were two other silverbacks in Group 4. The younger was named Uncle Bert, owing to a remarkable resemblance to a relative of mine. (I considered the epithet a compliment, but my uncle never quite forgave me.) The third silverback was eventually named Amok for his unstable temperament and his puzzling frequent screaming and running displays — aberrant behavior possibly caused by chronic illness. Amok's trails, usually found just outside Group 4's or within three hundred feet, were consistently spotted with diarrhetic, mucus-covered, blood-flecked dung deposits.

Amok, about twenty-five years of age, appeared too old to have been sired by Whinny, so I concluded that the cantankerous silverback was Whinny's half brother and that they had shared a common parent in the past. Uncle Bert, estimated at around fifteen years old, closely resembled Whinny and was well tolerated by the dominant silverback; this suggested that he had been sired by Whinny.

During the first several months of observation with Group 4, Whinny's health began to decline. Old Goat, who had no offspring when the group had been first contacted, assumed an increasing share of the leadership duties. I found her masculine physical and behavioral characteristics most unusual. It was Old Goat, rather than the young silverback Uncle Bert, who acted as Group 4's watchdog when Whinny lagged behind them.

Because of the female's behavior I often tried to maintain obscured contacts with Group 4 to avoid upsetting the remaining female members of the group: four young nulliparous females, two older multiparous females, who had already had offspring, and one infant female. Over a period of time, the young females, between the ages of six and eight years, came to be named Bravado, Maisie, Petula, and Macho; the two older females, Flossie and Mrs. X; and the infant, Papoose. The gorillas' names seemed to fit their personalities. Papoose was an adorable snuggly youngster whose mother disappeared not long after the study began. Mrs. X was always difficult to identify because of her shyness during the first several months. One of the young females had unusual wide, haunting eyes; her name Macho is a Swahili word meaning "eyes."

In mid-November 1967 I found the fourteen-member group on the opposite side of a wide ravine. Settling down behind heavy brush, I saw Old Goat and a second adult female, Flossie, traveling slowly together about a hundred feet below the rest of Group 4. Ponderously Flossie heaved her vast bulk down against the slope to pluck idly at *Galium,* revealing a shiny, black-headed, newly born infant squirming within the clutch of her left arm. The pink skin of the baby's palms and soles was in marked contrast to the shoe-wax sheen of its dorsal head hair as it rooted for Flossie's nipple without obvious assistance from the new mother.

As Old Goat awkwardly climbed the slope, Flossie moved farther uphill, walking quadrupedally with her infant clinging independently to her lower chest. On reaching Flossie's resting spot, Old Goat also sat and leaned against the bank, her ventral surface toward me. She relaxed her left arm and exposed a newborn infant from which hung about four inches of umbilical cord. Her baby's hands were tightly flexed against its wrists and its feet dangled limply. Old Goat looked down intently, almost quizzically, at her newborn before nuzzling her baby and drawing it closer to her.

While Old Goat rested, her baby's head hung to one side as though connected by rubber bands to its body. The instant Old Goat prepared to stand and move toward the rest of Group 4, the infant — almost in a spasm — tensed its entire body, straightened its head, and automatically clasped its mother's ventral body hair with its spiderish fingers. As Old Goat climbed, I noticed she walked with a mincing gait that assured her infant of partial support from her inner thighs, a typical travel pattern of experienced mothers. This suggested that Old Goat had previously given birth, though she did not associate closely with Group 4's two immatures Digit and Papoose, nor was she old enough to have been the mother of any of the young female adults. As I watched her rejoin the resting animals, I called her infant

Tiger, sure that any offspring of Old Goat's would live up to the name. At the same time I named Flossie's newborn Simba, a Swahili word meaning "lion."

* * *

Tiger and Simba were the first of 42 infants that were to be born among the 96 members of Karisoke's five study groups over the next years. Like most of those newborns, Flossie's and Old Goat's infants exhibited typical physical and behavioral characteristics.

The body skin color of a newly born gorilla is usually pinkish gray and may have pink concentrations of color on the ears, palms, or soles. The infant's body hair varies in color from medium brown to black and is sparsely distributed except on the dorsal surfaces of the body. The head hair is often jet black, short, and slick, and the face wizened, with a pronounced protrusion of the nasal region, giving a pig-snouted appearance. Like the nose, the ears are prominent, but the eyes are usually squinted or closed the first day following birth. The limbs are thin and spidery, and the digits typically remain tightly flexed when the baby's hands are not grasping the mother's abdominal hair. The extremities may exhibit a spastic type of involuntary thrusting movement, especially when searching for a nipple. Most of the time, however, a gorilla infant appears asleep.

The short durations of suckling observed for the first month of life seldom continue over fifty seconds. At that stage, suckling is usually accompanied by searching or rooting motions of the head. During the first year no marked breast preference is observed. As an infant ages, however, a choice for the left breast nearly doubles the time spent on the right breast. A newborn is always carried in a ventral position when the mother travels. When the mother is seated, the infant is cradled against her chest or held in her lap. I estimated a newborn's weight as approximately three and one-half pounds, an infinitesimal fraction of the weight an adult gorilla will attain.

As is typical following births, Group 4 traveled more slowly than usual and day-rested for prolonged periods. I was relieved, because Whinny had been having difficulties keeping up with his group even though they previously had slackened their pace for his benefit. By the time Tiger and Simba were two months old Whinny's body frequently was racked by prolonged coughing spasms that varied from high rasping gasps to deep harsh croaks. After such seizures, the old male would sit trembling, his eyes closed and his mouth twisted in a puckered, pained expression. Only Old Goat, on the alert as always, continued to sit nearby frequently looking at him with concern, though Whinny seemed unaware of much that was going on around himself.

One day I found Whinny sleeping alone, totally impervious to vegetation noises made by the tracker and myself while following Group 4's trail. I sat about fifteen feet away and watched the old silverback for nearly two hours as he dozed. He was lying in an unusual position on his stomach, with his head downhill to help his laborious breathing. Upon awakening he lethargically picked a few thistle leaves before weakly following Group 4's trail, which led up into the alpine zone — a high-altitude region soon denied to him because of his failing health.

By March 1968 Whinny no longer could make the effort to keep up with his group, now ranging on the mountain slopes. The old male remained alone in the relatively flat saddle area adjacent to Mt. Visoke, an area Group 4 had not been known to use. He wandered in ever narrowing circles, feeding little, spending most of the time resting. He seldom traveled more than fifty feet a day during the last month of his life, weaving from tree shelter to tree shelter, leaving vast pools of diarrhetic dung behind. The trackers and I made daily obscured contacts with Whinny in an effort to guard the dying silverback's last days against poachers who might stumble across his path. He was never aware of our presence nor did he indicate that he heard any of the distant intragroup outbreaks created by the increasingly frequent squabbles in Group 4.

On May 3, 1968, Whinny's emaciated body was found in his night nest, fourteen branches folded neatly around him. His was the second death among the Karisoke study groups. The men and I tied his body on to a sapling stretcher to carry down the mountain to Ruhengeri for an autopsy. Examination of the organs revealed that Whinny had advanced peritonitis, pleurisy, and pneumonia. Later, skeletal analysis showed extensive pitting on the right side of his skull, indicative of a spreading infection that suggested meningitis.

Following Whinny's death Old Goat assumed leadership by determining direction and speed of Group 4's movement in travel, settling intragroup squabbles, and even chestbeating or directing intimidating actions toward me when I attempted open contacts. At first Old Goat was supported only occasionally by Uncle Bert, who was about five years younger than she and seemed unsure of the responsibilities he had inherited upon Whinny's death. The amiable young silverback was more interested in cavorting with Group 4's younger members, all strongly attracted to him.

Once, from a hidden position, I watched Digit, about five and a half years old, tumble onto Uncle Bert's lap much like a puppy wanting attention. From a lazy sunning position, Uncle Bert had watched the youngster's approach, then quickly plucked a handful of white everlasting flowers (*Helichrysum*) to whisk back and forth against Digit's face as if trying to tickle the young male. The action evoked loud play chuckles and a big toothy grin from Digit, who rolled against Uncle Bert's body, clutching himself ecstatically before scampering off to playmates more his own size. I had been pleased to note that little Digit had undergone only a brief period of dejection following Whinny's absence and soon had formed close associations with other group members, particularly the four young females thought to be his half sisters.

Play, along with sexual behavior, is one of the first activities inhibited by the presence of an observer until the gorillas become fairly well habituated. This is particularly true with play behavior of small infants, whose mothers and fathers are extremely protective during the first two years of their lives. However, Digit and his half sisters were young adults when I first encountered them, thus were seldom guarded in their play activities. The freedom of their play depended to a great extent upon the type of contact I initiated with the group. During obscured contacts, when Group 4 did not know I was observing them, Digit and his young sisters engaged in prolonged wrestling and chasing sessions as far as fifty feet from day-nesting adults. The constant repetition of their actions seemed almost deliberate, simply to provoke a response from their partners. One by one the youngsters would wear themselves out before joining older members of the group for a resting period. During open contacts, when the group members knew I was present, a great part of the immatures' play behavior involved response reactions such as chestbeating, foliage whacking, or strutting. Each individual seemed to be trying to outdo the other in attention-getting actions. Their excitement was contagious and there were many days that I wanted to join in their escapades and could not do so until they lost their apprehension of my presence.

On one occasion, Group 4 were crossing a high grassy slope sup-

porting several rows of giant *Senecio,* the tall one-legged sentries of
the alpine zone. Led by Uncle Bert, the five young ones playfully
began a square-dance type of game, using the *Senecio* trees as "doh-
see-doh" partners. Loping from one tree to the next, each animal
extended its arms to grab a trunk for a quick twirl before repeating
the same maneuver with the next tree down the line. The gorillas,
spilling down the hill, resembled so many black, furry tumbleweeds
as their frolic resulted in a big pile-up of bouncing bodies and broken
branches. Time and time again, Uncle Bert playfully led the pack
back up the slope for another go-around with the splintered tree
remnants.

The first indication I had that the young silverback was aware of
life's more serious side immediately followed Whinny's death, when
Amok tried to rejoin Group 4 after several months of peripheral
travel. A violent outbreak of screams and roars was heard from Group
4, high on Visoke's slopes. Climbing toward the sounds, I saw Uncle
Bert rapidly run uphill toward a clustering of group members before
all fled from sight.

They had left a large area of trampled, blood-splattered vegetation
where Amok sat, slouched motionless under a tree, his head resting
on his chest. Minutes later Amok reached out for a few strands of
Galium, his face contorted into a grim expression of pain. Slowly
he began licking the forefinger of his right hand, passing it repeatedly
back and forth from his clavicle to his mouth. Only when he dropped
his arm could I see that his entire chest was covered with blood
coming from a deep, four-inch-long bite wound at the base of his
neck. For the rest of the afternoon Amok alternately rested or weakly
groomed his wound. Near dusk he laboriously built a night nest.
From that day onward, and for the next six years, the older ailing
silverback was never again observed mixing with the members of
Group 4.

* * *

Two weeks after this encounter, Group 4 had their first physical
interaction with Rafiki's Group 8. Uncle Bert was no match for the
bachelor band and showed his inexperience by excitable runs among
his own group members or incautious charges toward the males of
Group 8. The tyro leader had yet to become skilled in the more
subtle avoidance tactics — parallel strutting or chestbeating displays — and was fortunate in sustaining only minor wounds by the

end of the interaction. Group 4, with their inexperienced leader and four young adult females, all about to reach sexual maturity, was to become a frequent target for intergroup interactions, particularly with the all-male Group 8.

Group 4's third older adult female, Mrs. X, gave birth to Whinny's last offspring the same month in which the old silverback died. Simultaneously, Flossie lost her seven-month-old from unknown causes. Wanting to perpetuate the name Simba, I gave it to Mrs. X's infant female. Ten months later Flossie gave birth to the first offspring sired by Uncle Bert.

Like human mothers, gorilla mothers show a great variation in the treatment of their offspring. The contrasts were particularly marked between Old Goat and Flossie. Flossie was very casual in the handling, grooming, and support of both of her infants, whereas Old Goat was an exemplary parent.

Seven months old when Simba, Flossie's first baby, disappeared, Tiger was a thriving, strong-minded bundle of constant motion whose most remarkable feature was his long, wavy, reddish-brown head hair that stuck out in unruly curls around his face and hung in ringlets below his neck. His unusual flaming mane could be seen even at long distances, in marked contrast to Old Goat's jet-black body hair. Typical for his age, Tiger had the faint beginnings of a white tail tuft, his eyes were becoming the predominant feature of his face, and he weighed about eleven pounds. He was usually observed within arm reach of Old Goat and had begun traveling regularly on her back rather than in her arms. Tiger's own attempts to move around independently were still clumsy and uncoordinated. Like most gorilla infants around seven months of age, his main sustenance came from suckling. He was able to pluck at vegetation to eat but had not yet acquired the preparatory skills of stripping leaves or wadding vines. Tiger's increased dexterity in these activities was attributed to his intent scrutiny of the feeding methods of the older animals around him. Also, like other gorilla infants, Tiger never tried taking food away from another individual, though his mother often took dung or nonfood items, such as brightly colored flowers, away from him. He, like other free-living gorillas his age, obtained his first bits of food from remnants of vegetation or bark that had fallen onto his mother's lap.

In October 1969, when Flossie's second offspring approached seven months of age, the younger female Maisie, estimated at about nine and a half years, began showing increasing interest in Flossie's baby. Maisie groomed the older female a great deal, in a contrived effort to gain access to her young infant for grooming or cuddling purposes. Flossie, who strongly resembled Maisie, was extremely tolerant toward her; this led me to suspect that strong kinship ties existed between them. Flossie never objected when Maisie carried her infant off for entire day-resting periods to satisfy her maternal inclinations. This is often called "aunt behavior," a term that implies merely an affinitive relationship. Such behavior enables infants to become accustomed to adults other than their mothers and allows nulliparous females — who have not given birth — to gain maternal experience.

Uncle Bert reacted overprotectively toward his first offspring, and frequently tried to separate Maisie from Flossie by running between them or standing bipedally to whack at Maisie. Unbeknown to me, Maisie also had been impregnated by Uncle Bert, but she defended Flossie by pig-grunting or mockbiting the young silverback, who still had much to learn about handling his females.

After nearly a month of intense interest in Flossie's second baby, Maisie gave birth. The parturition apparently was a difficult one, because during her labor Maisie built four night nests spaced several feet apart from each other. Each nest contained an abnormal amount of blood, as did the intervening trail between them. Her infant was stillborn, its fetal body found intact in the last nest. The following day Maisie indicated no ill effects, but three years were to pass before she again gave birth.

Like most births, Maisie's was nocturnal. Gorilla groups are ordinarily stationary during the night and other group members are unlikely to interfere during the birth process. Experienced mothers usually give birth within a single night nest, its vegetation left saturated with blood and, occasionally, small bits of afterbirth. Females giving birth for the first time, or those having nonviable births as in Maisie's case, may build as many as five successive night nests adjacent to the main cluster of their group's nesting site.

Free-living gorillas giving birth to live offspring always eat most, if not all, of the placenta but leave afterbirth intact when producing stillborn infants. Possibly there are some dietary or even antibiotic benefits to be gained by wild mothers and their neonates from the parturients' consumption of afterbirth and, later, their infants' feces. This is suggested by the fact that captive gorilla mothers who live in artificial environments usually only lick or occasionally eat the afterbirth and have not been observed, to the best of my knowledge, ingesting the feces of their offspring.

Maisie was one of the first of many females to show that birth intervals for primiparous gorillas are longer than those for multiparous. This is due to the tendencies of younger females to transfer between silverbacks before settling down with a mate to whom they are ordinarily bonded for life. Such transfers make younger females three times more likely to lose their infants from infanticide than females remaining in one group during their childbearing years. In addition, transfer females need to undergo a period of bond formation with their newly acquired mate, somewhat like a courtship period, especially if the male already has an established harem or is in the process of obtaining more females.

* * *

At the time of Maisie's nonviable birth, Tiger was two years old and had developed his own distinct, engaging personality. Unlike most infants, who spend about 60 percent of their time at this age out of arms' reach from their mothers — a distance of some eight feet — Tiger delighted in Old Goat's indulgent company. His closest peers in Group 4 were Mrs. X's eighteen-month-old Simba, four-and-a-half-year-old Papoose, and Digit, by 1969 estimated at about seven years. On occasions when Tiger became involved in play with any of the three immatures and Old Goat wandered off to feed, Tiger immediately stopped playing. With a dismayed expression he tracked his mother's feeding trail through a maze of other trails — by means of scent only — to be hugged and cuddled even after a half-hour separation.

At two and a half years in mid 1970, Tiger was older than most infants who begin to practice day-nest building. Usually, by the time they are eighteen months old infants have been regularly observed clumsily patting down small stalks or attempting to arrange foliage stems around them during day-resting periods. In the first three years, practice nests require about six minutes to build because the activity is often interrupted by digressions of play with vegetation. (The youngest infant recorded to build and sleep consistently within his own night nest was thirty-four months old, and his mother was nearing parturition.) Typically, juveniles continue to build small night nests connected to those of their mothers for about a year following the arrival of their younger sibling. By then, regardless of age at their first attempts, young gorillas have had sufficient nesting practice to enable them to build their own independently rimmed nests near their mothers.

Tiger's first attempts at nest-building left a lot to be desired. One afternoon, when the adults of Group 4 had comfortably settled down in their leafy beds, the youngster confidently began bending long stalks of foliage one by one onto his lap. Standing on all fours, he tried to push the springy stems beneath him, then hurriedly attempted to reseat himself on top of them. The uncooperative *Senecio* vegetation naturally sprang back into an upright position; his small body could not control all the stalks he had succeeded in breaking. Tiger repeated the process four times before his confidence gave vent to utter frustration. He began whacking out at the remaining stems around him before jumping up and twirling rapidly in a circle to plop on his back, spreadeagled, in a last futile effort to hold down the unruly stalks. A few seconds later he grinned idiotically, slapped his belly as well as the mutilated leaves surrounding him, and kicked his feet up in the air as if riding a bicycle. He then jumped up and looked around inquisitively. Pulling out his lower lip, Tiger snapped it back and forth several times like a thick rubber band before running to join his mother, Old Goat, in her neatly structured day nest.

In August 1970, when Tiger was nearly three years old, there began a twelve-month period of gains and losses for Group 4 involving three births, three deaths, and three emigrations. Flossie lost Uncle Bert's first-conceived infant from unknown causes during its seventeenth month. Because of the infant's age and Flossie's tendencies to be a somewhat negligent mother, I felt the infant had probably succumbed to an accidental fate. Twenty-seven months earlier Flossie had lost a seven-month-old infant also from unknown causes. That disappearance and assumed death occurred at the time Uncle Bert had taken over the leadership of Group 4 following the natural death of the infant's sire, Whinny. Because infant loss may be strongly associated with new pair-bonding between sexually mature males and females, I felt that Flossie's first infant might well have been a victim of infanticide.

Also in August 1970 the primiparous female Petula gave birth to her first infant, a female named Augustus for the month in which she was born rather than for her sex. The newborn was Uncle Bert's third-conceived but only surviving offspring.

After the birth of Augustus to Petula, Uncle Bert was left with three older females (Old Goat, Flossie, and Mrs. X) and three younger females without infants (Bravado, Maisie, and Macho). Bravado was the first of the three to transfer. In January 1971, during an unobserved interaction, she emigrated to Group 5. This was a surprising move to me because the nulliparous female was entering an established group that contained a well-defined female hierarchy. As a young addition to Beethoven's harem of four older females, Bravado had no chance to improve her own status, nor that of any offspring she might conceive with Beethoven.

Uncle Bert's fourth-conceived infant was born to Old Goat in April 1971, but it did not survive and I thought it to have been stillborn. Several days before the birth, Old Goat and Tiger, then forty-one months old, began traveling into the saddle area west of Visoke for distances up to half a mile ahead of Group 4. Uncle Bert, followed by the rest of the group, patiently followed mother and son into an area which at that time Group 4 seldom used because of human encroachment. Far from Visoke's slopes, Old Goat gave birth. Her labor continued for three successive days and nights, marked both by night nests and long trails connecting the nests saturated with copious amounts of blood and tissue. Following the birth I caught only a brief glimpse of Old Goat dragging her infant's body. Moments afterward, a violent screaming outbreak occurred, and the entire group fled a mile back to Visoke's slopes.

For nearly a week, the camp staff, Bob Campbell, and I searched for the tiny corpse, but to no avail. Following the loss of the infant, the strong bonds between Tiger and Old Goat became even more pronounced. Tiger reverted to almost infantile behavior. Old Goat allowed him to suckle milk intended for her deceased infant, permitted him to resume traveling dorsally, although by now he weighed around forty-five pounds, and intensively groomed him for long periods. Her poignant need to be near Tiger, even when feeding or traveling, caused Old Goat to withdraw from interactions with other Group 4 members, most of whom were occupied with still another membership adjustment.

Within a day following Old Goat's unsuccessful parturition, when Group 4 were fleeing back toward the security of Visoke's slopes, they encountered the ailing Amok, who had been ranging alone mainly in the western saddle since his altercation with Uncle Bert nearly three years previously. When Group 4 returned to the mountain, old Mrs. X, who probably had been unable to keep up with the group, was found traveling with Amok. The consortship, which lasted only two months, seemed an odd but suitable pairing since both animals were ill, fed little, and traveled at a slow pace.

After Mrs. X's departure from Group 4, her thirty-seven-month-old daughter Simba changed suddenly from a happy, outgoing, sociable youngster into a pathetically withdrawn and sickly infant. She spent the days and nights huddled against Uncle Bert, rejected all play solicitations, and began eating her own dung. The young silver-back, now beginning his fourth year as Group 4's leader, responded fully to Simba's helplessness. In a maternal-like manner Uncle Bert groomed her, nested with her, and scrupulously protected her from other young gorillas seeking only to play with the despondent youngster.

In May 1971, old Mrs. X meandered back to Group 4, leaving Amok some 100 yards from the group exchanging chestbeats and roars with Uncle Bert for nearly a full day. Amok made no attempt to follow Group 4 to retrieve Mrs. X but returned to the saddle area. On and off for the next three years he was observed to be growing increasingly weaker and was always alone. When he finally disappeared, I thought he had died, though his body was never found in the vast saddle terrain where he had wandered during his final years.

Following the return of Mrs. X to Group 4, Simba reverted to her playful self and showed no psychological effects of the two-month period when she was deprived of her mother. Simba had been almost completely weaned at the time of the separation, so the main reason for her despondency was attributed to the lack of maternal body contact rather than to deprivation of suckling.

It was obvious when Mrs. X came back to Group 4 that she was terminally ill. After twenty-three days she disappeared. An intensive search for her body proved fruitless. Because of her long illness, I considered Mrs. X dead.

Although more than a decade has passed since the disappearances of some of the deceased animals mentioned, their skeletal remains have never been found despite the frequency with which numerous researchers and trackers have covered the entire study area. This is understandable when one considers the vastness of the forested ter-rain, the regenerative prowess of the vegetation, and the ruggedness of the Virungas. Moreover, dying gorillas often sought to conceal themselves in the hollow boles of *Hagenia* trees, thus compounding the difficulty of searching for their bodies.

With her mother's departure, Simba again withdrew into her shell, responding only to Uncle Bert, who immediately resumed his attentive guardianship. The silverback's persistent grooming sessions notwithstanding, the thirty-eight-month-old orphan's appearance showed lack of a mother's care. Her hair was matted, her once-white tail tuft badly soiled, and her eyes and nostrils frequently drained. Simba's most noticeable external signs of maternal deprivation were the pitted and nearly rotten appearance of her feet. Without her mother, Simba had no opportunities for dorsal travel. Neither Uncle Bert nor two other silverbacks among the study animals who took over the care of motherless three- and four-year-olds were ever observed carrying their charges dorsally, even during rapid travel periods. At such times, whenever Simba fell behind Group 4, only her half brother Digit waited for her to catch up with him; then both would rejoin the group together.

A year after her mother's death Simba began building her own nests by pathetically piling leaves into little mounds and making no attempts at rim construction. Her efforts, even when she was fifty months old, resembled the nest-building techniques of two-year-olds and offered no protection against the cold and dampness of the mountain. More often than not, she joined Uncle Bert at some point throughout the chilly nights. For well over a year, the young silverback's painstaking care continued without slackening. Simba's confidence under his doting attention grew to such an extent that she verged on being somewhat spoiled. If subjected to the slightest roughhousing when playing with Augustus, Tiger, or Papoose, Simba needed only to give the smallest squeak and Uncle Bert was quick to discipline Simba's bewildered playmates with pig-grunts or mock bites.

Simba's first shy attempts at play were gently encouraged by seven-year-old Papoose, three years Simba's senior. The orphan, however, remained hesitant about joining in paired play. She feigned her attraction by approaching the players to within a dozen feet, then sat and groomed herself intently. This activity enabled her to watch the other animals closely without obliging her to participate.

Uncle Bert used a similar wile toward human beings once Simba's confidence had been restored sufficiently to permit her to show curiosity toward observers. Seeking to intervene between his charge and humans, Uncle Bert would yawn — not from nervousness but to show lack of interest — feign feeding, and casually stroll directly to observers before giving a deliberated and loud *wraagh* that sent Simba scurrying to the background. Uncle Bert would then return to the group following our obedient retreat.

* * *

Indeed, the young silverback had come a long way in the assumption of his responsibilities toward the maintenance of Group 4's cohesion and the protection of its members. His maturity did not seem evident, however, when in June 1971 — the same month in which Simba's mother died — Group 4's two remaining young adult females, Maisie and Macho, immigrated into Rafiki's Group 8 during a violent physical interaction. At the time, it seemed to me that Uncle Bert had lost two capable breeding candidates. I did not take into consideration the fact that Group 8, then consisting of Rafiki, Samson, Geezer, and Peanuts, had gained two needed breeders. Moreover, the transfer enhanced the status of Macho and Maisie because the female hierarchy of Group 4 had been dominated by Old Goat and Flossie.

Over a four-year period of observations, Group 4 had survived many traumatic transitions. They had lost their silverback leader Whinny by a naturally caused death. Blood-bond reinforcement had enabled Whinny's oldest son, Uncle Bert, to keep the group together with the assistance of Group 4's most dominant female, Old Goat. That Group 4 was able to survive as an integral social unit was no miracle. It was simply a graphic demonstration of how kinship may function over time within gorilla society. Group 4's young, sexually maturing females had dispersed to breed with unrelated silverbacks; Group 4's inherent leader, Uncle Bert, was gaining experience in group control. Last, but far from least, the blackback Digit was beginning to learn, just as Uncle Bert had learned before him, the responsibilities of leadership. By June 1971, the future of Group 4 seemed well assured.

| ## Growing Family Stability: Group 4

WITH THE 1971 EMIGRATIONS of Bravado, Maisie, and Macho, Digit lost three half sisters, the peers with whom he had played during his transition from a juvenile to a blackback. Now nearly nine years old, he was too old to cavort freely with one-year-old Augustus, forty-month-old Simba, forty-five-month-old Tiger, or five-year-old Papoose; too young to associate closely with Group 4's older females, Old Goat, Flossie, and Petula. Perhaps for these reasons Digit became more strongly attracted to humans than did other young gorillas among the study groups who had siblings and peers.

I received the impression that Digit really looked forward to the daily contacts with Karisoke's observers as a source of entertainment. Eventually he showed that he could tell the difference between males and females by playfully charging and whacking men but behaving almost coyly with women. He was always the first member of Group 4 to come forward to see who had arrived on any particular day. He seemed pleased whenever I brought strangers along and would completely ignore me to investigate any newcomers by smelling or lightly touching their clothing and hair. If I was alone, he often invited play by flopping over on to his back, waving stumpy legs in the air, and looking at me smilingly as if to say, "How can you resist me?" At such times, I fear, my scientific detachment dissolved.

Like Puck of Group 5, Digit became fascinated by thermoses, notebooks, gloves, and camera equipment. He always examined, smelled, and handled everything gently, and occasionally even returned objects to their owners. His return of these items was not done from any sense of recognition of ownership but only because he did not like the clutter of human belongings around him.

One day I took a small hand mirror to Group 4 and stood it in the foliage where Digit could see it. Without hesitation he approached, lay propped up on his forearms, and sniffed the glass without touching it. As the young blackback viewed his image, his lips pursed, his head cocked quizzically, he gave a long sigh. Digit continued staring calmly at his reflection before reaching behind the glass to "feel" for the body of the figure before him. Finding nothing, he lay quietly gazing at himself for another five minutes before again sighing and moving away. Often I have puzzled over Digit's acceptance and apparent pleasure when gazing intently at his reflection. It would be presumptive for me to believe that he recognized himself. Perhaps the lack of scent clues informed him of the absence of another gorilla.

The Rwandese Office of Tourism, in an attempt to attract visitors to the Parc des Volcans, asked me for a photograph of a gorilla to use for a poster advertisement. Since I was a guest in their country, I complied with their request, just as several years previously I had provided a number of pictures of the forest and gorillas to the Rwandese postal service to serve as models for the first series of Rwandan stamps featuring the gorillas of the Parc des Volcans. The slide I selected for the tourist office was one of my lovable Digit. Shortly thereafter large color posters of Digit feeding on a piece of wood were scattered throughout Rwanda—in hotels, banks, the park office, the Kigali airport, and in travel bureaus throughout the world. In various languages the poster was captioned "Come and See Me in Rwanda!" I had very mixed feelings on first seeing the posters around Rwanda. Heretofore Digit had been an "unknown," only a young male maturing within his natal group. Suddenly his face was everywhere. I could not help feeling that our privacy was on the verge of being invaded. I certainly did not want the public flocking to Group 4, especially at a time when the group promised finally to become a secure and integral family unit.

In August 1971 Flossie gave birth to Cleo, sired by Uncle Bert, exactly one year after the disappearance of their first offspring. Cleo was born several hundred feet from Flossie's 1967 birth site of Simba, who had lived only seven months. The location, named Birth Gully, offers maximum visibility over surrounding terrain. Because the knoll protrudes into the middle of a wide and steep-sided ravine, access to the ledge is very difficult, at least for human beings.

Flossie showed maternal diligence for the first time with Cleo, Uncle Bert's second surviving offspring out of the five he had so far sired. I attributed Flossie's new attentiveness toward her infant to the growing stabilization of Group 4 under Uncle Bert's developing leadership as well as to the silverback's increased protectiveness and support of Group 4's infants. Although the maturing young leader was becoming steadily more experienced in handling relationships within the group, several years and many interactions were to pass before Uncle Bert had gained any degree of expertise during encounters with other social units.

In October 1971, on the southeastern edge of their range, Group 4 met Group 5. During the course of the two-day interaction, Bravado, who had transferred to Group 5 ten months before, became reacquainted with the members of her home group. All, other than the infants, instantly recognized her. Digit and Papoose played the most enthusiastically with Bravado and, as well, with Icarus, Pantsy, and Piper of Group 5. Their camaraderie gave me some idea of how young individuals of neighboring gorilla groups become acquainted during their formative years long before reaching sexual maturity.

Uncle Bert's reactions toward Beethoven were, unfortunately, impetuous. The younger silverback, often supported by Digit, directed numerous frenzied displays toward Group 5's dominant silverback. He neglected to keep Group 4 members away from Beethoven at a secure distance, and excitedly charged into the midst of his own group. This caused females and young to flee and imperiled the safety of his infants, five-month-old Augustus and two-month-old Cleo. Augustus, carried dorsally at the time, appeared terrified of the activity, flattened herself out on Petula's back, clutching tightly and crying loudly whenever the disorganized Group 4 members retreated from Beethoven's charges. Cleo, carried ventrally by Flossie, slept through most of the commotion. Old Goat, with the dourest of expressions, finally led Group 4 youngsters and females away from the two displaying silverbacks. On the afternoon of the second day of the interaction Beethoven firmly herded Bravado away from Group 4 and back to Group 5. In a futile attempt to have the last word, Uncle Bert strutted after Beethoven and was followed by Tiger,

Simba, and Papoose, who amusingly mimicked the young silverback's exaggerated strutting walk.

Over the course of these two days, I couldn't help wondering at the differences in behavior between the two silverback leaders. At one time, many years before I entered the scene, Beethoven himself must have been every bit as awkward in handling Group 5 members as Uncle Bert was during this encounter with his own Group 4. Age, coupled with experience, had given Beethoven the practice and ability to control such a situation without needless, overt aggression. I felt confident that only time would enable Uncle Bert to deal with other gorilla groups or lone silverbacks as effectively as Beethoven. For now, however, Uncle Bert had much to learn, just as did both groups' supporting males — Digit of Group 4 and Icarus of Group 5.

Although still a blackback, or sexually immature male, Digit had begun assuming responsibilities to the best of his ability. As the second-oldest male in Group 4, he had supported Uncle Bert during the interaction with Group 5, though it was obvious that he was intimidated by Beethoven while at the same time very excited by the opportunity to renew his ties with Bravado. Although it was not clear to me whether it was Digit or Old Goat who began seeking the other's company, it suddenly became apparent that the adult female was extremely tolerant of Digit's proximity to her near the edge of Group 4. The developing situation was obviously beneficial to all Group 4's members. Old Goat had someone with whom to share the duties of group defense; Digit acquired a function within his natal group; and Uncle Bert had two "watchdogs" rather than one to ensure greater familial security.

Digit's new responsibilities became significantly obvious when, in early 1972, Group 4 began to use the saddle extensively. The new area afforded a means of getting away from the overlapping ranges shared with Groups 8 and 9 and also provided a variety and abundance of the saddle terrain's vegetation and trees. Unlike the mountain slopes, the saddle, because of its relative flatness, restricted visibility of the land surrounding Group 4's travel and resting periods; thus Digit and Old Goat maintained an even closer allegiance to the protection and defense of their group.

Digit's support of Uncle Bert during intergroup interactions carried the risk of his being wounded by older and more experienced males, specifically those of Group 8. In March 1972, after an unobserved interaction with Group 8, Digit received several serious bite wounds on his face and neck. For more than four years the deep neck injury drained a foul-smelling exudate. The position of the wound prevented the blackback from cleaning it orally. The best Digit could do was to lick the drainage from his forefinger, which undoubtedly explained why the infection became systemic. For nearly two years his entire body conveyed an unusual sour odor, he frequently expelled copious amounts of gas, and was often heard retching. In addition to his becoming listless and apathetic, Digit's body slowly assumed a humped, angular appearance as though he were on the verge of sitting down at any moment.

Digit showed his usual alertness only on the two or three days of each month when seven-year-old Papoose started showing estrous cycles. Uncle Bert never interfered with Digit's mountings of the sexually immature female, but he would not tolerate Digit's proximity within the group when Old Goat returned to estrus in mid 1972. At these times Digit could usually be found alone, peripheral to the group, rocking from side to side, a motion suggestive of masturbation, although this was never verified.

During several of Group 4's interactions with Group 8, when the two groups were separated by no more than 100 feet, Papoose and Simba left Group 4 for brief periods to play excitedly with Peanuts and Macho of Group 8, much to Digit's concern. Uncle Bert and Rafiki showed indifference to the movements of the two young females. I attributed this to Papoose and Simba being far from sexually mature as well as to Rafiki's advanced age, which made him less interested in obtaining new females.

Digit's increasing maturity meant that he had much less interest in human observers because of his roles as a Group 4 sentry and as a potential breeder. This was a considerable relief, for I had worried about his becoming too people-oriented. As it happened, just about this time Digit became world famous from the tourist bureau poster.

* * *

Like Digit, Tiger also was taking his place in Group 4's growing cohesiveness. By the age of five, Tiger was surrounded by playmates his own age, a loving mother, and a protective group leader. He was a contented and well-adjusted individual whose zest for living was almost contagious for the other animals of his group. His sense of well-being was often expressed by a characteristic facial "grimace." Resembling that of a person preparing to blow a wad of bubble gum, it resulted in nearly unbelievable stretch contortions of Tiger's mouth, wrinkling his nose and squinting or even closing his eyes. Unlike Digit, Tiger seldom was inclined to seek the company of observers. Only when his seemingly endless supply of energy wore out his peers did he even acknowledge the presence of humans. Always more interested in action than in curious investigation, he delighted in tug-of-war with foliage stalks that he would try to pull out of observers' hands, apparently enjoying the give-and-take sensations, especially at the moment the human would let go of the stem and send Tiger rolling over backward chuckling heartily before he returned to repeat the game.

Tiger's favored play companion was seven-and-a-half-year-old Papoose, who was entering the throes of adolescence. Roughhouse play with Tiger evoked her tomboy inclinations; little Simba prompted Papoose's maternal instincts. Her sexual interactions with Digit were beginning to take precedence over other interests. Papoose also sought the company of Petula, the lowest-ranking female of Group 4. Because of their physical resemblance and close association it was highly possible the two females were half sisters.

Petula reminded me very much of Liza of Group 5: both were at the bottom of their group's female hierarchy, their proximity was not well tolerated by older females of their respective groups, and both had obstreperous offspring. Petula was inconsistent in her maternal behavior. Often, after squabbling with Flossie or Old Goat, she would direct frustrations to her daughter Augustus. Whenever her mother pig-grunted at her or mock-bit her for no obvious reason, Augustus' face wrinkled into an expressive moue and she began whimpering. When whimpers grew into whines, Augustus would be chastised again by Petula.

During her first year, probably because of her mother's irregular care and the rejection of older companions, Augustus developed an unusual repertoire of solo play activities in trees. Her jungle-gym acrobatics all but devastated stands of young *Vernonia* saplings. She was the only gorilla I have observed who frequently used the heights of trees to locate other animals, particularly Petula, when her mother was obscured by tall foliage.

When Augustus was eighteen months old she discovered that by clapping her hands together she could create a unique sound. She continued hand-clapping until the age of five years. I have never observed hand-clapping by any other free-living gorilla, though it is not uncommon among captives. Judging from the startled expressions of the animals in Group 4, the clapping noise was also novel to them. Augustus sometimes tended to overdo her act. For up to a minute at a time she could sit and hand-clap, wearing a rather foolish grin of self-accomplishment. (Some wild youngsters often slap the soles of their feet, but this action possibly is provoked for tactile rather than for auditory reasons.)

Cleo at the age of six months had developed into a spritely, inquisitive infant who spent long day-resting periods crab-crawling within arms' reach of her mother, Flossie. At this time Cleo received a serious eye injury from unknown causes. The wound drained for nearly two years before healing. Cleo did not seem bothered by it, nor was Flossie ever observed grooming the injury.

Repeating her pattern with the two earlier infants, Flossie again became casual in her maternal behavior. I was therefore surprised one day when she rushed to Cleo's side and threw, overhand, two freshly deposited lobes of Petula's dung out of the infant's reach. I wondered if Flossie's actions were maternally inclined or an expression of dominance over Petula, the lowest-ranking female.

Cleo was the last infant to be born in Group 4 for a three-year period. By the end of 1973 all three adult females were coming into regular estrous cycles and soliciting Uncle Bert for copulations. The young silverback showed great interest in Old Goat, dutiful interest in Flossie, and virtually no interest in Petula. Old Goat might take as long as fourteen minutes between the initiation of her slow-motion approach to Uncle Bert before hesitantly reaching his side to be mounted by him. Their copulations were of greater intensity and longevity than those he completed with Flossie or Petula. Uncle Bert mounted Flossie more enthusiastically on days when her periods of receptivity occurred simultaneously or overlapped with those of Petula. Petula displayed more coquetry than Flossie when soliciting Uncle Bert, but he usually responded with grooming rather than by mounting behavior. In fact, Augustus was forty months old before Uncle Bert resumed copulations with Petula, and these were very casual.

It is frequently the case that when an adult female comes into receptivity, there is a great deal of vicarious sexual play within her group. Flossie usually mounted a most impassive young Tiger, and

Petula was accepted benignly whenever she chose Flossie or Old Goat to mount. Only Old Goat was never observed soliciting other females. Tiger was six years old when his mother came into estrus regularly in 1973. He showed great interest in Old Goat's sexual activities but never attempted to interfere when Uncle Bert was seriously copulating with her. It was at this age that Tiger began expressing interest in mounting Simba or Papoose, but only when Digit, now around eleven years old, was not in the immediate area.

When Simba was nearly six, Digit began mounting her as well as Papoose without interference from Uncle Bert on the days when one of the three adult females was receptive. The age and size discrepancies between them made the mountings somewhat ludicrous. Simba's bland look was a comical contrast to Digit's serious puckered-lip facial expression.

* * *

In January 1974 Uncle Bert began following Group 8. Old Rafiki's group then consisted of Peanuts, Macho, the seven-month-old daughter Thor, whom he had sired by Macho, and Maisie, who had joined Group 8 in mid 1971. Experience made Uncle Bert more of a strategist now and he succeeded in regaining Maisie from Rafiki. During the young female's first few weeks after her return to Group 4, Uncle Bert directed numerous runs and whacks at her. When a female transfers into a new group, she commonly becomes a target for excitable displays from the group's silverback as he enforces his dominance over her. Maisie's case was slightly different because Group 4 had been her natal group, and she was well acquainted with all the members except two-and-a-half-year-old Cleo, born after her departure in June 1971. Cleo, more than any other Group 4 member, did not welcome Maisie's return, and constantly expressed antagonism by pig-grunting or giving infantile bluff charges toward the immigrant. Augustus, on the other hand, had been only eleven months old when Maisie had emigrated. The forty-one-month-old youngster avidly sought attention from the new transferee for grooming and play.

Maisie remained with Group 4 only five months. She did not integrate well, as judged by the number of pig-grunts directed toward her and the wide distance she maintained from the group's adult females — Old Goat, Flossie, and Petula. In June 1974 Maisie began a series of transfers between Group 4, Group 8, and the lone silverback Samson, formerly of Group 8, with whom she eventually settled.

Upon Rafiki's death in April 1974 Uncle Bert spent increasing amounts of time adjacent to or within Group 8's range. A month later, the three remaining members of Group 8, Peanuts led by Macho carrying Thor ventrally, had a violent physical interaction with Group 4. Thor was killed by Uncle Bert. Like most cases of infanticide this one also occurred after the death of the silverback leader of the infant's group and was followed by the mother's transfer to the group of the infanticidal male. By killing Thor, Uncle Bert eliminated an unrelated offspring that had been sired by a male competitor, added a female to his own group, and shortened his waiting time toward insemination of the female, Macho. This case differed from most infanticide incidents in that Uncle Bert did not take Macho for another five months after killing Thor.

There were two likely factors responsible for Uncle Bert's delay in regaining Macho. Rafiki's son Peanuts, then around twelve years old, was not yet sexually mature — thus not a breeding competitor. Also, within the southern part of their range Group 4 was being subjected to multiple interactions with two lone silverbacks, Samson and an older silverback, Nunkie, who first had been encountered within the study area two years before. Uncle Bert was no match for Nunkie and probably would have lost Macho to Nunkie, if not to Samson, had he taken her from Peanuts at the time of the infanticide. One month following Thor's killing, the totally unexpected happened.

Petula and Papoose emigrated from Group 4 to join Nunkie, marking the formation of a new Visoke group. Papoose probably left Group 4 because of her long-term close association with Petula, thought to be her half sister, and because there were no immediate breeding opportunities available to her in Group 4.

Petula's emigration was rather surprising. Like Liza of Group 5, the low-ranking female had left her four-year-old offspring behind with the youngster's father. Petula, like Liza, improved her status by removing herself from the established female hierarchy of her group. The most important similarity between the two adult females was that they had been nursing their four-year-old juveniles beyond the normal weaning age, suggesting that prolonged lactation inhibits or postpones conception. Neither Petula nor Liza had been able to provoke intensive copulation from their silverback breeding partners.

It was difficult to compare the behavioral adaptations of Group 4's two motherless juveniles, Simba and Augustus. Augustus had two advantages over Simba: she was a year older when she lost her mother, and her father, Uncle Bert, was present within her group. Augustus, unlike Simba, never became withdrawn, though her play with other youngsters was considerably reduced. She sought to be close to Uncle Bert during daily feeding and resting periods, and made her own poorly structured night nests adjacent to his following her mother's departure.

Augustus was never mollycoddled by Uncle Bert as Simba had been, and she matured far more rapidly than she might have done had her mother remained within the group. She became very protective toward Cleo and also groomed all other Group 4 individuals with the exception of Flossie. Without low-ranking Petula, Augustus was able to spend more time near her father, thus further strengthening her bonds within the group.

In August 1974, exactly three years following the birth of her daughter Cleo, Flossie gave birth to a male infant, Titus. Over the years of research the average interbirth interval for thirteen females was 39.1 months. When only viable births (those in which consecutive infants survived) are considered, the average interval for ten females was 46.8 months. Flossie's birth interval remains the shortest yet recorded between viable births. When only nonviable births (those in which infants died or disappeared) are considered, the average interbirth interval for seven females was 22.8 months. The birth of Titus came as a complete surprise. Although Flossie had been observed copulating with Uncle Bert eight and a half months previously, she did not appear pregnant, nor had there been a significant reduction in the number of mounting solicitations she directed toward Uncle Bert or other females.

Behavior associated with estrus often occurred during pregnancy, especially in the later stages, and was seen even the day before Flossie gave birth. Nearly all females regularly observed before labor mounted both dominant and subordinate adult males of their groups, as well as other females. The recipients of such mountings initiated by the pregnant females were usually passive. However, the mounted nonpregnant females appeared to respond actively to the attention, as shown by their copulatory vocalizations or thrusting responses. The more dominant the mounting female, the more likely was the mounted female to respond. I suspect that such behavior helps to strengthen the pregnant females' social bonds within the group before parturition.

Titus was Flossie's second offspring to survive beyond infancy, and like her other newborns, this baby too seemed underdeveloped and spindly. In addition, he had serious breathing difficulties. With his mouth wide open, he inhaled air in loud, gasping breaths followed by sneeze-like motions of his head. The baby's acute respiratory distress continued for nearly eight months. I became increasingly concerned over Flossie's apparent indifference to Titus, especially during travel, when the infant's head dangled without support over the arm Flossie chanced to be using to grasp him to her ventral surface.

Flossie had aged considerably during the three years between births. It seemed as if the elderly female had about expended herself on the successful rearing of Cleo and could only provide Titus with the basic necessities of maternal attention. She either ignored or thwarted her son by pig-grunting or mock-biting him whenever he tried to play on her body.

Flossie groomed Titus in a perfunctory manner. He seldom attempted to wiggle free, kick, or whack — commonly observed infant protests to grooming. Unlike most infants, for whom grooming is almost always initiated by the mother and terminated by the infant, Titus thrived on his mother's attention. He was no different from any other baby, though, when it came to suckling, which is always initiated by the infant and terminated by the mother. Flossie was more authoritative with Titus than she had been with Cleo. Her intolerance was probably the main reason for her son's quick acquiescence to end suckling bouts and thereby avoid his mother's irritability.

At the time of Flossie's parturition thirty-eight months had passed since Old Goat's nonviable birth, thus her pregnancy was long overdue. She appeared to be having regular cycles, as judged by the consistency of her monthly solicitations of Uncle Bert. Other than a six-week period of illness in September 1973, Old Goat seemed in excellent health and was as attentive and loving with Tiger as ever.

On a lovely warm day in October 1974, I was contentedly sharing the sun and the peacefulness in the midst of Group 4. Old Goat bemusedly lay down on her side near me to watch seven-year-old Tiger play. Gleeful, he picked handfuls of foliage to beat into shreds on the ground, against his head and Old Goat's side, all the while wearing his ecstatic facial expression. As I watched the pair, I marveled again at the cohesiveness of gorilla family bonds.

The next day I discovered Group 4 had had a violent interaction with Peanuts and Samson. During the encounter Samson took Maisie from Peanuts and Uncle Bert reclaimed Macho from Peanuts, leaving this young silverback a lone traveler. The unobserved interaction was determined as having been violent because of the vast blood-spattered area filled with tufts of silverback hair and permeated with pungent silverback odor. From the encounter site, Group 4 fled almost four miles into the distant western saddle area and onto Visoke's southern-facing slopes. Peanuts had pursued Group 4 the entire way in an attempt to retrieve Macho from Uncle Bert. For the next month he remained between one and ten yards from Group 4, prompting countless physical interactions of varying degrees of severity. Gone were the months of calm that Group 4 had enjoyed.

Gone also was Old Goat. After first making sure she was not with Samson, and that no other fringe groups had been in the area, the camp staff and I began the grim search for her body. A month passed without finding a trace of her in the vast one-square-mile saddle area encompassing Group 4's flee route from the interaction site. Then, late in November, one of the searchers looking for Old Goat had to climb a tree to avoid a herd of buffalo. Aware of the putrid odor of death, he looked down and found the old female's decaying body nestled in the hollow of the huge *Hagenia* tree and nearly hidden by vines.

Her corpse was so decomposed that I had to take specimens of her organs on the spot for later histological examination. Cutting into the body of the noble female was an indescribably loathsome job.* Many months were to pass before I could get over the emptiness felt upon contacting Group 4 without its indomitable character, Old Goat.

I was surprised that Tiger showed no emotional slump following the death of his mother and attributed his absence of distress to the pressure created by the month-long presence of the lone male Peanuts. Together, Digit and Tiger succeeded in keeping Peanuts from entering the group. As they strutted and swaggered in front of the young silverback, they reminded me of two little boys playing

---

* Medical experts later diagnosed the cause of death as viral hepatitis. They also found that Old Goat was not pregnant at the time of her death.

soldiers. Digit, now about twelve years old, was not yet capable of emitting a mature silverback's hootseries vocalization, but he often positioned his lips before chestbeating as if expecting the sound to emerge. It never did. Seven-year-old Tiger imitated all of Digit's exaggerated stances, strutting displays, and bluff charges in the first encounter in which he was observed directly confronting a silverback. Uncle Bert was quite aware of the defensive maneuvers of the three males. Keeping himself between Peanuts and Macho, he was also occupied in copulating with his newly acquired female.

At the end of November 1974 a dejected, weary, and wounded Peanuts finally gave up to return to the northern Visoke slopes where he had earlier ranged. After his departure, Digit seemed a young silverback without a cause. From his usual rear-guard position to Group 4, he sat morosely for prolonged periods staring off into space in the direction where Old Goat's body had been found. Neither the presence of human observers nor Uncle Bert's active sex play with Macho could arouse Digit's interest. His brooding was reminiscent of that of Samson following the death of Coco, Group 8's old female. The effects of Digit's neck wound, inflicted thirty-two months before, were still obvious. His body growth appeared not to have kept up with his head growth and this resulted in an ungainly and disproportionate appearance. The young male's present mood of deep dejection and his pathetic appearance were characteristics totally unlike the lively, curious Digit whom I had known as a youngster.

*　*　*

The return of Macho to Group 4 prompted some unusual spatial changes within the group. Motherless Simba and Augustus sought to be near the dominant older female Flossie for security even though she totally ignored them. Flossie, with three-month-old Titus clinging to her ventral surface, badgered Macho whenever Uncle Bert was not in sight. Soon Simba, Augustus, and Cleo seemed to duplicate Flossie's antagonistic behavior, making Macho's first month in Group 4 a difficult one. Flossie's aggressiveness was identical to the behavior that she had directed toward Macho back in 1969 when the young female was approaching sexual maturity. I felt this was attributable to a lack of kinship ties between the two.

Although Flossie became the dominant female of Group 4 after Old Goat's death, she was ostracized from proximity to Uncle Bert when he acquired the sexually receptive female Macho. Titus was only four months old, therefore at least two and a half years would pass before Flossie, Uncle Bert's inherited, rather than acquired, mate, would return to estrus. The old female redirected her antipathy toward her infant son, disciplining him for even the mildest offense. Uncle Bert seemed unaware of Flossie's behavior. The usually sedate group leader now spent a great deal of his time copulating with Macho or playfully cavorting with Tiger, Simba, Augustus, and Cleo. Their tickling and wrestling sessions were periodically interrupted while the silverback, who wore his gentle facial expression, cuddled and groomed the older youngsters.

Once Macho was impregnated by Uncle Bert he totally ignored her. Flossie and Group 4's three young females resumed their harassment in the silverback's presence, forcing her to retreat to the outer edge of the group, often near Digit. Macho conveyed the impression of wanting to be close to her new mate, but her apprehension when near other group members was extreme. She often walked as though treading on eggshells whenever approaching them.

Kweli was born in July 1975, Macho's second- and Uncle Bert's seventh-conceived offspring. This tangible proof of Uncle Bert's investment with her made Macho more self-assertive. Even Flossie exchanged contented belch vocalizations with the new mother and patiently shared Uncle Bert's proximity with Macho.

One day when Kweli was about three months old, Macho, for no discernible reason, ran pig-grunting at eight-year-old Tiger, who fled screaming in alarm. Thereupon Uncle Bert charged Macho, who crouched submissively as he stood over her vocalizing harshly. Uncle Bert threateningly tried three times to grab Kweli, clutched in Macho's arms, before the cowed female slowly crept away. I wondered if Macho recalled Uncle Bert's killing of Thor, her first and Rafiki's last infant, the preceding year. I have never known a silverback to kill his own offspring; no reproductive advantage would be gained by this strategy.

When Kweli was five months old, the youngsters of the group began ganging up on Macho, who still carried her son ventrally. They herded her to the edge of the group away from an impassive Uncle Bert. It was distressing to watch the young mother become an introverted scapegoat. Sharing the group's periphery with Digit, Macho developed a nervous tic. She would turn her head rapidly, briefly look at Uncle Bert, then instantly lower her eyes and bite her lips. Her expressions indicated keen apprehension, though the silverback seemed unaware of her. Whenever she behaved in this manner, five-month-old Kweli, an unusually alert and well-coordinated baby, flinched and gazed at his mother in distress, his eyes as wide and penetrating as those of Macho.

A year passed without external interventions to Group 4. Then, in January 1976, another violent interaction occurred with Peanuts. After nearly thirteen months of lone travel, the young silverback had acquired an unknown adult gorilla companion on the northern slopes of Visoke. Uncle Bert succeeded in taking the individual, estimated as about ten years old, away from Peanuts, who made no effort to recover his consort as he had so futilely attempted to recover Macho.

My perplexity over the identity and the sex of the emigrant prompted the name Beetsme. The animal appeared scruffy and undernourished, a condition often seen in gorillas from Visoke's northern slopes, which lie adjacent to a relatively bleak vegetation saddle zone. The individual also indicated some degree of habituation, thus might have been an infant from Geronimo's Group 9. Although Beetsme physically resembled a blackback, the immigrant was mounted enthusiastically by Uncle Bert and began spending a large part of his time grooming and cuddling seventeen-month-old Titus, who relished the unaccustomed attention.

Eventually our Karisoke records designated Beetsme as a male — the first and, thus far, only recorded transfer male to be accepted into an established gorilla group. I could not understand why Uncle Bert had resorted to violence to obtain another male at a time when Group 4's male to female ratio was 1 : 1. Perhaps, by taking Beetsme away from Peanuts, Uncle Bert was reducing Peanuts' strength, therefore lessening the younger silverback's opportunities to form his own group. This speculation might erroneously imply a degree of forethought on Uncle Bert's part. However, the strategy, much like infanticide, should only be considered as an evolved mechanism functioning for genetic perpetuation.

It became apparent within a month that Beetsme was a born rabblerouser. He gave the impression that life was one big summer vacation punctuated by occasional sabbaticals. His rough wrestling and chasing play with Tiger, combined with his unruly sexual advances toward Simba and Cleo, created much unrest in Group 4 and demanded increasing discipline from Uncle Bert. Tiger, for the first time in his life, had a male near his own age with whom to play, and subsequently he abandonded the watchdog duties he had shared with Digit following Old Goat's death. The new young male's disruptive behavior gave Macho a second opportunity to intermingle with group members without having to resort to her former obsequious approach.

Macho's new-found security resulted in an abrupt transformation in Kweli's personality. The year-old infant became vibrantly alive, playfully squiggling his way through each day as if he could not get enough of all that life had to offer. Soon, Kweli's physical and social development paralleled, then surpassed that of Titus, one year older.

Observations of the widening variations in development between the two infants led me to believe that Titus had been born prematurely and was trying to catch up even during his second year of life. Beetsme's grooming and gentle play with Titus might possibly have been a contrived, yet efficient, means of enhancing his immigrant position in Group 4. Certainly his attentions stimulated Titus' social development and enabled the infant to mix more freely with his peers than he might otherwise have done without Beetsme's intervention on his behalf.

When almost three years old Titus discovered that by whacking both hands rapidly under his relaxed chin he could create a rhythmic, *clackety-clack, clatter* noise between his upper and lower teeth. The sound was as unusual as that produced by Augustus' hand-clapping seven years earlier. Perhaps some gorillas, lacking opportunities for normal social interactions, are prone to develop atypical behavior patterns as a substitute for social stimulation. Young Digit's rocking behavior may have been prompted for the same reason, just as is the idiosyncratic behavior of many captive gorillas.

After her mother's emigration from Group 4 Augustus seldom hand-clapped; however, she resumed the activity when Titus started chin-slapping. Together the two sounded like a mini-minstrel band. On sunny, relaxing days, their claps and slaps could prompt playful spinning pirouettes from Simba, Cleo, and little Kweli. After several months of closely scrutinizing Titus' attention-getting feat, Kweli began chin-slapping whenever he was without play partners.

Titus was two years old when his mother Flossie resumed soliciting Uncle Bert for mounting. The old female was ignored even though the Group 4 leader had no other receptive mates in his group. At this time, August 1976, Simba, eight years and eight months, was showing regular monthly receptivity. Uncle Bert still expressed no sexual interest in his orphaned charge. It was Digit, now approximately fourteen and verging on sexual maturity, whom the young female solicited for mountings. Digit responded enthusiastically toward Simba's flirtatious invitations and showed a revived interest in life.

Simba's cyclical condition affected most Group 4 members, particularly Beetsme, who vicariously began mounting Cleo, Augustus, and Titus. Neither he nor Tiger was allowed to mount Simba on the days she was in estrus. On those days Digit actively protected his potential breeding rights with the young female by staying near her in the group and preventing Tiger or Beetsme from approaching. The two young males often feigned interest in Simba by boisterously roughhousing within her sight, but watched Digit's movements surreptitiously. Cleo, five and a half years old by the end of 1976, was very inquisitive about the change in status of her playmate Simba; from a sideline position she excitedly watched the capricious actions of the others and sometimes tried to draw Tiger's and Beetsme's attention to herself.

On Simba's receptive days old Flossie was able to solicit mountings from an unenthusiastic Uncle Bert, or even from Macho, who, along with her son Kweli, was now well integrated into Group 4. Eighteen-month-old Kweli was the only member of the group not influenced by Simba's impending maturity. Kweli had developed into an unusually independent youngster whose actions drew overt, affectionate responses from his father.

One day, when Simba was the center of attention, Kweli went off to the sidelines to feed with Uncle Bert, who paused to urinate. Fascinated, Kweli immediately cupped his hands under the steady stream to try to catch the urine. Uncle Bert, wearing a comically irritated expression, reached back to swish at his young son as if the infant were a pesky fly. Kweli reluctantly withdrew a few feet, sat down sulkily, and continued staring intently at his father. The silverback then reached back to catch two lobes of his dung before they hit the ground and sat down to eat both with lip-smacking gusto. To young Kweli this seemed far more interesting than all the frenzied sexual behavior taking place nearby. (Coprophagy allows the absorption of nutrients not available in plant matter.)

On the days Simba was not sexually receptive, Digit could be found at the edge of Group 4 maintaining his guard position. Tiger and Beetsme were usually on the opposite side of the group pulverizing the forest's vegetation in extremely rough chasing and wrestling play.

Contacting Group 4 one horrible, cold, rainy day, I resisted the urge to join Digit, who was huddled against the downpour and about thirty feet apart from the other animals. It had been many months since he had shown any interest in observers, and I did not want to disrupt his growing independence. Leaving him to his solitude, I settled several yards from the group, a cluster of humped forms barely visible in the heavy mist. After a few minutes, I felt an arm around my shoulders. I looked up into Digit's warm, gentle brown eyes. He stood pensively gazing down at me before patting my head and plopping down by my side. I lay my head on Digit's lap, a position that provided welcome warmth as well as an ideal vantage point from which to observe his four-year-old neck injury. The wound was no longer draining but had left a deep one-inch scar surrounded by numerous seams spidering out in all directions along his neck.

Slowly I took out my camera to take a picture of the scar. It was too close to focus on. About half an hour later the drizzle let up, and, without much warning, Digit stretched back his head and yawned widely. Quickly I snapped the shutter. The resultant photograph shows my gentle Digit as a King Kong monster because his wide-mouthed yawn displayed his massive canines in a most impressive manner.

Not long thereafter, Digit's canines were viewed in an entirely different context. In December 1976 a tracker, Nemeye, and I spent five futile hours in a downpour searching for Group 4 in the western saddle area, now an intrinsic part of the group's range. Since camp was hours away, we gave up the search and were heading back along a wide open trail that eight years earlier had been known as the Cattle Trail.

Nemeye was trudging about ten feet ahead of me when, through a momentary lifting of the fog curtain, I could just see the humped backs of Group 4's members huddled together in the rain near the base of Visoke's slopes about 130 feet to the left of our trail. After deliberating the pros and cons of attempting to make a contact so late in the afternoon in such weather, I decided against it. As I

was about to catch up with Nemeye, Digit came running out of
the thick brush on the right side of the trail and unexpectedly encountered the African tracker nearly head-on. Both came to a horrified
halt. Digit stood bipedally, gave two terrified screams, exposing all
of his canines, and released a gagging fear odor. The young silverback
seemed undecided as to whether to flee or charge. He had not yet
seen me. Running forward, I pushed Nemeye behind me. Upon
recognizing me Digit dropped to all fours and fled toward Group
4, who, led by Uncle Bert, were already running for the security
of Visoke's slopes. With the abrupt cessation of Digit's vocalizations,
the group nervously halted until they discerned the reason for Digit's
alarm. The unfortunate incident graphically revealed the value of a
peripheral sentry to a gorilla group's safety.

*   *   *

Over the years Digit and Uncle Bert had grown into a cooperative,
defensive team and relied strongly on one another for support during
both intra- and intergroup disturbances. Theirs could not be termed
a close relationship, such as Uncle Bert had with Tiger, but it was
harmonious because of mutual concern for the cohesion of the family
group.

The silverbacks' reliance on one another was more evident when
Group 4 was traveling in the saddle area, which was never completely
freed of poachers. One day in the beginning of 1977 I was about
to contact Group 4 far west of Visoke's slopes when I heard a prolonged outbreak of alarmed *wraaghs* from Uncle Bert. In dread I
ran toward the vocalizations and found the group settled down peacefully for their day-resting period. Only Uncle Bert remained tensely
sitting upright and alert as if on guard duty.

Some fifteen minutes passed. Still the silverback sat, rigid, wearing
a facial expression of fear. All at once a pair of ravens, which had
been cawing nearby, flew over the group and dive-bombed toward
Uncle Bert's head. *Wraagh!* Quickly he leaned over, cringing, with
his hands covering his head. For nearly an hour at irregular intervals
the ravens continued their mortifying teasing of the majestic silverback, who was totally ignored by the other members of Group 4.
I was embarrassed for my noble friend.

Once the ravens left, Uncle Bert, again the dignified Group 4
leader, led his family off to feed. Thinking the gorillas had gone, I
slowly stood up to determine their direction for the next day's contact.

Above: Flossie directs mild pig-grunts, a disciplinary vocalization, at juveniles wanting to play with her six-month-old daughter, Cleo. (*Dian Fossey*)

Above: A young adult male, Ziz, succeeds in eliciting a roughhouse play session from his old father, Beethoven, during a day-resting period, a time when social interactions between group members are at their highest. (*Dian Fossey*)

Opposite: Nine-month-old Poppy, the fifth offspring of Group 5's Effie, jubilantly chestbeats while jumping up and down on her mother's back. (*Dian Fossey*)

One of my first rules to visitors was "Never touch the gorillas." This rule was occasionally broken once I learned how much gorillas loved to be tickled. (*H. von Rompaey*)

OPPOSITE

Top: The gorillas enjoyed play sessions with people they knew and trusted, although this type of behavior sometimes interfered with observations of normal gorilla behavior. (*Dian Fossey*)

Bottom: When I approached a study group for a contact it was important for me to remain at the gorillas' level and announce my approach by giving gorilla contentment vocalizations. The animals then would continue with ordinary behavior and not be alarmed by the sudden appearance of an observer. (*Warren and Jenny Garst/Tom Stack Associates*)

I labeled Pablo a "dirty data stealer" because of his continually mischievous way of stealing notes or film accumulated after hours of observation. (*Dian Fossey*)

In an effort to distract Puck's attention from my camera, I offered her a *National Geographic* magazine and was amazed at her interest in large color photographs. (*Dian Fossey*)

Above: I often felt that the *Hagenia* tree must have been made just for gorillas, especially when watching Coco and Pucker play gorilla chase-and-grapple games on the strong moss-covered limbs. (*Dian Fossey*)

Opposite: Exhausted by a morning of strenuous play, the two orphans snooze in the nook of a *Hagenia* tree comfortably cushioned by soft moss in the warm sun. (*Dian Fossey*)

Left: Eleven years after the departure of Coco and Pucker, three-year-old Bonne Année (also captured from the wild and confiscated from poachers) came to Karisoke for a three-month recuperation period. During our walks in the forest Bonne Année was attracted to the gorilla graveyard of victims killed by poachers. (*Dian Fossey*)

Right: Unlike Coco or Pucker, Bonne Année was reintroduced to the wild. When she had completely recovered from poachers' neglect, she was unsuccessfully introduced to Group 5, who rejected her because of their own close kinship ties. On that day, as seen here, the orphan huddled against aged Beethoven's back for protection against heavy rain and also for security against the silverback's progeny who sought to kill her. Some six weeks later, when Bonne Année had recovered from the bite wounds inflicted by members of Group 5, she was successfully introduced to a heterogeneous group—one without strong blood ties. (*Dian Fossey*)

Bonne Année was finally free to live among her own kind. (*Elisabeth Escher*)

Come to
meet him in Rwanda

Office Rwandais du Tourisme et des Parcs Nationaux, B. P. 905, Kigali

Above: On reaching sexual maturity, Digit remained with his natal group because breeding opportunities were available for him. After the death in 1974 of Old Goat, Group 4's dominant female who had shared Digit's sentry duties, the young silverback took on a haunted look. (*Dian Fossey © National Geographic Society*)

Opposite: By 1972, aged eleven, Digit's trust in the people he knew had become famous. This poster from one of the author's photographs appeared in tourist bureaus around the world.

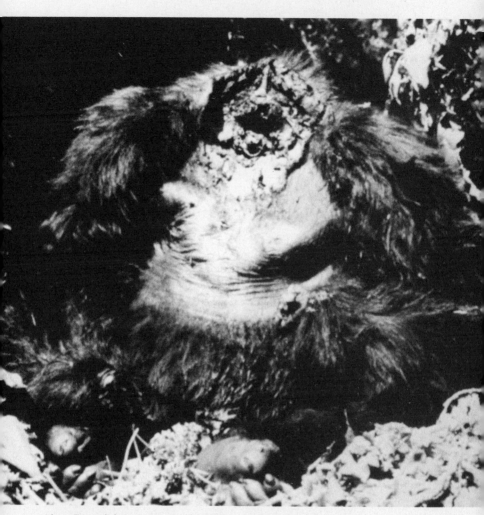

Above: Digit did not live to see his only sired infant. On New Year's Eve day in 1977 he was slain by poachers. Digit gave his life so his family group might survive for the perpetuation of his kind. (*Dian Fossey*)

Opposite: Unaware that this would be the last picture of Digit, the author took this photograph of him on sentry duty in early December 1977 as he was sitting in the shade apart from his group. Digit and Simba's infant was due to be born in four months. (*Dian Fossey*)

Overleaf: In April 1978 Digit's first and only offspring, named Mwelu, an African word meaning "a touch of brightness and light," was born in Group 4 to Simba. (*Dian Fossey*)

Suddenly I heard a noise in the foliage by my side and looked directly into the beautifully trusting face of Macho, who stood gazing up at me. She had left her group to come to me. On perceiving the softness, tranquillity, and trust conveyed by Macho's eyes, I was overwhelmed by the extraordinary depth of our rapport. The poignancy of her gift will never diminish.

11 | Decimation by Poachers: Group 4

BY JANUARY 1977 Uncle Bert had matured into an authoritative group leader who commanded the respect of all of his family members. It had taken nearly eight years for the young inexperienced silverback to be thus transformed. Uncle Bert had learned through a series of interactions with other groups and lone silverbacks, by settling disputes whenever they arose within his own group, and by acquiring increased responsibility for the young he had sired as well as those of his deceased father, Whinny. By the tenth year of the Karisoke study, Group 4 consisted of eleven members. They had had eight deaths, five emigrations of young females, six births, and two immigrations, one a female, the second a male.

The immigration of the blackback Beetsme, who was about ten years old, puzzled me. Beetsme was not well tolerated by Group 4 members except by the immatures, particularly eight-year-old Tiger, son of the group's previously dominant female and former silverback leader. Before Beetsme's arrival, Tiger and Digit, now a sexually mature silverback, had been the Group 4 sentries, maintaining guard positions at the group's periphery on watch for either human intruders or other gorillas seeking interactions with Group 4. Beetsme was never inclined to take on the responsibilities of sentry duty. He shared no blood ties with any of Group 4's members and appeared to contribute nothing toward group cohesion or defense.

Macho, the adult female who had left and subsequently rejoined Group 4, was now an integral part of the group. Her son Kweli, eighteen months old in early 1977, was well favored by his father, Uncle Bert, and was one of the liveliest gorilla infants I had yet to meet. Much like Poppy in Group 5, Kweli was accepted by all Group 4 members for grooming or playing activities.

One bright warm morning I met the group sunning in a small meadow surrounded by hills in the saddle area. The sound of my approach caused Uncle Bert to sit up abruptly. Recognizing me, he gave a soft belch vocalization greeting and lay back down in the hot sun, wearing what I called his "smaltzy face," an expression of ultimate contentment and well-being. Macho strolled over, gazed at me with her wide, trusting, gentle eyes, and lay down by her mate's side. Spirited Kweli was far too playful to settle quietly with his parents. Using a caterpillar type of crawl, Kweli approached me on bent elbows, his white-tailed rump protruding upward. Within seconds his button-bright eyes were looking into my own, his whiskers tickling my face as he smelled my hair. Inquisitively he plucked at my clothing and knapsack before kicking up his heels in delight and rolling over backward to tumble against Uncle Bert. The sprightly infant next somersaulted onto Macho's body for a brief suckle. Mother and son gently grappled together and gave soft play chuckles, while wearing smiles of lazy contentment.

Tiger, who had been at the group's edge playing roughly with Beetsme, joined the family, clustered as usual around Uncle Bert. Tiger's kinship ties with Uncle Bert over time were stronger than those of any other Group 4 member; a very deep social bond existed between them as well. Frequently the probable half brothers indulged in long play and grooming sessions. This particular January day was too warm for strenuous activity. After a mild exchange of tickling and grooming, both males lay back, wearing smiles, and dozed off. By now all the family — with the exception of Digit, on the group's edge in his customary sentry position — were gently snoring in a compact little nesting circle. At that moment I could not think of any place in the world where I would rather be than sitting in the midst of Group 4, as satisfied with the sun and the seclusion as the gorillas were.

About half a drowsy bee-droning hour passed before I thought I heard a whistle from the top of the nearest hill. Uncle Bert, who had been snoozing so heavily that his lower lip was hanging down to his collarbone, immediately sat up and stared in the direction from which the sound came. The silverback's eyes, ears, and nose seemed as keen as antenna receivers. His body tensed for about five minutes. Digit, who had been resting on the slope above the group, slowly started climbing toward the source of the sound. Tiger, now serious and alert, left Uncle Bert to follow Digit up the hill. During the next hour nothing further was heard. Uncle Bert relaxed but led his group off to feed in the opposite direction from the noise.

The confidence of the group's movements assured me that I could begin the long trek back to camp. About twenty minutes away from the contact site, I saw a poacher running through a wide open meadow carrying his spear, bow, and arrows high above his head. Like an antelope the man literally skimmed across the meadow and into the thick forest, where other poachers and their dogs were waiting for him. Running as fast as I could, I took off after them. Once in the forest, I hid and imitated the poacher's whistles to draw the scattered men and dogs to me. However, upon seeing "Nyiramachabelli," all fled.

Returning to camp after a futile chase, I asked Ian Redmond and the excellent tracker Rwelekana to take up the poachers' trail where I had left it. Meanwhile I went to Group 4 to check on their safety. Backtracking the poachers' trail showed that the men I had met were responsible for the whistling heard while I was with the gorillas. They had been following a newly set trapline that ended on the hilltop above Group 4's day-resting site. They also had just speared a duiker to death and had been in the process of cutting up its body when I came onto the meadow. This explained the unusual flaunting openness of the poacher as he attempted, and succeeded, in diverting my interest from the slain antelope. After checking to confirm that the gorillas were all right, I broke the poachers' traps and went back to camp. Ian and Rwelekana eventually returned with the remains of six duiker victims, spears, bows, arrows, and hashish pipes confiscated from the poacher party.

The summer months of 1977 were nearly idyllic for Group 4 as they peacefully meandered throughout the western saddle area disturbed neither by poachers nor by other gorilla groups. Their contented, harmonious days were replete with sunning, playing, and feeding. Between August and September one of Digit's intensive copulations with Simba resulted in conception. The young female abruptly ceased her mounting solicitations, socialized little with others, and spent more time feeding — behavior typical of an impregnated female. Digit then returned full-time to his watchdog position, sometimes as far as one hundred feet away from Group 4.

That entire year Ian Redmond, the camp staff, and I increased our antipoacher patrols in the saddle area and also our periods of contact with Group 4, especially since the family had moved farther away from the security of Visoke's slopes.

On December 8, 1977, when approaching the group, I first encountered Digit some distance from the others sitting alone in a hunched position and looking totally dejected. I felt compelled to spend some time with him to exchange belch vocalizations. Since his impregnation of Simba, the young silverback once again seemed an animal without a cause. I was prompted to snap a few pictures, even though he was in the shade and appeared very morose. After a while Digit began feeding. As he left me, he momentarily assumed his mischievous face and whacked some foliage down on my back — his old way of saying, "Goodbye."

I went on to contact Group 4 and found Uncle Bert, resembling a great black Buddha, surrounded by his two mates Macho and Flossie and frolicking offspring. Augustus, nearest to the family leader, was cheerfully clapping the soles of her feet. Kweli, in one of his drunken-sailor moods, wobbled back and forth on his hind legs between the group and myself with squinted eyes and a lopsided grin. The peaceful aura of Group 4's togetherness would have been complete if only Digit had been in their midst.

* * *

The holiday season with its annual threat of increased encroachment within the park was approaching. My usual dread of this time of the year was somewhat alleviated because our patrols had been quite successful in confiscating poachers' weapons and in demolishing traps. However, limited staff and funds meant we could cover only so much of the extensive saddle area at any one time. We therefore alternated the regions patrolled on a regular basis.

January 1, 1978, Nemeye returned to camp very late in the day to announce that he had not been able to find Group 4. Their trail had merged into numerous buffalo, elephant, poacher, and dog trails. Fearful, he added that he had also found a great deal of blood along the trails, as well as diarrhetic gorilla dung. In spite of the evidence of poachers and dogs, Nemeye had shown a commendable amount of courage in tracking Group 4 along their two-mile flee route back to Visoke's slopes. The following day, four of us — Ian Redmond, accompanied by Nemeye, and myself, accompanied by Kanyaragana, the houseman — left camp at dawn to begin searching for whatever the vast saddle area might reveal.

It was Ian who found Digit's mutilated corpse lying in the corner of a blood-soaked area of flattened vegetation. Digit's head and hands had been hacked off; his body bore multiple spear wounds. Ian and Nemeye left the corpse to search for me and Kanyaragana, patrolling in another section. They wanted to tell us of the catastrophe so that I would not discover Digit's body myself.

There are times when one cannot accept facts for fear of shattering one's being. As I listened to Ian's news all of Digit's life, since my first meeting with him as a playful little ball of black fluff ten years earlier, passed through my mind. From that moment on, I came to live within an insulated part of myself.

Digit, long vital to his group as a sentry, was killed in this service by poachers on December 31, 1977. That day Digit took five mortal spear wounds into his body, held off six poachers and their dogs in order to allow his family members, including his mate Simba and their unborn infant, to flee to the safety of Visoke's slopes. Digit's last battle had been a lonely and courageous one. During his valiant struggle he managed to kill one of the poachers' dogs before dying. I have tried not to allow myself to think of Digit's anguish, pain, and the total comprehension he must have suffered in knowing what humans were doing to him.

Porters carried Digit's body back to camp, where it was buried several dozen yards in front of my cabin. To bury his body was not to bury his memory. That evening Ian Redmond and I debated two alternatives: bury Digit and retain the news of his slaughter or publicize his death to gain additional support for active conservation in the Parc des Volcans through regular and frequent patrols to rid the park of encroachers.

Ian, relatively new to the field, held optimistic views about all that could be gained by publicizing Digit's death. He felt that a public outcry over the needless slaughter would exert pressure on Rwandese government officials to imprison poachers for prolonged periods. He also believed that the incident might compel a greater degree of cooperation between Rwanda and Zaire so that the contiguous sections of the Virungas could function as one.

I did not share Ian's optimism. By the time of Digit's death I had been working eleven years in the Virungas. I had met only a handful of park guards or park officials who had not fallen victim to the general inertia and malaise of their poor and overpopulated countries. Indeed, one of the greatest drawbacks of the Virungas is that it is shared by three countries, each of which has problems far more urgent than the protection of wild animals. I did concur with Ian that an indignant public outcry might well result in large sums of unchanneled conservation money pouring into Rwanda, but little would be intended for active antipoacher patrols. Following the captures of Coco and Pucker, both funding and a new Land-Rover had been obtained by the Rwandese officials connected with the park at that time; however, neither financing nor Land-Rover was applied toward the park interests. I had come to believe that self-motivation must necessarily accompany monetary assistance if long-term goals were to be accomplished. My greatest fear was that the world would climb evangelistically onto a "save the gorilla" bandwagon upon hearing of Digit's death. Was Digit going to be the first sacrificial victim from the study groups if monetary rewards were to follow the news of his slaughter? This was my line of reasoning as Ian and I continued to discuss the pros and cons of publicizing Digit's death.

The black night skies faded into those of a gray-misted dawn when I realized that, like Ian, I did not want Digit to have died in vain. I decided to launch a Digit Fund to support active conservation of gorillas, the money only to be used to expand antipoacher foot patrols within the park. This would involve the recruitment, training, outfitting, and remuneration of Africans willing to work long tedious hours cutting down traplines and confiscating poacher weapons such as spears, bows, and arrows. I would have preferred to employ park guards for such work. Cooperation with the government is essential, especially when one is a guest in a foreign country. The park guards, not I, have the legal right to capture poachers, and they could also have used the extra income to supplement their monthly salaries of about sixty dollars. However, the park guards ostensibly work for the Parc des Volcans' Conservator, who is in turn employed by the Director of the Rwandese national parks and located in Kigali. The guards routinely receive their salaries whether they go into the park or not; thus this aspect of motivation had never worked successfully. For many years, I had returned to Rwanda after brief trips to America laden with boxes of new boots, uniforms, knapsacks, and tents for the guards. I had tried countless times to encourage the men to participate actively in antipoacher patrols conducted from

Karisoke in the heartland of the park. The uniforms and boots, of course, were eagerly received, likewise the extra money paid and the food eaten at camp, but the guards' brief efforts were meaningless. The men wanted only to return as rapidly as possible to their villages and the local *pombe* bars, where most of the boots were sold to more affluent Rwandese in order to buy more *pombe*. My naïve attitude eventually ceased once I came to realize that the long-standing poachers of the park are on extremely good terms with the guards, whom they regularly pay either in francs or meat for permission to hunt in the park. I also learned that the men the guards pretended to have captured during their stays at Karisoke were, in actual fact, friends or relatives who always managed to "escape" when being escorted to prison. I had made the mistake of paying the guards extra for each poacher caught instead of paying a flat salary for each day's work. This was a mistake I never repeated when later working with non-park-related recruitments, the only men I could personally motivate to work honestly and effectively. It was equally futile to pay a bonus for each trap snare brought to camp, since this practice only encouraged the manufacture of nooses to be brought to camp in exchange for the reward.

For several days Ian, the camp staff, and I backtracked the poachers' trail to and from Digit's death site and maintained brief contacts with Group 4 — well secluded on Visoke's slopes — while mulling over the decision we had to make. We found that Digit had not been killed for the trophies of his head and hands as we originally had thought. Six poachers had been working their trapline before inadvertently running into Group 4 at the line's end. Digit's body lay only 80 feet from the last trap and some 260 feet from Group 4's day-nesting site where he had been alone on sentry duty.

Tracking along the poachers' trail showed that the men had been in the park killing antelope and setting traps for two days before encountering Group 4. They then fled back to the notorious poacher Munyarukiko's village of Kidengezi, adjacent to Karisimbi's eastern slopes. The poachers took Digit's head and hands as afterthoughts, because such items had previously been sold profitably to Europeans. Our original mistaken belief that Digit had been killed only for trophies is regretted. It continues to capture the public's imagination far more than the actual truth: Digit was not killed as an intended victim of slaughter by trophy hunters; he gave his life to save his family — which, tragically, had been in the wrong place at the wrong time, especially on New Year's Eve. If Digit's death proved to be an economic boon to the park system, I wondered just how long Group 4 could survive — a month? six months? a year? I awakened each morning wondering who would be next.

Ian and I finally decided to publicize Digit's slaughter. A few days later North American television viewers heard Walter Cronkite announce Digit's death on the "CBS Evening News." The Rwandese park Conservator was invited to camp to see Digit's body before it was buried. He arrived with Paulin Nkubili. The Chef des Brigades was genuinely horrified at the mutilated corpse and promised to do all that he could to apprehend any known poachers his men might encounter in Ruhengeri. Active law enforcement is what active conservation is all about.

Six days following Digit's slaying, I was typing in my cabin when I heard the woodman scream, *"Bawindagi! Bawindagi!"* — "Poacher! Poacher!" The four Rwandese working at camp immediately took up the chase of an unknown man who had crept into camp attempting to kill one of the many antelope now thriving in the security Karisoke offered. After a long chase, the poacher was caught and brought back to my cabin. The man was wearing a yellow T-shirt extensively stained and splayed with dried blood. He was also carrying a blood-stained bow and five arrows. Questioning revealed that this was indeed the blood of Digit.

The Chef des Brigades again climbed to camp with armed commandos to take legal custody of the poacher, who was later tried, convicted, and sentenced to prison in Ruhengeri. While at Karisoke, Nkubili questioned the man and obtained the names of the other five poachers responsible for killing Digit. Within a week two of them had been captured. Three leading poachers of the Virungas — Munyarukiko, Sebahutu, and Gashabizi — eluded capture by hiding in the forest.

I resumed my contacts with Group 4 but, for countless weeks unable to accept the finality of Digit's death, I found myself looking toward the periphery of the group for the courageous young silverback. The gorillas allowed me to share their proximity as before. This was a privilege that I felt I no longer deserved.

Tiger and Beetsme attempted to take Digit's place as Group 4 sentries. However, the two young males frequently were diverted by their roughhouse wrestling games, leaving Uncle Bert to bear full responsibility for group security. Shortly after their flight to Visoke's slopes, Group 4 were harried by interaction attempts from Nunkie. Uncle Bert led his family back to the saddle area to get away from the persistence of the older silverback, who appeared to be trying to obtain Simba. Group 4 went directly to Digit's death site, circling the area for days as if looking for Digit, whose slaying, of course, they had not witnessed. Their actions were surprising to me. Over the previous ten years of research, whenever gorillas had encountered cattle herds, traps, or poachers, they usually avoided immediately returning to the areas that had threatened danger.

* * *

Because of all the terror the animals had been through, I was reluctant to consider the highly traumatic process of herding Group 4 back onto the relative security of Visoke's slopes, away from the jeopardy of traps and poachers remaining in the saddle. My indecision dissolved when I observed Tiger with a fresh wire-snare injury on his right wrist.

The herding day was as terrifying for Group 4 as it was repugnant to me. The frenzied gorillas had no way of knowing that they would not be harmed by their unseen pursuers, that their flee route had been cleared of traps, and that they purposely were being driven toward their favored section of Visoke currently unoccupied by Nunkie's Group. Only this knowledge made it possible for the staff and me to endure Group 4's screams of terror as they fled toward the mountain, led by Uncle Bert, flanked by Tiger and Beetsme.

Twenty-four hours later Group 4 showed no obvious effects of their earlier harassment. If anything, they seemed calmer than they had for a long time, though, understandably, much of their calmness was attributed to fatigue.

For nearly six months Group 4 contentedly remained on or adjacent to Visoke's slopes without encountering other gorilla groups or poachers. Simba, carrying Digit's first and only offspring, was showing increasing signs of her impregnation. Group 4's immatures were highly attracted to the young female, but Simba continued to prefer to spend most of her time alone feeding voraciously. She allowed Beetsme to mount her several times, and Flossie, also nearing her term, solicited numerous mountings from Uncle Bert. Flossie was carrying her fifth infant to be born within Group 4 since the Karisoke study began, only two of them currently alive. This would be Uncle Bert's seventh-conceived offspring, four of which were still living.

Flossie showed more obvious physical signs of her impending birth than did Simba. Both were cantankerous, not only toward one another, but toward Macho in particular. The mother of Kweli, who was now thirty-two months old, evidenced no return to estrus and once again retreated to the sidelines of the group to avoid antagonistic encounters with Flossie and Simba.

Another sure sign of Flossie's condition, typically seen in females about to give birth, was the increased amount of grooming attention she gave to her youngest, three-and-a-half-year-old Titus. The young male thrived on her unaccustomed maternal care just as his older sister Cleo had relished similar attention from Flossie before Titus' birth. Play sessions were consequently reduced for Kweli, who was undergoing his most intensive period of weaning. That stress, combined with the curtailment of play with Titus, transformed Kweli into a whining, pouting infant, though he was, as always, receiving consistent and loving affection from Macho.

Three months and seven days following Digit's killing, a diminutive part of Digit came into the world when Simba gave birth to Mwelu, an African word meaning "a touch of brightness and light." Digit's perpetuation was an extraordinarily beautiful female infant with long fluttering eyelashes that framed bright sparkling eyes. Simba proved herself tenaciously uncompromising about her infant's protection, since, like most newborn gorillas, Mwelu was an object of great curiosity among the group's younger members.

Throughout the entire study period, Simba was the second female to give birth within her group without having the support of her offspring's sire. Ironically enough, Simba herself had been the first born following the death of her father. As an infant she had relied only on her aged mother, Mrs. X, and, when fully orphaned, upon Uncle Bert, then Group 4's new leader. Whenever other Group 4 members grew overzealous in curiosity about the newborn, once again Uncle Bert became a protector for Simba as well as for tiny Mwelu. Mwelu was forty-five days old when Flossie gave birth to a female infant named Frito.

When Frito was nearly a month old in mid-July 1978, Uncle Bert led his group from Visoke's slopes back to the saddle area, where no signs of poachers had been seen for several months because of increased patrol work. For a week the adults of Group 4 sunbathed blissfully while the young unleashed boundless energy climbing and chasing one another in huge *Hagenia* trees. All the animals were taking advantage of the saddle's wealth of lush and varied vegetation. They were living each day to the fullest upon their return into favored terrain.

On the morning of July 24, 1978, one of four students then at Karisoke knocked on my door. I was surprised to see him because he had left camp only an hour earlier to contact Group 4. One look at his face told me another disaster had struck.

"Poachers." I stated rather than asked.

The student replied, "Uncle Bert has been shot through the heart and decapitated."

That morning the student had proved himself quite intrepid. Even though alone, he had searched the area surrounding Uncle Bert's still-warm corpse looking for Group 4, and was willing to return to the slaughter site accompanied by staff members. The men spent hours sorting out a maze of flee trails leading from Uncle Bert's body back to Visoke's slopes. There they found an unhabituated fringe group of thirteen animals excitedly confronting the remaining ten members of Group 4, who were being bravely protected by ten-and-a-half-year-old Tiger. Upon seeing human observers, the fringe group fled and Group 4 gathered around Tiger, their new leader. As the only living son of Whinny, probably the founder of Group 4, and the only surviving offspring of the group's previously dominant female, Old Goat, Tiger had been groomed throughout his short life for the role of Group 4's next leader. Because the young male so closely resembled Uncle Bert and had also shared strong social bonds with the silverback, I considered it likely that Tiger's newly acquired position was one that had been enhanced by strong kinship bonds to the group's probable original leader, Whinny.

Kweli was found whining pitifully, and Macho was missing. It was assumed that the fringe group had succeeded in taking her off with them. That night Tiger nested with his half cousin, three-year-old Kweli, who had never yet slept alone. Tiger's protective behavior exactly duplicated that of Uncle Bert, who, seven years previously as Group 4's new leader, had given consistent care toward safeguarding his half sister Simba when she had become orphaned.

All the horror and shock of Digit's murder returned. Even more abhorrent details were to be revealed concerning the newest slaying. Following poachers' footprints, two trackers and I backtracked from Uncle Bert's body to find a still-smoldering fire where the killers had spent the night only two hours from Group 4's July 23 nesting site. The poachers' trail, both toward and away from Group 4, led directly to Kidengezi, Munyarukiko's village. Trail evidence also revealed that the poachers had been interrupted by the student's arrival before they could further mutilate Uncle Bert's body; thus the silverback's hands had been left intact. Unlike the slow, agonizing killing of Digit by arrows, spears, and dogs, Uncle Bert had been murdered by a single bullet piercing his heart. He had probably comprehended no more than a brief moment of terror before dying.

Uncle Bert was killed very near Group 4's night nests at the beginning of what should have been another sun-filled day of sharing activities and contentment among mates and offspring. Yet, unwittingly, the poachers gave one small thread of consolation by letting me know that Uncle Bert had not endured Digit's suffering. Flagrant in conveying the threat of their new power to kill gorillas with guns, they left the single bullet hole above Uncle Bert's heart intact. They had used knives and *pangas* to slash open the right side of the noble silverback's chest and remove the damaging evidence of the bullet. We carried Uncle Bert's body back to camp and buried it next to Digit's.

I went to Ruhengeri to report the newest slaughter to Paulin Nkubili. The Chef des Brigades promptly organized a commando force to raid Munyarukiko's village and invited me to go with them. With the element of surprise in their favor, the commandos were able to surround the village that night and quickly move in to search the small thatched huts. Within an hour they had confiscated a pile of spears, bows, arrows, and hashish pipes. More important, under a bed in one of the huts the soldiers found the third most notorious poacher of the Virungas. Gashabizi was later proven to have been involved with Digit's death and also the latest slaughter. Although Munyarukiko again had escaped, Gashabizi's capture made the long night's effort worthwhile. Eventually he was tried and convicted to ten years' imprisonment in the Ruhengeri prison.

The following morning Chef Nkubili conducted another surprise raid in a small village where Sebahutu, a second treacherous poacher, lived with his seven wives and numerous children. Repeating the routine of the previous night, the commandos surrounded the huts and carefully searched them one by one. The results were gruesomely rewarding. Piles of spears, bows, arrows, and hashish pipes were accumulated and placed in the center of the compound. Then Sebahutu's sodden, blood-stained clothing, several knives, and *pangas* gummed with sticky blood were discovered hidden under straw bed matting.

The presence of the incriminating evidence sent Sebahutu's wives into loud wails proclaiming the innocence of their common husband. Just at that moment, a man in a bright red sweater ran from behind the compound hedge and down an open trail through the village. The commandos caught him and returned him to the cluster of huts for questioning. Sebahutu, the poacher who, we later learned, had fired the bullet into Uncle Bert's heart, was captured. A court tribunal subsequently pronounced him guilty and sentenced Sebahutu to the Ruhengeri prison. Now only Munyarukiko remained at large.

Ready to climb back to camp, I met a porter waiting for me with a note at the base of Visoke. The body of Macho had been found 150 feet from the spot where Uncle Bert had been slain. Macho had also been shot by a single bullet that had passed through her rib cage and splintered her spinal column before exiting. As in the case of Uncle Bert, the poachers had retrieved the bullet casing.

Dazed, disbelieving, I drove back to Ruhengeri thinking of the day Macho had walked up to my side to gaze into my face with her wide trusting eyes and the tenderness she had always lavished on Kweli. How would the three-year-old survive without his mother or father?

Nkubili's reaction upon hearing of yet another killing was intense anger. He immediately planned a third patrol and ordered *all* suspected poachers be brought in for questioning. The next day I drove my VW bus filled with armed soldiers and a police inspector to a small village adjacent to the Parc des Volcans. I parked out of sight of the village and the soldiers poured from the car. Carrying their guns high over their heads and moving as if they were marines making a beach landing, the men quickly surrounded the marketplace to confine several hundred people inside the square. This was the first of five surprise raids made that day on villages around the park. The raids resulted in the capture of fourteen poachers, all of whom were detained in the Ruhengeri prison to await trial.

Driving back to Visoke's base, I saw the Rwandese park Conservator walking along the road and offered him a lift to his headquarters. During the short drive he spoke curtly in rapid Kinyarwanda to the camp staff members in the back of the bus while ignoring me totally. After we dropped off the Conservator, the men translated his conversation. The main reason for his anger, they explained, was that some Rwandese, long known as poachers, had been imprisoned. The Conservator had stated that he was going to demand the men's immediate release because Uncle Bert and Macho had been killed within the Zairoise sector of the Parc des Virungas. He therefore concluded that their deaths had to be considered a responsibility of Zaire and not of Rwanda.

The staff members and I all too vividly recalled that Digit also had been slain in Zaire, very near the spot of the latest killings, and that all the poachers connected with Digit's slaying had been proved to be Rwandese. Since neither poachers nor gorillas carry visas, I was amazed that the Conservator could impute blame to the Zairoise, who seldom ventured at all near the Rwandese border where the killings had occurred.

While I drove back along the jarring lava-rock road to Visoke's base, the men continued translating the rest of the Conservator's remarks. He had told them that he was just returning from a three-day stay in Gisenyi, where he had gone to collect a young newly captured gorilla. Not finding the victim, he had to return to the park headquarters, which lay about halfway between Ruhengeri and the base of the mountains.

As the men and I began the trail for the long dismal climb back to Karisoke, I found myself constantly remembering the captures of Coco and Pucker, used as barter items nine years earlier between Germany and Rwanda. I became increasingly puzzled about what motives might lie behind the present Conservator's abrupt departure to Gisenyi to pick up a young gorilla on the same day that Uncle Bert and Macho were killed. In an effort to find a link between his trip and the latest slayings, I asked several long-trusted Rwandese assistants to undertake a quiet investigation. They knew how to gather information discreetly from villagers and other informants knowledgeable about poacher activities.

Two days after the killings of Uncle Bert and Macho, a European conservation team, accompanied by a reporter, had arrived in Kigali. Their long-planned visit had been greatly anticipated by park officials, who were to receive additional financial assistance, equipment, and supplies from the consortium of gorilla-preservation agencies that had sprung up following Digit's much publicized killing. The conservationists were met at the Kigali airport by the park Director, his assistants, and a Belgian aide. Immediately, of course, they were informed of the latest gorilla slaughters. The reporter was able to telephone his story to London, datelined Kigali.

The team spent two days more in Kigali before going on to Ruhengeri, where I encountered them while organizing legally conducted village searches for poachers. I was muddy, hungry, exhausted, and more depressed than I had ever been at any point of the eleven-year research. The Europeans' chauffeured van pulled up alongside my bus. The reporter nimbly jumped out, tape recorder in hand, wanting an on-the-spot interview about the happenings of the past few days. My mind flashed back to the long deliberation between Ian Redmond and myself the night six and a half months earlier following the killing of Digit. Since Digit's killing had proved so profitable to the Rwandese park officials, could there possibly be a connection between the first tragedy and the latest timely slaughters? I refused the reporter's demands to climb to Karisoke to photograph either the bodies of the gorillas or their gravesites because I did not want to foster more publicity. So far it seemed only to have brought disastrous effects on the known animals of the study groups.

The next day, five days after Uncle Bert and Macho had been killed, the European conservation mission left Rwanda. From an article subsequently published in a British conservation journal, I read that the group had been extremely gratified by the timeliness of their visit, the financial assistance they had given and pledged to the Parc des Volcans, and the wide attention that the reporter's newsbreaking articles had received from a sympathetic public.

Even now, four years later, most people are under the impression that gorillas are being slaughtered for their heads, a practice Paulin Nkubili had ended years before the killings of Digit, Uncle Bert, and Macho. That gorillas will die defending their family members is, to many media hounds, not as newsworthy as the inaccurate story of gory mutilations for trophy purposes.

Six days after the killings of Uncle Bert and Macho the reason for Kweli's incessant whining became evident. The infant had been shot through the top of his right shoulder at the time of his parents' murders. The bullet had chipped Kweli's clavicle before exiting through the musculature of his scapular region. Since both Sebahutu and Munyarukiko were crack shots, I found it odd that they had missed killing young Kweli, if indeed this had been their reason for meeting Group 4 on that fateful July morning.

Using the same procedures of extensive backtracking that

had proved so revealing following the killing of Digit, the Rwandese assistants, several students, and myself unraveled the sequence of events involved in the latest killings. Kweli had been the first victim. He was shot from a tree by Sebahutu shortly after Group 4 had arisen from their night nests and spread out to feed in their customary morning routine. Macho was shot as she ran in from a side feeding position in a futile effort to protect her son. Uncle Bert had been in the lead of his group fleeing toward Visoke's slopes when the screams of Kweli and Macho caused him to return to and charge the poachers head-on in defense of his mate and son. The silverback had to have been standing upright for the fatal bullet to strike him as it did. Dead before he hit the ground, his heart shattered. Only the intervention of Kweli's parents made it possible for the juvenile to flee with Group 4. Both Uncle Bert and Macho could have escaped had they not instinctively tried to protect their son. They gave their lives so that Kweli might live.

Several days following my knowledge of Kweli's bullet wound, the men whom I had asked to collect information from villagers living adjacent to the park returned to camp. They withdrew from their pockets hand-scrawled notes of places, dates, times, names of people, and lists of events possibly connected with the killings. From my informants I learned that two Africans, both strangers and thought locally to have been "Congo-men," or Zairoise, had visited the Conservator the day before Uncle Bert and Macho had been killed. They and the Conservator spent several hours together before the strangers left and the Conservator informed his staff, for the first time, that he would be leaving to go to Gisenyi to collect a newly captive gorilla youngster. The Conservator then asked his staff to reinforce an enclosure near his offices for the young gorilla's temporary stay.

It was only at this moment — eleven days after the killings — that I realized the full implications of the Conservator's conversation with my staff at the back of the bus on the day I gave him a lift following his return from Gisenyi. It seemed entirely likely that Kweli had been his intended capture victim, but that neither he nor the poachers had anticipated the lengths to which gorillas will go in defense of their own kind. However, they did realize that the capture of a gorilla *outside* the study area would not generate the profitable publicity to attract monetary aid for gorilla conservation from foreign countries. Furthermore, for the protection of the Rwandese connected

with the sector of the Virungas on which Karisoke study animals
ranged, any gorilla capture had to be attempted only from the Zairoise
section of the Virungas. For several years both Groups 4 and 5 alternated their travel between the two countries, but only Group 4 was
ranging in Zaire at the time the European conservation team arrived.

* * *

The newly orphaned Kweli, deprived of his mother, Macho, and
his father, Uncle Bert, and bearing a bullet wound himself, came
to rely only on Tiger for grooming the wound, cuddling, and sharing
warmth in nightly nests. Wearing concerned facial expressions, Tiger
stayed near the three-year-old, responding to his cries with comforting
belch vocalizations. As Group 4's new young leader, Tiger regulated
the animals' feeding and travel pace whenever Kweli fell behind.
Despondency alone seemed to pose the most critical threat to Kweli's
survival during August 1978.

Beetsme, having no kinship ties within the group, was a significant
menace to what remained of Group 4's solidarity. The immigrant,
approximately two years older than Tiger and finding himself the
oldest male within the group led by a younger animal, quickly developed an unruly desire to dominate. Although still sexually immature,
Beetsme took advantage of his age and size to begin severely torment-
ing old Flossie three days after Uncle Bert's death. Beetsme's aggression was particularly threatening to Uncle Bert's last offspring, Frito.
By killing Frito, Beetsme would be destroying an infant sired by a
competitor, and Flossie would again become fertile.

Neither young Tiger nor the aging female was any match against
Beetsme. Twenty-two days after Uncle Bert's killing, Beetsme succeeded in killing fifty-four-day-old Frito even with the unfailing efforts
of Tiger and the other Group 4 members to defend the mother
and infant. Flossie carried Frito's corpse for two days before being
forced to drop it in self-defense during another attack by Beetsme.
The tiny body was buried next to that of her father in the graveyard
in front of my cabin. Frito's death provided more evidence, however
indirect, of the devastation poachers create by killing the leader of
a gorilla group.

Two days after Frito's death Flossie was observed soliciting copulations from Beetsme, not for sexual or even reproductive reasons —
she had not yet returned to cyclicity and Beetsme still was sexually

immature. Undoubtedly her invitations were conciliatory measures aimed at reducing his continuing physical harassment. I found myself strongly disliking Beetsme as I watched his discord destroy what remained of all that Uncle Bert had succeeded in creating and defending over the past ten years.

A week after Frito's death, Flossie had her first opportunity to emigrate from Group 4 during a violent interaction with Nunkie's Group. She transferred accompanied by her seven-year-old daughter Cleo and Petula's eight-year-old daughter Augustus. Flossie's four-year-old son Titus remained behind in Group 4. Although the emigrations meant a further fragmentation of Group 4, I was relieved when Flossie left. It seemed unlikely that she could have endured much more physical abuse from Beetsme.

Flossie and Cleo spent only nineteen days with Nunkie before transferring into a small fringe group of four animals, again at the first opportunity given for an additional emigration. As there was just one other female in this group, Flossie was allowed an opportunity to become higher-ranking than she would have been in Nunkie's established group of four females and young. Sadly, the second transfer of Flossie and Cleo meant a sort of farewell to them; the group they joined was not often found within the Karisoke study area. Known as the Suza Group, the animals usually traveled beyond the far side of the Suza River on Karisimbi's distant slopes. During the ensuing months, however, we made greater efforts than usual to keep track of the Suza Group. With considerable delight we observed that, eleven months following her transfer, Flossie was carrying a newborn infant, who must have been sired by John Philip, the Suza Group's dominant silverback. Nearly forty months later, Flossie became a grandmother for the first time when Cleo gave birth in December 1981.

Augustus had not transferred with Flossie and Cleo because her mother, Petula, had been Nunkie's first female acquired four years earlier. She therefore occupied the highest rank among his four mates. By remaining in Nunkie's Group, Augustus could possibly share the status of her mother and her three-year-old half sister Lee, sired by Nunkie.

Flossie's departure from what was left of Group 4 — Beetsme, Tiger, Titus, Kweli, Simba, and her infant, Mwelu — noticeably reduced aggression within the group, though Beetsme occasionally directed threatening runs or whacks at Titus. Like Titus, Simba and

Kweli counted heavily on Tiger for protection. I allowed myself the faintest glimmer of hope that Simba's four-month-old, Digit's only offspring, might survive the odds against her.

I also became increasingly concerned about Kweli, who had been, only a few months previously, Group 4's most vivacious and frolicsome infant. The three-year-old's lethargy and depression were increasing daily even though Tiger tried to be both mother and father to the orphan.

Three months following his gunshot wound and the loss of both parents, Kweli gave up the will to survive. On the morning of his death he was found breathing shallowly in the night nest he had shared with Tiger. Kweli had only enough strength to give faint screams and whines when the group slowly fed away from him. Responsive to Kweli's sounds of distress, the gorillas returned to his side repeatedly throughout the day to comfort him with belch vocalizations or gentle touches. Once Beetsme even tried to pull Kweli into a sitting position, as if to induce the nearly dead baby to get up and follow them. Every animal seemed to want to help but could do nothing. After spending their day-resting period near the rapidly failing youngster, each member of the group went to Kweli individually to stare solemnly at his face for several seconds before silently moving off to feed. It was as though the gorillas knew Kweli's life was nearly over.

Late that afternoon the body of young Kweli, the poachers' intended capture victim, was brought back to camp to be buried between his mother and father, Macho and Uncle Bert. All that now remained of Group 4 were Simba, her six-month-old daughter Mwelu, and the three males, Titus, Tiger, and Beetsme. It was difficult to think of Beetsme as an integral member of Group 4 because of his continual abuse of the others in futile efforts to establish domination, particularly over the indomitable Tiger.

Four months after the shootings, in December 1978, Simba followed Flossie's example by taking advantage of her first opportunity to transfer into Nunkie's Group. To my deepest sorrow, Mwelu, Digit's only offspring, was killed by Nunkie during the interaction. The bright and shining light was extinguished.

The decimation of Group 4 left the three young males restlessly meandering on Visoke's slopes for six weeks after Simba's loss. Tiger helped maintain cohesiveness by "mothering" Titus and subduing

Beetsme's rowdiness. Because of Tiger's influence and the immaturity of all three males, they remained together. In January 1979 they joined the older Peanuts, who was still traveling alone. Under the leadership of the young silverback, the all-male band left Visoke's slopes to go into the saddle area for the first time since the Group 4 killings six months previously.

Reinforced by three males, Peanuts traveled more broadly and acquired two adult males from fringe groups during unobserved interactions in the northwestern saddle area. The two young strangers, named Ahab and Pattie (the latter first thought to have been a female), provided new opportunities for play interactions for the remnant members of Group 4. The possibility that Group 4 under Peanuts' leadership might have the ability to form a new group was an encouraging thought. Of even more consolation at this time was the news that the gorillas of the Virungas had gained a slight edge on survival. The infamous poacher, Munyarukiko, had died.

* * *

The 1979 holiday season around Karisoke and adjacent areas passed peacefully among the study and fringe groups. The Digit Fund had been paying for the expansion of patrol work conducted from Karisoke for nearly eighteen months. This fund and donations from the Humane Society of America made it possible for my men finally to have waterproof boots, raingear, lightweight bivouac tents, warm clothing, and gloves. After each day in the field cutting traps or confiscating poachers' weapons, the men could return to Karisoke for a hot meal and warm bedding.

Whenever the men became apprehensive about patrolling for more than four or five hours away from camp — because the distance increased the possibilities of running into encroachers armed with rifles — Ian Redmond or myself would accompany them to bolster their confidence. Their willingness to work under difficult circumstances was combined with intrepid honesty and convictions that they had a personal stake in protecting the remaining wildlife of the Virungas. However, the men were not legally park guards but, in effect, implementers of Karisoke Research Centre's conservation policies.

Every week after three days of work the six-man patrol returned to their villages with wages in their pockets. They left their sodden boots and clothing behind at camp to be cleaned and dried out in preparation for the next week's work. This not only ensured long life for the gear but avoided any risk of losing the equipment while the men were off the mountain.

During the first year and a half of these Digit Fund patrols, nearly 4000 poacher traps had been destroyed. The combined costs of food and salaries per day, per man, totaled $6.00.

While Karisoke's active-conservation patrols were cutting traps and confiscating poacher weapons in the heartland of the Virungas, other gorilla conservation organizations were also trying to save the mountain gorillas. These agencies had received substantial contributions from the public in response to the news of the deaths of Digit, Uncle Bert, and Macho. Working outside the Parc des Volcans or near the edge of the park boundary, the groups emphasized expansion of tourism, acquisition of new vehicles and equipment for the park staff, and educational programs to increase the Rwandese people's knowledge of and interest in the gorillas. Because such activities enhanced the image of the Parc des Volcans, they were loudly applauded by Rwandese park officials. In spite of my dismay at the amount of money spent on nonactive conservation efforts, I was nonetheless pleased that Rwandese officials were content and that Karisoke was allowed to continue active-conservation patrols unhampered.

* * *

On the morning of January 1, 1980, someone knocked loudly on my cabin door. I opened it to find one of my food porters carrying a bulging potato basket on his head. I was about to tell him that I had not ordered any potatoes, when he excitedly exclaimed, *"Iko ngagi!"* — "It's a gorilla!" My heart sank. We put the basket into a large seldom-used room. I slowly opened it. Out crawled a pathetically weak female infant of about three years, Kweli's age.

She had been taken from Zairoise poachers attempting to sell her on New Year's Day to a French physician in Ruhengeri for the equivalent of $1000. Only because of the cleverness of Dr. Vimont was the captive acquired from the poachers, who were subsequently jailed. Again, active law enforcement is active conservation at work. I never learned how many group members had been killed for the infant's capture. All I could definitely discover was that she had been held for about six weeks in a damp, dark potato shed near the park boundaries below Mt. Karisimbi and fed bread and local fruits. Like all other gorilla capture victims, she was badly dehydrated and had developed severe lung congestion. Terrified of the presence of humans, the baby immediately hid under a bed when she saw me. For two days whenever anyone entered the room she retreated there. Fresh food vegetation and nesting materials were brought to her. I was pleased when she finally began eating and using the nests I built for her to sleep in at night.

Six weeks of care were required before Bonne Année was well enough to play in the meadows surrounding camp. An additional six weeks were needed for the infant to regain tree-climbing dexterity and food-preparation skills, such as peeling celery stalks, stripping thistles, and wadding *Galium.* The transformation of the sickly captive into a typically lively young gorilla was a joy to observe. Bonne Année's recovery was aided by Cindy, who cared for the infant exactly as she had watched over Coco and Pucker eleven years earlier. Although considerably aged, Cindy provided Bonne Année with cuddling or body warmth whenever the baby wanted to rest. The dog also participated in mild wrestling or chasing games during the gorilla's two-month convalescence.

The Rwandese park Director in Kigali had been informed of Bonne Année's arrival as well as of my intention to introduce her into a free-living group once she recovered from the trauma of capture and poacher confinement. I was pleased with his acceptance of my decision. The enforcement of legislation, both in Rwanda and abroad, had come a long way since 1969, when Coco and Pucker were exploited as pawns traded between Rwanda and Germany.

I felt that Bonne Année's best survival chances lay with the newly forming, heterogeneous Group 4, the only study group without strong blood ties or infants. The major drawback to the choice of Group 4 was Peanuts' inclination to lead his five followers into the saddle area west of Visoke, where poachers and traps were still being encountered. Would Bonne Année be freed only to fall victim to poachers once again?

By March Bonne Année's health was completely restored. The time for her release could no longer be put off. It was first necessary to "wean" her from Karisoke's commodities such as treat food, the cabin's warmth, and the numerous camp residents, including Cindy and visitors who provided the infant with constant attention and play opportunities. To accomplish this, a bivouac camp, consisting of only a small tent and sleeping bags, was set up in Group 4's range far from Karisoke. There, for four days and nights, Bonne

Année learned to "rough it" along with a helpful student, John Fowler, and an African assistant.

The day of Bonne Année's attempted reintroduction to the wild was jinxed from the start. Not only was it pouring rain, but Group 4 had traveled far from the bivouac camp and were involved in an aggressive interaction with an unidentified fringe group. Because of their highly excitable state, it seemed unlikely that Group 4 would have accepted Bonne Année at that time. While returning to the bivouac camp, we reluctantly decided to attempt introducing the infant to Group 5 the next day. Group 5 could offer Bonne Année a relatively poacher-free range, although their strong blood ties could jeopardize acceptance of an infant from outside their own gene pool.

My greatest apprehensions about Bonne Année's introduction to Group 5 were centered on the dominant female Effie, who had the most to lose if alien bloodlines entered the group. Unknown to me, she was three months away from parturition (her sixth infant to be born into the group during the years of the study). I was also concerned about the provocative behavior of Effie's daughter Tuck, coming into regular cyclicity and being mounted frequently by Icarus. A third factor strengthening the kinship ties within Group 5 was the conception of Icarus' second offspring with Pantsy during the same month as the attempted release of Bonne Année. At this time, early spring 1980, I had yet to gain any evidence concerning the extreme lengths to which a silverback will go to protect the integrity of his familial bloodline, other than incidents that had occurred during interactions with extragroup silverbacks. For this reason, I did not consider Icarus a potential threat to Bonne Année.

As John and I took Bonne Année out to Group 5, my misgivings grew. I was surprised that my feelings were not conveyed to the youngster, happily riding piggyback on John during the long morning trek. On reaching Group 5, clustered together under drizzly skies during a day-resting period on Visoke's southern slopes, we were relieved to note that no other groups or lone silverbacks were around. Perhaps this second reintroduction attempt had a chance after all.

It was my intention to find a tree near the group. Being in a treed position would give Bonne Année the option to remain with us if afraid or to return to us if she was not accepted. The three of us climbed into a tall and gently sloped *Hypericum* about fifty feet away from the resting group. Five minutes passed before Beethoven, after a startled, "doubletake" stare, gave a short alarmed scream.

He looked at Bonne Année quizzically, as if trying to determine whether she was one of his group. The infant matter-of-factly returned his gaze as though she had known the old silverback all her life. It was hard to believe that Beethoven was the first gorilla Bonne Année had seen in three months.

Beethoven's vocalization alerted the other members of Group 5 to our presence. Immediately, Tuck left the group and strutted to the base of the tree wearing compressed lips and nervously whacking at vegetation during her approach. She was followed by her mother. Effie was also walking stiffly and wearing a none-too-pleasant facial expression.

Humans ceased to exist for Bonne Année. The baby slowly left John's arms and descended the tree to rejoin her own kind. As she passed me I reached for her almost as instinctively as a mother reaches out to protect her child from danger. Then, fully realizing that it was not for me to interfere with the infant's decision, I withdrew my arm. Bonne Année climbed down to Tuck. Both gorillas gently embraced. John and I beamed at each other, disregarding all our previous fears and doubts about the captive's acceptance into Group 5. It was the last time either of us smiled for the rest of the day.

All that I had feared came to pass. Effie strutted over to Tuck. The two females began fighting for possession of the baby, pulling at her extremities, dragging her away from one another and biting her. Bonne Année screamed with pain and fright. After ten minutes, I could not take any more. My intentions to remain a detached scientific observer dissolved. Yelling, "Get out of here! Get out of here!" I climbed down to rescue the baby. I passed her up to John, who then climbed higher above me. Effie and Tuck returned to the tree base after their momentary fright at my interference and glared up at us threateningly, as if intending to climb the *Hypericum* and retrieve Bonne Année.

Then, to the complete surprise of John and myself, the baby again left the protection of his arms and returned to Tuck and Effie. This time I made no move to stop her. Bonne Année was clearly determined to become a free-living gorilla.

Tuck and Effie instantly resumed their cat-and-mouse game of torture. Bonne Année's screams began anew. The brutality of Group 5's two females was agonizing to watch, the infant's cries unbearable to hear. The sounds prompted Beethoven to charge, roaring, to the base of the tree and caused Effie and Tuck to flee. Shaken, Bonne

Année went directly to the old silverback, who smelled her with mild interest but did not open his arms to her pathetic attempts to be held or cuddled by him. As it started raining heavily, Beethoven turned his back on the infant to gain protection from the downpour in a dense clump of vegetation. Little Bonne Année huddled against his massive silvered back, drenched and shivering.

When the rain let up slightly, other Group 5 members came to investigate and smell the small stranger. The presence of young animals seemed to bolster Bonne Année's confidence. She moved in among them, sat down, and calmly began feeding. She was almost lost from our view as the gorillas crowded and displayed around her, strutting, and chestbeating as if trying to gain some kind of response from the infant. Suddenly Icarus entered into their midst, scattering the youngsters by strutting in a threatening manner, lips compressed. He made a direct run at Bonne Année and dragged her by one arm through the foliage. Effie and Tuck ran to join the young silverback in a combined attack on the baby, knocking her over whenever she tried to stand up. Their abuse became far rougher than before, because of Icarus' involvement. Ruthlessly, he grabbed Bonne Année away from the females and ran some sixteen feet, carrying her in his teeth. The infant screamed in terror. The noise brought Beethoven and other individuals charging toward Icarus, who hastily dropped the baby and fled.

My gratitude for Beethoven's intervention was short-lived. After about a minute the old silverback left and went downhill to feed by himself. It was as though physical limitations brought on by advanced age prevented him from active chastisement of Icarus. Also reproductive opportunities seemed at an end for Beethoven, whose remaining two females, Marchessa and Effie, were not likely to be sexually receptive again for at least three or four years. The future generations of Group 5, Beethoven's progeny, were now the responsibility of Icarus.

With Beethoven's departure, Icarus returned and, along with Tuck, resumed even more concentrated harassment of Bonne Année. It seemed to John and me as if they wanted to extend her suffering as long as possible for their own sport. Finally, the infant gave up her weak attempts at self-defense. She lay down, remained absolutely still, and gave no more vocalizations. This was a sign of total defeat.

For the last time Icarus chestbeat, grabbed her, and violently dragged her downhill, throwing her to the side before completing his display with a run and another chestbeat. Miraculously, Bonne Année managed to crawl back to the base of our tree, but she lacked the strength to climb up to us. Stunned at the extent of Group 5's xenophobic brutality, I was almost too slow to move to her rescue. However, I did manage to retrieve Bonne Année and hand her up to John just before Icarus and Tuck returned to the tree base to glare at us belligerently. John hid the baby under his rain jacket. We could only pray that she would not vocalize her usual dislike at being restrained. I had little doubt that the sound of her cries would provoke Icarus to climb the tree and forcibly retrieve her.

Four of Group 5's younger animals now surged around the tree before climbing up the trunk in an attempt to locate Bonne Année. As they tried to pass me, I had to pinch, lightly kick, or shove them away when Icarus was not looking. Puzzled by this unaccustomed treatment from me, they backed down, yet did not lose their interest in the newcomer. I was grateful that the shielded infant was neither moving nor vocalizing. Icarus scattered the youngsters, then he began climbing the tree. I'll never forget the feeling of the young silverback's hot breath penetrating my sodden boots, his head inches from my feet. It was only because John and I were positioned above him — and there were two of us — that Icarus did not continue his pursuit of Bonne Année.

During the next hour Icarus and Tuck maintained a hostile vigil at the tree base, barking or pig-grunting harshly at the slightest movement either John or I made. The male's head hair stood erect and he emitted a pungent silverback odor indicative of his tension. Both animals yawned repeatedly, exposing all of their teeth while shaking their heads rapidly from side to side. I received the impression that they wanted to attack, but neither had sufficient courage to risk climbing up the trunk with two human beings above them. I do not recall ever feeling so helpless.

Several times Icarus unleashed his stress by charging violently down the slope, chestbeating, and tearing at vegetation. To our dismay, he always returned to resume his guard position. After nearly an hour the rest of Group 5 moved away to feed. This prompted Icarus and Tuck to follow, but they soon strutted back to stare at us suspiciously. Only when the pair moved out of sight from us, some thirty feet below into dense foliage, was it safe for John to leap out of the tree and run uphill still carrying the silent baby gorilla under his jacket. Five minutes later I followed, expecting to be pounced on from behind at any minute. That evening John told me that he had had the same fear. Neither of us was at all relaxed until we were well over half an hour away from the group. The irrational behavior of Tuck, and especially of Icarus, had severed all ordinary bonds of rapport and communication. It was impossible to know if what certainly appeared to be xenophobia had rendered Icarus' behavior as unpredictable to himself as it was to us.

Once back at camp, Bonne Année was dried and put into her sleeping cage along with her treat box of fruit. Her wounds proved not to be serious and she seemed content to be back in familiar surroundings.

Twenty days later, when the injuries incurred from her encounter with Group 5 had healed, Bonne Année was successfully introduced to Group 4. Group 4's lack of strong kinship bonds permitted instant acceptance of the baby into the heterogeneous group of two young adult animals from a fringe group, the three males Beetsme, Tiger, and Titus of the old Group 4, and the group's young silverback leader, Peanuts, about eighteen years old at this time. Within an hour after meeting her foster family, Bonne Année was playing with five-and-a-half-year-old Titus. Bonne Année had become a gorilla of the mountains at last.

For a year Bonne Année was an integral part of Peanuts' Group 4, protected and cuddled by all the members. In May 1981 she succumbed to pneumonia after a prolonged period of heavy rains and hail.

Bonne Année's death raises the question as to whether she should have been released to the wild in the first place. My answer remains yes. A putative argument would claim that, if there are only some two hundred and forty mountain gorillas left in the wild, shouldn't Bonne Année have been preserved in a zoo? This line of reasoning would ask that the captive be exhibited simply for exhibition's sake because of the rarity of her kind. There are no mountain gorillas surviving in zoos, therefore even if Bonne Année had endured the adjustment she would never have had any opportunities to breed for the perpetuation of her species. She had that chance in the wild with others of her own kind. Bonne Année, at least, died free.

All too vividly I remember photographs taken of Coco and Pucker during the years of their confinement in the Cologne Zoo, photographs that revealed their depression even though they had one another for comfort and companionship. I would never have wanted

Bonne Année to experience Coco and Pucker's trauma of captivity only to gain a few more years of sterile existence. By her successful reentry to a free-living group, where she thrived for a year, Bonne Année proved that it is possible to reintroduce captive gorillas to their natural habitat if a receptive free-living gorilla group is available. I believe that the benefits involved — especially the perpetuation of the species — outweigh the risks. In Bonne Année's case the threat was extremely bad weather, though poachers had been the most feared risk.

* * *

Because there were no reproductive opportunities available to thirteen-and-a-half-year-old Tiger, he emigrated from Peanuts' Group 4 in February 1981, shortly before Bonne Année's death. Although a third individual, about eight years of age, had joined the group, it too was thought to be a male. Born with the potential of inheriting Group 4 just as his half brother Uncle Bert had inherited the group's leadership from their father, Whinny, Tiger had become instead a member of another silverback's group that contained no females.

How brutally a single bullet had deprived Tiger of his legacy and had altered his life! At this writing he has spent nearly two years wandering alone on Visoke's slopes, gaining experience in interactions with other groups, mostly the group led by Nunkie, a far older and very authoritative silverback. It is possible that Tiger might have to extend his search for mates on still another mountain within the Virungas. Wherever he needs to wander in the perpetuation of his species, it is my fervent hope that Tiger may succeed to form and lead a gorilla family just as other lone silverbacks have before him.

Formation of a
New Family Lends Hope:
Nunkie's Group

A SIGNIFICANT NEW PERSONALITY entered Karisoke's study area for the first time in November 1972. Encountered ranging in a portion of Group 5's southern core area in the hills that separated Visoke from Karisimbi, the grizzled lone silverback, estimated in his mid to late thirties, was named Nunkie. All trail evidence indicated that Nunkie had come from Mt. Karisimbi, the most heavily encroached area within the entire Virungas.

Nunkie's noseprint pattern matched none of the sketches or photographs of any formerly identified silverbacks, nor did the somewhat garrulous male behave like a habituated animal. Nunkie thoroughly distrusted human beings. I could only speculate as to the circumstances responsible for this older silverback's wandering without a family group.

It seemed unlikely that Nunkie could have been the leader of an established group decimated by poachers, since the protective nature of silverbacks compels them to give up their lives defending their family members. I tried to account for the new silverback by comparing him to silverbacks of his age in the study groups. Amok, Group 4's silverback who had ended up traveling for nearly six years alone, was probably unable to form a group because of ill health. Nunkie's sole apparent physical abnormality was a four-toed webbing of his right foot, leaving only his big toe unfused, and a left foot webbing of the two middle toes — all characteristics suggesting inbreeding.

I could not compare the then-mysterious silverback to Peanuts, originally of Group 8, a young silverback inept at acquiring and keeping females. Nunkie soon proved himself extremely capable during interactions with other groups in his unrelenting pursuit of females for his own harem. The mature silverback's behavior suggested that he had led a group, though Karisoke observers could never find what had become of it. I finally concluded that Nunkie's past must have been similar to that of Rafiki, the aged leader of Group 8. Perhaps, like Rafiki, Nunkie had lost the males of his group through emigration due to lack of breeding opportunities within his group; perhaps his females, like those in Group 8, had died from natural causes.

For nearly two years after Nunkie's arrival at the Karisoke study area, the silverback seemed a phantom wanderer as he interacted and attempted to obtain females from Groups 4 and 5 and several fringe groups. During this period, in addition to familiarizing himself with the individuals he met in each group, Nunkie was gaining knowl-.dge of the terrain on and adjacent to Mt. Visoke. Unlike any other lone silverback within the study area, Nunkie had no inherent range, a necessity for a silverback's establishment of his own group.

Nunkie needed to find an area not preempted by the established Visoke Groups 4, 5, 8, and 9. He tended to wander erratically until finally he spent much of his time high on Visoke's slopes. This was not an ideal choice of terrain; the region near the subalpine zone lacks the lushness and variety of the richer vegetation in the saddle area or on Visoke's lower slopes. However, it was relatively free of other groups and was the area to which Nunkie retreated in June 1974 after acquiring his first two females, Petula and Papoose, from Group 4. These females were to remain with Nunkie permanently.

The females' transfers were complex and involved a series of multiple interactions. Out of her home Group 4, Petula then left Nunkie to travel alone over a forty-eight-hour period back toward Group 4 and her first offspring, forty-six-month-old Augustus. Instead of rejoining Group 4, Petula united with the lone silverback Samson, Rafiki's son, formerly of Group 8. Nunkie, accompanied by Papoose, followed Petula. Instead of regaining Petula, Nunkie lost Papoose to Samson, who had developed into a powerful silverback rival. The two females remained with the younger silverback for only three weeks before transferring back to Nunkie. I suppose their choice of the older and scruffy Nunkie stemmed from the probability that he had had experience leading a group and therefore could offer them more protection and security than could the less-experienced Samson.

Petula had been the lowest-ranking adult female in Group 4. Since the dominance order of females usually depends upon their order of acquisition by the silverback, Petula became the top-ranking female in the hierarchy that was eventually to form among Nunkie's mates. She has remained, as of this writing, the top female in Nunkie's Group.

In Group 4 Petula had not been receiving sexual attention from Uncle Bert even though it had been forty-six months since she had given birth to her daughter Augustus, whom she was still occasionally nursing. Petula was one of two females whose ability to conceive was possibly inhibited by prolonged lactation. Within eleven months after her transfer to Nunkie, Petula bore his first Visoke infant. I named her Lee, after a very close friend, wildlife photographer Lee Lyon, who was killed by an elephant in Rwanda at the time of the birth. In May 1976, ten months after Lee's birth, Papoose produced a male named N'Gee, also sired by Nunkie and named for the National Geographic Society. At nine years and ten months, Papoose remains the youngest gorilla mother thus far observed during the study period. A mean age of ten years at first birth has been recorded for nine primiparous females.

Like Thor, Macho and Rafiki's offspring, Lee was at first the sole infant in her group. Even though deprived of the benefit of playmates, she developed normal motor skills during the ten months preceding the birth of N'Gee. Lee showed the same type of originality in solo play as had Augustus, her half sister in Group 4. When about a year old, Lee began chin-slapping, producing as unusual a sound, by the clattering together of her upper and lower molars, as Augustus' hand-clapping activity. (Chin-clapping by Titus of Group 4 when he was nearly three years old had been the first observed.) Neither in Group 4 nor in Nunkie's Group did other youngsters adopt the attention-getting practice. Lee's development of the unusual activity again suggested that the absence of siblings or close peers might prompt original means of self-entertainment.

*   *   *

For nearly a year — a period of bond formation between Nunkie and his first two females — the silverback did not try to obtain additional mates. Then, one month following the birth of Lee, Nunkie acquired a sexually immature female from a fringe group. She remained with him just seventeen months before transferring to another small fringe group. After Papoose had been impregnated, Nunkie actively resumed interactions with other groups. One month before N'Gee's birth Nunkie obtained two more adult females. One remained with him only ten months. The other, Fuddle, came from Group 6, as did a third. Pandora joined Nunkie's growing harem in August 1976, four months after Fuddle. Because of noseprint similarities and a close rapport, I consider Fuddle and Pandora to be either full or half sisters.

Pandora was more like Group 4's Old Goat than any gorilla I ever met. Her plucky character and strong sense of responsibility in supporting Nunkie during interactions with other groups were just two character traits endearing her to me. Her most significant physical feature was her hands. She had only a thumb on her right hand, which ended in a stump. Her left was claw-shaped, with atrophied and twisted fingers. The backs of both hands bore old scars and suggested that past wounds, rather than birth defects, were responsible for her deformities. Undoubtedly Pandora had been a poacher's trap victim — one of the lucky ones who had, at least, escaped with her life.

Pandora took somewhat longer to feed than did other gorillas. Considering her handicaps, she was extremely adept in pulling, stripping, and wadding vegetation. She carried out daily activities with a skill and dexterity that had developed over years of compensation, having learned to substitute her toes and mouth for her missing fingers.

The intervals between Nunkie's acquisition of Fuddle and Pandora and the births of their first offspring by Nunkie were twenty-six and twenty-seven months, respectively. For unknown reasons, these spans between transfers and births were far lengthier than those for Petula and Papoose, eleven and twenty-one months. Finally, in June 1978, Fuddle gave birth to a male, Bilbo, Nunkie's third-conceived offspring, and Pandora bore his fourth in December 1978. Pandora's son was named Sanduku, a Swahili word meaning "box" after the object entrusted to the Pandora of Greek mythology. I had often wondered to what extent the hand deformities might impair Pandora's maternal capabilities. I was delighted to see that Pandora was a capable and conscientious mother. During Sanduku's early months, he had to cling unaided to his mother's stomach more often than did other infants, but showed no signs of neglect. Pandora managed to groom him with her left hand while cradling him in the nook of her right arm.

Bilbo and Sanduku provided young Lee with two more half siblings to play with. By March 1979 Lee and N'Gee had developed into lively, socially active youngsters, both highly favored by Nunkie, perhaps because of having been the first he had sired.

In August 1978, Nunkie acquired the aging Flossie, her seven-year-old daughter Cleo, and Petula's eight-year-old daughter Augustus from the decimated Group 4, then led by a very young Tiger. Flossie and Cleo soon immigrated to the Suza Group, the small fringe group that offered the females greater opportunities for the advancement of their status than Nunkie's four females allowed. Augustus remained with her mother. Two years later, in August 1980, she gave birth to an infant named Ginseng, Nunkie's seventh since 1972. It was almost impossible for me to think of the little hand-clapping infant, now ten years old, as a mother.

Group 4's last remaining female, Simba, transferred to Nunkie in December 1978 during a violent interaction that cost the life of her infant, Mwelu, Digit's sole progeny, and left Group 4 with only three young males. Thirty-two months following her transfer into Nunkie's Group, Simba gave birth for a second time, in August 1981, to Jenny, Nunkie's sixth living offspring.

As the size of Nunkie's Group increased, so did the group's terrain. By early 1979, following the decimation of Group 4, Nunkie had begun using a large portion of Uncle Bert's old range both on Visoke's slopes and in the western saddle terrain.

On the morning of March 3, 1979, the student who had found Uncle Bert's decapitated body went out to contact Nunkie's Group, then ranging in the saddle about half a mile away from the death site of Uncle Bert and Macho. The young man returned to camp far earlier than expected. He told how he had left Nunkie's Group hysterically displaying around Lee, whose left foot was snared in the wire noose of a trap. Because of the frenzied state of Nunkie's Group, it was impossible for an observer to approach Lee and remove the wire.

Using mostly the same trails that had led to Group 4's slaughter site, and with the same sense of dread, staff members and I followed the student back to the scene of the newest poacher encroachment.

Nunkie's Group, including Lee, had fled. They had left behind a large area of broken trees and trampled foliage around the trap pole, which — much to our distress — bore no sign of the wire that had snared Lee.

Days passed before we spotted the wire, which could be seen embedding itself deeper and deeper into the tissue and bone of the forty-six-month-old female's ankle. Although both Lee and her mother, Petula, frequently groomed the wound, neither they nor Nunkie was ever observed trying to remove the gripping wire. Adult gorillas are averse to touching alien objects. Even if this were not the case, it seems unlikely that Lee's parents could have had any comprehension of the wire's lethal hold on their offspring. Nunkie could thus only adjust the group's travel and feeding routine to meet his daughter's declining physical condition. For three months, Lee suffered a painful and lingering death as the wire injury became gangrenous and she developed pneumonia. On May 9, 1979, Lee, weighing thirty-three pounds, half the normal weight for her age, became the seventh poacher-related victim in twelve years of study.

Seven months later Nunkie's second-born, N'Gee, Lee's constant companion during their short lives, disappeared. Trail evidence showed that Nunkie's Group had encountered poachers, dogs, and traps in the western saddle area very near the spot where Lee had been trapped. The loss of forty-three-month-old N'Gee was the impetus needed to send Nunkie and his family of ten back to the security of Visoke's slopes; there he spent most of the next three years expanding his group's range.

* * *

As expected, Nunkie's 1979 return to Visoke's slopes exerted a domino effect on other gorilla groups also unable to venture freely into the encroached saddle. The group least affected by the human presence in the western saddle area was Group 5, whose portion of saddle range lay to the south and southeast of Visoke's slopes. This is where the most concentrated bamboo zone of nearly three miles exists along the park's eastern boundary. Relatively safe in this area, Group 5 continued to divide their time on and off the slopes except during bamboo shooting months, when they remained below Visoke to feast on bamboo and related vegetation highly favored by the gorillas.

Visoke's surface can only maintain so many animals, however. With the addition of Nunkie's Group, the consecutive abutment of one group's range to another gave way to increased overlapping around the entire mountain. The situation became much like that which had existed on my arrival in 1967, when gorilla groups suffered crowded conditions on Visoke's slopes because of human encroachment in the saddle terrain.

Because Nunkie was now responsible for a large familial group rather than just two mates and offspring, he sought a range area offering greater variety and quality of vegetation than what the higher Visoke altitudes offered. He began using the entire mountain at varying altitudes, making use of Group 5's herbaceous-range section when they were off in the bamboo zone, and even going down into the bamboo fringing Visoke's northeastern slopes, a region he previously had never been known to use. At least five other groups, primarily fringe groups, were affected by Nunkie's movements in and out of their ranges. Although encounters between groups naturally increased because of Nunkie's extensive travels, the majority of the interactions were excitable in nature rather than violent, possibly because the size of Nunkie's group precluded the necessity of his acquiring additional females.

*　*　*

By December 1982 Nunkie's Group had grown to 16 individuals. From Group 4 he had obtained four adult females — Petula, Papoose, Simba, and Augustus. With them he had sired seven offspring. The first two, Lee and N'Gee, born to Petula and Papoose, became poacher victims in 1979. Within a month following the loss of Lee, Petula gave birth to a female infant named Darby, and forty-two months later to a male named Hodari. Five months after the loss of N'Gee, Papoose gave birth to a female infant named Shangaza. With Group 4's former females Simba and Augustus, Nunkie sired Jenny and Ginseng, both thought to be females, born within a twelve-month period between August 1980 and August 1981. The progeny produced by Nunkie with Group 4's former females represented a perpetuation of the bloodline of Whinny, the deceased leader of Group 4, by four grandchildren and one great-grandchild.

From the fringe Group 6 Nunkie had obtained two adult females, Fuddle and Pandora. With them he had sired four infants as of March 1982 — Bilbo and Mwingu with Fuddle, Sanduku and Kazi with

Pandora. The four offspring, all thought to be males, mark Nunkie's contribution toward the extension of Group 6's bloodlines.

Nunkie's seventh acquired adult female was one named Umushitsi, a local word for "witch doctor," who was thought to have come from yet another Visoke fringe group during an unobserved interaction around the beginning of 1981. In May 1982 Umushitsi gave birth to Nunkie's twelfth offspring. For reasons unknown, Umushitsi and her infant disappeared from Nunkie's Group shortly thereafter.

Nunkie's formation and subsequent expansion of a new gorilla group is something of a phenomenal success story. It well illustrates many of the requirements that enable a silverback to create, maintain, and enlarge his family group — the distinctive cohesive gorilla structure enhanced by the bonds of strong kinship ties. Which gorilla will emigrate from, or which will immigrate into, a gorilla group is largely determined by what is beneficial to both the group and the transferring individual.

We have seen how Groups 8 and 9 disintegrated because of the natural deaths of their silverbacks Rafiki and Geronimo. No gorilla group can exist without its unifying force, the silverback leader. In the case of Group 8, while Rafiki's aged mate Coco survived, the family unit did remain together for a brief period but not as a reproductive unit. Old Coco's death left a group of five males — the leader Rafiki, three males verging on sexual maturity (Pugnacious, Samson, and Geezer), and Coco and Rafiki's youngest son, Peanuts. Employing a basically sound reproductive strategy, the three oldest males emigrated and began lone travel in order to acquire females of their own. Peanuts stayed with his father, who subsequently began rebuilding Group 8 by securing females from other social units. If Rafiki had had another decade of life, his daughter Thor would now be nearing sexual maturity at nearly nine years of age. Thor would probably have remained within her natal group, Group 8, to breed eventually with her half brother Peanuts, instead of becoming a victim of infanticide following Rafiki's death.

The silverback leader of Group 5, Beethoven, has been allowed a long and productive life. Beethoven's son Icarus has remained within his natal group because breeding opportunities were available with his half and full sisters. Beethoven, at this writing, perhaps no longer sexually mature, has sired at least nineteen offspring and perpetuated his gene flow both within Group 5, among the daughters who remained, and also outside Group 5, among the daughters who emigrated. Beethoven is now taking a backseat to his silverback son, who after the aged male's death assuredly will seek, for purposes of exogamy, females from other groups for Group 5's genetic expansion.

Group 4, a group with all the potential and promise of Group 5, disintegrated upon the slaughter by poachers of their noble silverback, Uncle Bert — a young leader who was able to sire only eight offspring before dying in the prime of his life. Not more than three of Uncle Bert's offspring survive. Two of them, Augustus and Cleo, have produced two of his grandchildren within two different groups. Had it not been for the annihilation by poachers of Uncle Bert and Group 4's courageous supportive silverback, Digit, future generations of mountain gorillas would be perpetuating the bloodlines of Uncle Bert and Digit. The younger silverback's sole progeny, Mwelu, was a victim of infanticide when killed by Nunkie during the interaction that occurred at the time Simba transferred into Nunkie's Group. Actually, the eight-month-old female's death was caused by poacher encroachment into the supposed sanctity of the Virungas, the last stronghold of the mountain gorillas.

* * *

The gorillas' evolved strategies of transfers between groups, or even infanticide, may serve as means of dissemination of genetic variability throughout the small remaining population. Indeed, the formation of Nunkie's Group was possible only because of these techniques. What now of Nunkie's Group? How will the females he has so diligently obtained and protected throughout the past ten years fare during the next ten years? Will they or their present offspring suffer the fate of Nunkie's first two, Lee and N'Gee, because of continued human pressure in their man-locked mountain island?

Thirty-two percent of the Virunga protected area lies within Rwanda, which has an annual population growth of 3.8 percent. The mountain gorilla population, at least since George Schaller's study in 1960, has been declining at a rate of 3 percent each year because of poaching activities within the Virungas and land encroachment of gorilla habitat. Also since Schaller's time, 4050 acres have been excised from the Parc des Volcans; this almost halved Rwanda's sector of the park and reduced by one fifth the total protected area, a land reduction that alone could account for a 60 percent drop in the gorilla population during the past twenty-odd years. The 30,000 remaining acres of the Parc des Volcans occupy only ½ of 1 percent of Rwanda's soil. Rwanda, with its 5,000,000 inhabitants, has the highest population density of any African country south of the Sahara. About 95 percent of the people earn a meager living on small 2½ acre plots, and the land fringing the boundaries of the Parc des Volcans supports 780 inhabitants. Each year another 23,000 Rwandese families need additional land for cultivation. Even if the entire park were given over for agricultural purposes, it would provide land for only one quarter of one year's increase in Rwanda's population. This cultivation would, of course, mean total annihilation of the mountain gorilla and other wildlife now trying to survive there, as well as the destruction of the rain forest upon which people and wildlife in Rwanda depend in order to live. The real prospect of this destruction is as frightening as the fact that every year some 445,500 acres of the planet's rain forest are destroyed, at a rate of 49.2 acres per minute.

Foreigners cannot expect the average Rwandan living near the boundaries of the Parc des Volcans and raising pyrethrum for the equivalent of four cents a pound to look around at the towering volcanoes, consider their majestic beauty, and express concern about an endangered animal species living in those misted mountains. Much as a European might see a mirage when stranded in a desert, a Rwandan sees rows upon rows of potatoes, beans, peas, corn, and tobacco in place of the massive *Hagenia* trees. He justifiably resents being refused access to parkland for realization of his vision.

American and European concepts of conservation, especially preservation of wildlife, are not relevant to African farmers already living above the carrying capacity of their land. Instead, local people need to be educated about the absolute necessity of maintaining the mountains as a water catchment area. The farmers need to know, not so much about what foreigners think about gorillas, but rather that 10 percent of all rain that falls on Rwanda is caught by the Virungas and is slowly released to irrigate the crops below. Each farming family's personal survival depends upon the survival of the Parc des Volcans. Cultivation of this vital catchment area would mean virtually the end of both present and future crops. If the importance of the ecosystem to the lives of the populace becomes a prime local priority, which is not now the case, the rain forest might stand a chance to survive, and with it the animals it contains and the people who rely on it. As a country with much to gain both sociologically and economically from active preservation of land resources, Rwanda could well serve as a monumental example to Zaire and Uganda to gain fuller cooperation in safeguarding the future of the Virungas that all three nations share.

There also should be an internationally coordinated policy of stringent law enforcement against any type of human encroacher within the Parc des Volcans of Rwanda, the Parc des Virungas of Zaire, and the Kigezi Gorilla Sanctuary of Uganda. Prolonged imprisonment of convicted encroachers, such as that bravely advocated and enforced by Paulin Nkubili, should be the rule rather than the exception. By acting impartially toward both African and European intruders in the Rwandese park domain, the Chef des Brigades set an example for others — Rwandese, Zairoise, and Ugandans — to follow and perhaps even extend.

Karisoke Research Centre's study of the mountain gorilla from 1967 to 1983 covers a tiny segment of time. Centuries ago a decade and a half was just a passing moment in the life span of a species. At present, however, the next two decades are estimated to see the extinction of twenty species of animals. Human beings must decide now whether or not the mountain gorilla will become one of them, a species discovered and extinct within the same century. The gorillas' destiny lies in the hands of those who share their communal inheritance, the land of Africa, the home of the mountain gorilla.

Epilogue

THERE IS A GROWING TREND among conservationists, economists, sociologists, and journalists to approach the compounding problems of Third World countries more realistically than was advocated by the theorists of the past. This welcome trend is becoming increasingly widespread in conservation efforts in Africa, where until recently anachronistic conservation policies all but ignored the complexities of any country's bureaucracy, the basic needs of the human population, and the varying degrees of corruption among officials at a local level. Such disregard of reality has resulted in a form of short-lived innocuous diplomacy inhibiting resourcefulness and self-motivation among those most directly connected with the future of their country's wildlife. The self-eulogizing attempts of expatriates to impose the notion of wildlife as a treasured legacy overlook the reality that to most of a local impoverished and inert populace wildlife is considered an obstacle — tolerated only as long as it proves economically valuable on a practical basis in the form of tusks, meat, or skins.

On the other hand, promotion of tourism, if properly directed, might well prove profitable on a nationwide basis and thus compel the one-to-one reapers of wildlife proceeds to give way to the rule of the majority. This aim might be accomplished in Africa — a continent where tribalism, nepotism, and distinct class systems have evolved — only by consistent, uncompromising individuals able to consider the needs of the animals before their own.

Certainly, endangered species live on a day-to-day basis — be they the 242 mountain gorillas remaining in Africa, the 1000 giant pandas remaining in China, or the 187 grizzly bears remaining in America.

The survival chances of these species are little improved by tourism compared with more expedient actions that could be taken on their behalf. Active conservation includes frequent patrols in wildlife areas to destroy poacher equipment and weapons, firm and prompt law enforcement, census counts in regions of breeding and ranging concentration, and strong safeguards for the limited habitat the animals occupy. Such inglorious activities monetarily benefit no person, yet offer the diminishing number of animals in the forests an opportunity to survive into the future.

Methods of active conservation must necessarily be supplemented by longer-term projects. However, just as A comes before B, arduous daily conservation efforts must precede the Z's of theoretical conservation. In the case of the mountain gorilla, this means cutting traps and imprisoning convicted poachers; in the case of the giant pandas, this means thorough investigation of the availability of food resources; in the case of grizzly bears, this means strict enforcement of penalties against poachers and stringent surveillance of legal park boundaries.

I would like to express my heartfelt gratitude to the many intrepid Rwandese and Zairoise who have assisted me in realizing the expediency of today's aims toward the preservation of the mountain gorilla. For Digit, Uncle Bert, Macho, Lee, N'Gee, and so many other gorillas, I sorrow that I was too late to change the quixotic ways of many Europeans and Africans who, in hoping for a brighter dawn tomorrow, have yet to realize that avoidance of very basic conservation issues may ultimately push Beethoven, Icarus, Nunkie, their mates, and their progeny into the mountain mist of times past.

Appendixes

Bibliography

Index

APPENDIX A

Food Vegetation Types Utilized by Gorillas of Study Groups 4, 5, 8, and Nunkie's Group

Fern Type
Polypodium sp.

Grass—Sedge Type
Arundinaria alpina
Carex petitiana
Cyperus sp.
Pennisetum purpureum

Herbaceous Type
Carduus afromontanus
Carex petitiana
Chaerefolium silvestre
Cineraria grandiflora
Conyza gigantea
Cynoglossum amplifolium
Cynoglossum geometricum
Helichrysum formosissimum
Helichrysum guilelmii
Laportea alatipes
Leucas deflexa
Lobelia giberroa
Lobelia wollastonii
Peucedanum kerstenii
Peucedanum linderi
Rumex ruwenzoriensis
Senecio maranguensis
Senecio trichopterygius
Solanum nigrum
Thalictrum rynchocarpum
Urtica massaica

Shrub Type
Pycnostachys goetzenii
Rubus apetalus
Rubus runssorensis

Tree Type
Afrocrania volkensii
Erica arborea
Hagenia abyssinica
Hypericum revolutum
Pygeum africanum
Rapanea pulchra
Senecio alticola
Senecio erici-rosenii
Vernonia adolfi-friderici
Xymalos monospora

Parasitic Type
Ganoderma applanatum
Loranthus luteo-aurantiacus

Vine Type
Basella alba
Clematis simensis
Crassocephalum bojeri
Droguetia iners
Galium spurium
Mikania cordata
Piper capense
Stephania abyssinica
Stephania dinklagei
Urera hypselendron

APPENDIX B

Data for 1967–1982
on Which Birth Frequency Graph
Is Based

Month	Births	Names	Month	Births	Names
January	1	Ziz	July	5	Kweli
February	1	Shinda			Quince
March	4	Muraha			Anjin
		Simba I			Flob
		Query			Lee
		Mwingu	August	7	Cleo
April	4	Unnamed			Titus
		(Old Goat's)			Augustus
		Mwelu			Pablo
		Simba II			Ginseng
		(Mrs. X's)			Jenny
		Poppy			Curry
May	4	Tuck	September	0	
		N'Gee	October	3	Unnamed
		Shangaza			(Maisie's)
		Unnamed			Banjo
		(Umushitsi's)			Kazi
June	5	Frito	November	3	Unnamed
		Thor			(Flossie's)
		Unnamed			Tiger
		(Maidenform's)			Cantsbee
		Bilbo	December	5	Puck
		Darby			Unnamed
					(Marchessa's)
					Sanduku
					Safari
					Hodari

TOTAL 42

Birth Frequency: Frequency Distribution in Which
42 Births Occurred Over a 15-year Period
(*September 1967–December 1982*)

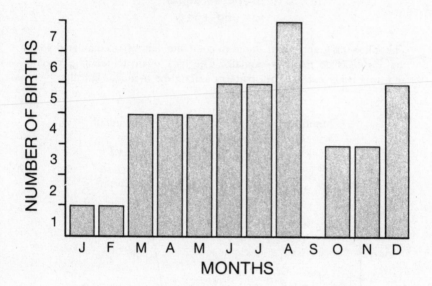

APPENDIX C

Rainfall Distribution
in the Virungas,
1969–1979

The following graph charts the most complete data yet accumulated within the Virungas on monthly rainfalls. The data relate to feeding, ranging, and, to a lesser extent, birth patterns among the mountain gorillas.

Monthly Rainfall: Variations (mm) in Rainfall
Over a 10-year Period, 1969–1979
(*no. of complete records shown for each month*)

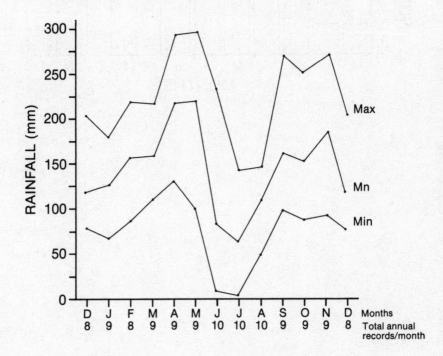

APPENDIX D

1981 Census Findings

Between 1959 and 1960 George B. Schaller estimated there were 400–500 mountain gorillas remaining within the Virungas. Schaller's count showed 169 individuals in ten groups around Kabara (22 animals were not included in the above count because of not being consistently around Kabara throughout the study period). In 1967 I found only three groups at Kabara; they totaled 52 individuals, although these three groups, considered to be the same as Schaller's, contained 62 animals during his period of study.

The map (on next page) plots the distribution of the Virunga mountain gorilla population as found during the 1981 census work. Counts obtained from two camps, Kabara and Karisoke, are included in the counts of the camps' mountains, Mikeno and Visoke respectively. The 1981 census revealed a total population of 242 free-living mountain gorillas, approximately a 50 percent reduction in the population over a 22-year period.

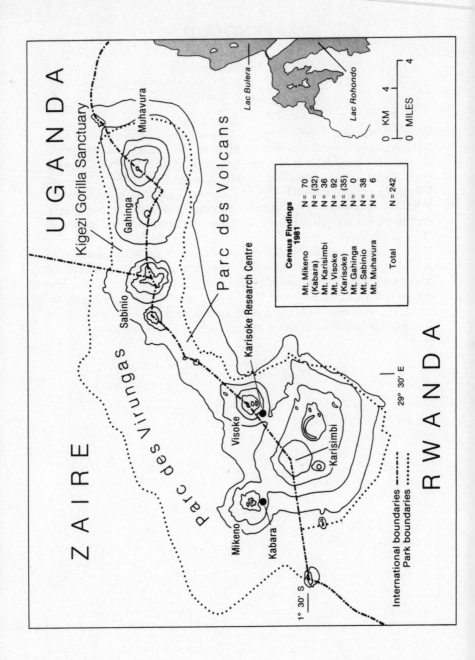

ZAIRE

UGANDA

Kigezi Gorilla Sanctuary

Parc des Virungas

Mikeno

Kabara

Visoke

Sabinio

Gahinga

Muhavura

Karisimbi

Parc des Volcans

Karisoke Research Centre

Lac Bulera

Lac Rohondo

RWANDA

1° 30' S

29° 30' E

Census Findings
1981

Mt. Mikeno	N = 70
(Kabara)	N = (32)
Mt. Karisimbi	N = 36
Mt. Visoke	N = 92
(Karisoke)	N = (35)
Mt. Gahinga	N = 0
Mt. Sabinio	N = 38
Mt. Muhavura	N = 6
Total	N = 242

International boundaries ━ · ━ · ━
Park boundaries · · · · · · · ·

0   KM   4
0   MILES   4

APPENDIX E

Common Gorilla Vocalizations from Main Study Groups and Coco and Pucker, the Captive Juveniles

Aggressive Call

Figure 1. Roar: This monosyllabic loud outburst of low-pitched harsh sound lasted from .20 to .65 second, beginning and ending abruptly. As may be noted in Fig. 1, there were individual differences in the frequency concentrations of a roar. Roars were heard only from silverbacks in situations of stress or threat and were primarily directed at human beings, although occasionally at buffalo herds. The vocalization was always followed, on the part of the emitter, with varying degrees of display, ranging from bluff charges to small forward lunges.

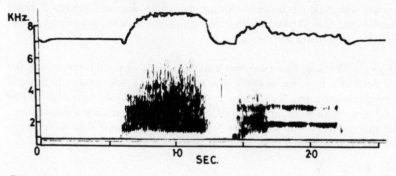

Figure 1

Alarm Calls

Figure 2. Screams: These shrill and prolonged emissions of extremely loud sound could last up to 2.13 seconds and be repeated as often as ten times. Unlike with the roar, individual differences in screams could not be denoted, either spectrographically* or subjectively. Screams were heard from gorillas

---

* Spectrograms are permanent records of sounds tape-recorded and subsequently played in conjunction with a vibralizer (a spectrographic machine), which graphically defines measurements of tonality, rhythm, amplitude, quality, and frequencies. Spectrogram grid lines are spaced at ½-second intervals. Spectrogram verticals record the number of sound waves occurring per second or kilocycle intervals.

251

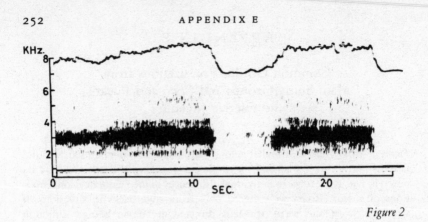

Figure 2

of all ages and sex classes, but more frequently from silverbacks. The vocalization was most often heard during intragroup disputes, though it could be directed toward human beings or even ravens if alarm, rather than threat, was the motivation for the call.

Figure 3. Wraagh: This explosive monosyllabic loud vocal outburst was not as deep as a roar nor as shrill as a scream. Like roars, *wraagh* began and ended abruptly and lasted between .20 and .80 second. As may be noted in Fig. 3, there were individual differences in the frequency concentrations of the sound, which was more harmonically structured than were roars. *Wraaghs* were heard from all adult gorillas but far more frequently from

Figure 3

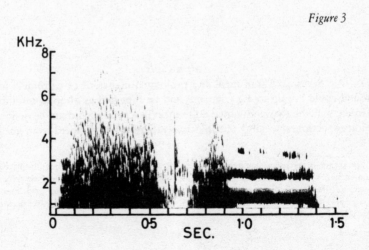

silverbacks. They were usually precipitated by sudden situations of stress—the unexpected arrival of an observer, duiker alarm calls, rockslides, thunderclaps, or loud wind noises. The vocalization was most effective in scattering group members and, unlike the roar, was never accompanied by aggressive display behavior.

Figure 4. Question Bark: This vocalization is best described by its characteristic composition (both subjective and spectrographic) of three notes with the 1st and 3rd lower than the middle, as if asking the question "Who are you?" The sound was short, lasting between .20 and .30 second, and was heard more from silverbacks than from gorillas of any other age or sex. It was usually given in situations of mild alarm or curiosity and was a common response to discovery of an obscured observer or to branch-breaking noises by gorillas not readily visible to other group members.

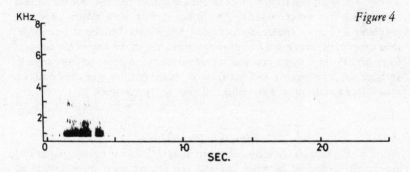

Figure 4

Figure 5. Cries: These sounds, resembling wails of human infants, could build up into shrieks much like human beings' temper tantrums. (Figure 5a shows cries going into shrieks; in 5b they subside.) They were emitted between .03 and .05 second apart and could last for nearly 19 seconds at

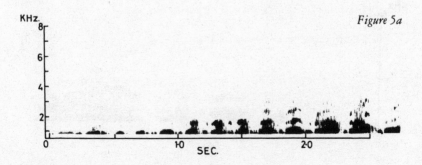

Figure 5a

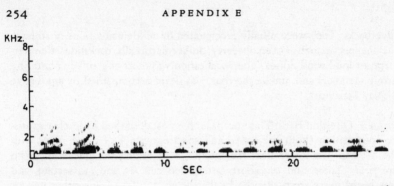

Figure 5b

a time. The wails had four distinct frequency concentrations, but the shrieks were much less structured. Cries were heard only from infants or young juveniles and most frequently occurred when they had been left alone, thus temporarily separated from their mothers, or, in the case of the captives Coco and Pucker, when one was separated from the other or from myself. In both the free-ranging and the captive young gorillas, the cries built up into temper tantrums if a stressful situation was prolonged.

Coordination Vocalizations

Figure 6. Pig-Grunts: A series of short, rough, guttural noises, pig-grunts are usually delivered between .15 and .40 second apart in sequences of nine or ten outbursts. The sounds, resembling the grunting of pigs feeding at a sty, tended to become louder and more closely spaced if prolonged.

Figure 6

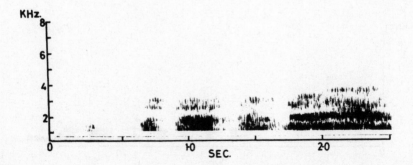

Pig-grunts were most frequently heard when individuals were traveling, for this was when trail disputes and altercations over limited food resources were more apt to occur. On such occasions pig-grunts were effective rebuttal vocalizations and also served as disciplinary enforcements between adults and young.

Figure 7. Belch Vocalizations: These sounds resemble deep, prolonged rumbles (*naoom, naoom, naoom*) rather like throat-clearing utterances. The belch vocalization is undoubtedly one of the most complex of the entire gorilla vocabulary because of its multiple intergradations and functional variations. The sound, as recorded from both free-ranging and captive gorillas, had two frequency concentrations and gradated into croons, purrs, hums, moans, wails, and howls if prolonged in situations of maximum contentment. It was most frequently heard from stationary gorillas at the end of a long, sunny resting period or when in a lush feeding site. Variations of the belch vocalization were heard from gorillas of all age and sex classes.

Silverbacks tended to be heard more frequently when launching belch vocalizing exchanges with other group members as a means of establishing location in dense vegetation. Other than for expressing contentment or verifying position, a slightly shortened belch vocalization was often used

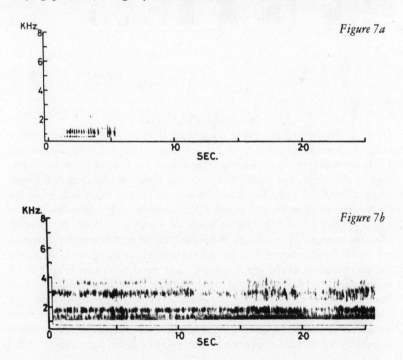

Figure 7a

Figure 7b

by adults when mildly disciplining gorilla young or human observers. I found the belch vocalization to be the most useful gorilla vocalization in allaying the gorillas' apprehension during the habituation process.

Figure 8. Chuckles: These raspy expirations of noise verge in intensity dependent on the degree of play activity involved (wrestling, tickling, or chasing). They were difficult to record, both from free-ranging young and the captive infants, because of the interference of background noise resulting from the activity itself. Chuckles were irregularly spaced spurts of sound varying in length from .02 to .10 second with a low frequency concentration. No individual differences were noted.

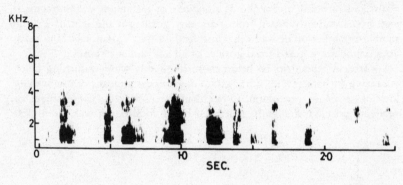

Figure 8

Intergroup Vocalization

Figure 9. Hootseries preceding chestbeats: The hootseries, given with or without a terminating chestbeat, consists of prolonged distinct *hoo-hoo-hoos*. These were low-pitched, often undetectable to the human ear at the beginning of the series, but usually built up into plaintive-sounding and longer hoots toward the end. The lengthier the series, the more individual the fluctuations in harmony and phasing. Frequencies ranged between 1.4 and 1.8 kilocycles per second for as many as eighty-four hoots per second. Silverbacks were, by far, the most frequent emitters of hootseries. Depending on the distance between silverbacks of distinct familial groups, the gorilla leaders would or would not terminate their vocalizations with mechanical noises such as chestbeating, ground thumps, branch-breaking, or runs through thick foliage. From impressions received in the field, it seemed that the closer the distance between callers (some twenty feet), the more frequently mechanical noises were utilized, because visual proximity between interacting silverbacks appeared to necessitate display punctuation of the

hootseries. The farther the distance between callers—the hootseries may
travel for roughly a mile—the less frequently the intergroup vocalization
was terminated by mechanical noises. Hootseries thus functioned well as
a type of vocal probing that did not reveal the precise location of a group
or of a particular gorilla. There was individual variation in the pattern of
delivery of hootseries between silverbacks, though all primarily utilized
the hootseries in long-range communication.

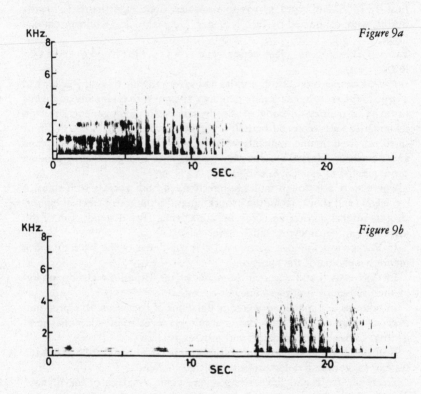

Figure 9a

Figure 9b

APPENDIX F

Summaries of Findings from
Autopsies of 14 *Gorilla Gorilla Beringei*

COCO AND PUCKER: Two captive females, approximately 14 and 15 years old respectively, of the Cologne Zoo; date of respective deaths, June 1, 1978, and April 1, 1978. Autopsies, done eight hours following deaths, were conducted by Drs. Krueger, Neumann, and Kullmann of Cologne.

External Examinations. Respective weights = 155.2 lb (70.0 kg) and 135.5 lb (61.1 kg).

Internal Examinations. Both gorillas had severe thymic lesions. Pucker had a large multicystic thymic tumor weighing 385 gm with essentially no thymic tissue. The multicystic body of the thymus was dysontogenetic since no thymic tissue was seen and Hassall's corpuscles were absent. Coco had four small pea-sized thymic remnants weighing a total of 15 gm, cystic dysplasia and occasional Hassall's corpuscles representing a degeneration product consequential of extensive involution.

There were also found multiple infections in both gorillas with signs of septicemia and shock, including "shock lungs," numerous petechial hemorrhages, adrenal cortical necroses, hepatic edema, fatty degeneration of the myocardium, hepatocytes, and Kupffer cells.

In Pucker was found a transitional cell papilloma of the bladder and a pseudolymphoma of the appendix.

In Coco was found a gastric adenoma of the Brunner's glands, unpigmented substantia nigra and nucleus coeruleus.

In both, tissues contained mixed populations of bacteria with a predominance of Gram-negative rods. Bacterial cultures revealed pseudomonas aerogenosa, klebsiella pneumoniae, and proteus vulgaris.

Both also had marked atrophy of thymic-dependent parts of the lymphatic system, suggesting a defect in cell-mediated immunity.

Conclusions. Coco and Pucker had severe cystic dysplasia of the thymus. Combined morphological, histochemical, and immunological studies suggested a defect in cell-mediated immunity. They died of Gram-negative septicemia and shock.

CURRY: 9-month-old male infant of Bravado and Beethoven of Group 5. Date of death, April 14, 1973; body recovered April 15; autopsy conducted April 16 by Dr. Klein of the Ruhengeri Hospital, report written by Ric Elliott.

External Examination. Two small puncture wounds on the right side of neck, each 2 to 3 mm in diameter, one 2½ cm above the other. Clean puncture wound into the subclavian vein 1 cm above level of junction of internal and external jugulars, resulting in heavy bleeding. Puncture wound above right armpit and below clavicle 1 cm in diameter, 3 to 3½ cm in depth, resulting in heavy bleeding. A large surface wound below left rib cage on ventral surface about 1½ to 2 cm in diameter. A 7½ cm long, deep wound that ruptured the colon some 7 to 8 cm from its junction with the rectum. Two puncture wounds, ½ cm apart, of 1 and 2 cm diameter, respectively, near the base of the large colon wound. A surface wound of 2 cm in diameter on the right ventral surface of the abdomen opposite the large colon wound. A small surface wound near the lower end of the left femur.

Internal Examination. Recent bruising found on the fascia overlying the skull, each about 3 cm in diameter; one overlying the front portion of the frontal lobe, the second sagitally over the junction of the central and lateral frontal sutures. The left femur was totally fractured 2.2 cm from the acetabulum with its splintered end still attached to the acetabulum and all ligaments intact.

Cause of Death. Infanticide.

DIGIT: Silverback of Group 4 approximately 15 years old. Killed by poachers December 31, 1977; body recovered January 3, 1978; autopsy conducted by Dr. Desseaux of Ruhengeri Hospital, Ian Redmond, and the author.

External Examination. Five spear wounds, any one of which could have been fatal, into ventral and dorsal body surfaces; decapitated and hands hacked off. No external parasites found.

Internal Examination. All internal organs appeared healthy except for a 3 cm cyst in spleen. A large trematode parasite found in left lung measuring 3.2 cm in length and 1.9 cm in width. The trematode had a median ventral sucker 1.4 cm long, 0.6 cm wide; its dorsal surface was smooth, coming to a slight point at the anterior symmetrical end; the posterior end was not symmetrical. Consistent with findings of fecal examinations throughout 1977, no cestodes were found in Digit's dung, which, two days prior to death, contained strongyle eggs and nematodes.

FRITO: 3-month-old female offspring of Flossie and Uncle Bert of Group 4; date of death August 14, 1978; body recovered August 16 by David Watts; autopsy conducted by the author.

External Examination. Fatal wound was a 3 cm bite into the right brachial plexus which splintered the humerus and severed the right brachial artery and vein. No external parasites found. Weight = 5.1 lb (2.3 kg). Accompanying sketch shows body measurements.

Cause of Death. Infanticide.

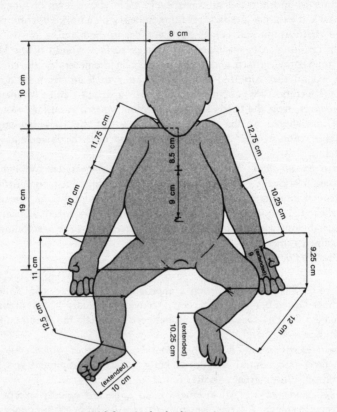

Frito, 3-month-old female; body dimensions

IDANO: Adult female of Group 5 approximately 30 to 35 years old. Estimated date of death May 1, 1973; body recovered May 2 by Ric Elliott; autopsy conducted by Dr. M. Philippot and Dr. F. Vounderick of the University of Butare.

External Examination. Aspecific granulomatous reaction (4 cm in diameter) of recent date on dorsal surface of the radius and ulna overlying the muscle 3 cm from junction with carpals. Right hand flexed abnormally. Milk could be squeezed from nipples; the female had not been known to have an infant since July 1970.

Internal Examination.   Stomach and colon empty. Genital organs normal except for inflammation of head of uterus; no corpus luteum cells found. Heart showed moderate edema in pericardium. Left lung normal; right lung considerably atrophied, especially in median lobe. Upper lobe showed many fibrous adhesions suggestive of pleurisy of long duration. Brain normal; weight = 378 gm; length (frontal-occipital poles) = 120 mm; width (across temporal poles) = 95 mm. Colon indicated enteritic condition and had strong fibrin attachments between it and abdominal wall suggestive of peritonitis of 1 to 2 years' duration. *Anoplocephala gorillae* found in large intestine; *Murshidia devians* in small intestine and *Paralibyostrongylus hebrenicus* in liver. Weight before autopsy — 120 lb (54.4 kg).

Microscopic Examination.   Liver: Lobular architecture intact. Heavy cholostasis all over the lobules; sinuses had many polymorphonuclear and mononuclear white blood cells. Kupffer cells heavily reactive. Several focal spots of necrosis. In contrast, interlobular spaces only slightly infiltrated with inflammatory elements. Turnbull reaction on iron was negative. Many dinucleated hepatic cells with bizarre, large nuclei. Bile ductuli normal. No fatty necrosis. Spleen: Polymorphonuclear W.B. in great number. Pancreas: Diffuse edema; moderate polymorphonuclear infiltration around vessels and between acini; rare spots of necrosis or microabscesses. Kidney: All glomeruli show thickening of the flocular mesangium and increased cellularity; slight thickening of the Bowman capsule; no interstitial reaction of fibrosis. Ovary and Endometrium: Deposits of bilirubin evident. Myocardium: Uneven nuclei surrounded by granules of lipofuscin, as in humans of older age. Cerebrum: Lipofuscin pigment in some neurons; corpora amylaceae in white matter.

Conclusions.   Animal died of bacterial hepatitis. The fairly deep granulomata on the feet and the involvement of the pancreas strongly suggest a bacterial process where the liver lesion plays a major role.

K W E L I :   39-month-old male offspring of Macho and Uncle Bert of Group 4. Shot through right clavicle July 24, 1978, by poachers when both parents killed. Date of death, October 26, 1978; body recovered same day by David Watts; autopsy conducted by Ian Redmond and the author. *External Examination.*   Victim in moribund condition when received and failed to respond to resuscitation. A large lump had developed around the right clavicle, which was still draining pus at time of death. The animal was heavily infested with lice, particularly around the groin and axillary regions. (*Pthirus gorillae;* see 100× magnification sketch.) Weight = 38.8 lb (17.5 kg). Accompanying sketch shows body measurements.

Internal Examination.   Lungs found extremely discolored; fragments of shattered clavicle fused together; nasal sinuses filled with thick green mucus; stomach and small intestine empty, with only a few lobes of digested vegetation remaining in large intestine. All other organs appeared normal; no internal parasites found.

Histology Results. Done by D. L. Graham, D.V.M. Mild subacute congestion and edema accompanied by increased numbers of alveolar macrophages; occasional microfoci of acute interstitial and intra-alveolar hemorrhages.

Conclusions. Kweli died of subacute pulmonary congestion, edema, and microfocal hemorrhages compatible with heart failure associated with bacterial septicemia.

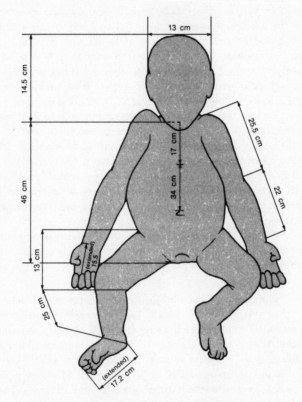

Kweli, 39-month-old male; body dimensions

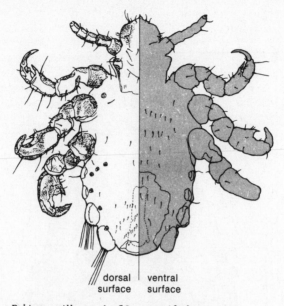

dorsal | ventral
surface | surface

Pthirus gorillae, male, 50x magnified

LEE: 46-month-old female juvenile offspring of Petula and Nunkie of Nunkie's Group. Date of death, May 9, 1979, resulting from a wire-snare trap injury received March 3, 1979; body recovered May 10, 1979. Autopsy conducted by Drs. Vimont and Poppoff of the Ruhengeri Hospital and the author at Karisoke.

External Examination. Moribund condition shown by eye orbits indicating conjunctivalis muquese; advanced anemia indicated by pale pink gums; injured left leg had caused atrophy of cervical and lumbar back muscles. Weight = 33.3 lb (15 kg). Comparative measurements of left leg and foot, snared extremity, with those of right, uninjured extremity; see below.

	Left	*Right*
Upper femur shaft at groin:	20 cm (diameter)	23 cm
Knee joint:	21 cm	20 cm
Midsection of foot:	11 cm	15 cm
Foot length from 4th toe to ankle:	14 cm (length)	16¾ cm
Groin to knee joint:	18 cm	18 cm
Knee joint to ankle:	16½ cm	17 cm

Accompanying sketch shows other possibly related abnormalities caused by snare-injured left leg; i.e., left-hand length, with fingers extended, was 15 cm compared to right hand, with fingers extended, of 14 cm. Ankylosis

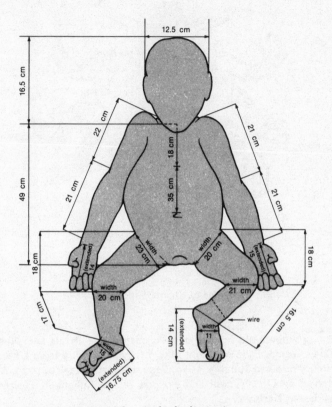

Lee, 46-month-old female; body dimensions

was thought to have developed in left knee joint, which Lee had used as a substitute foot because of having kept her left lower leg in a flexed position to avoid touching the injured foot to the ground; there was a large callus forming on the ventral surface of her left big toe, seeming to indicate that weight was put on the injured foot when necessary. Following death, and before rigor mortis, the right leg could be fully extended; the left leg had a 45 percent degree of contraction. Another difference between the left (injured) extremity and the right was that the tissue substance of the left foot was granulated, had collapsed blood vessels, and a grayish, mucous substance surrounding the ankle bone. There were minor sores visible on the palmar surfaces of both feet and hands revealing an epidermic breakdown. Lice eggs (no live specimens) were found in abundance around the groin area. There appeared to be an abnormality in the ability of the sphincter muscles to contract.

Internal Examination. A 2 cm piece of white fatty tissue was found within the heart; the aorta was thickened and fibrous; all lung lobes, particularly dorsal sections, discolored and pale; stomach and small intestines contained no vegetation deposits though ¾ of large intestine was filled with apparently well-digested food remnants; appendix measure 17½ cm.

Conclusions. Lee died of pneumonia coupled with heart failure and gangrene.

MWELU: 8-month-old female offspring of Simba and Digit of Group 4; date of death December 6, 1978; body recovered same day. Autopsy conducted by Ian Redmond.

External Examination. A deep puncture wound behind and above the left ear that had bled heavily and had deposited bits of pink brain tissue outside the cranial cavity; second deep puncture wound over left hip near groin; third deep puncture wound on inner surface of left thigh. Lesser wounds resulting in torn or broken skin found above right ear, outer corner of left eye. Lower left ribs concaved into chest cavity. No lice or lice eggs were found, but numerous soft-bodied mites (*Pangorillalges gorillae*) were found on the top of the head and around the groin and axillary regions. Weight = 8.6 lb (3.87 kg). Accompanying sketch shows body measurements.

Cause of Death. Infanticide.

OLD GOAT: Adult female of Group 4 approximately 30 to 35 years old. Estimated date of death October 28, 1974; body recovered November 25, 1974. Autopsy conducted by the author.

External Condition. Grossly deteriorated. Mammary glands in inactive phase.

Internal Gross Examination. No pleural or abdominal adhesions; lungs appeared healthy. Stomach and intestines full of well-digested matter; no parasites were visible. Only abnormality apparent to the eye was a yellow-orange slimy deposit lining the right pleural cavity that was not evident on the left side and which could not be attributed to position in which body was lying after death. Liver, spleen, kidneys, pancreas, lungs, heart (kept intact), mammaries sectioned for histology.

Histology Results. No lesions found in either the heart or lungs; however, moderate autolysis had occurred. Necrotic suppurations found in the hepatic parenchyma, but it could not be determined if the etiology of the hepatitis was bacterial or parasitical. Blood group O, rhesus negative.

QUINCE: 8-year-3-month-old female offspring of Liza and Beethoven of Group 5; date of death October 20, 1978, following 24 days of increasing weakness, diarrhetic, bloody dung deposits; body recovered same day; autopsy done by Drs. Poppoff and Preciado of Ruhengeri Hospital and the author at Karisoke.

External Examination.   No external parasites found; vulva unswollen; heavy residue and strong odor (resembled silverback odor) noted around right axillary region, though total lack of residue or odor from left axillary; both nipples very protruberant (breasts not swollen) and, when squeezed, produced a great deal of milk.

Internal Examination.   Right and left lungs extremely discolored and hardened, indicative of pneumonia; all other organs normal and ovaries totally inactive. Dilation of stomach and intestines initially suggested poisoning; contents of pancreas and intestines were mustard-colored, extremely foul-smelling liquid; beginnings of peritonitis noted on dorsal side of large intestines in form of abnormal "rubberband" type of adhesions to pleura; liver, spleen, kidney, and appendix (13 cm long) appeared normal. Left kidney weighed 195 gm; right kidney 180 gm. No food remnants found in stomach, small or large intestines. In the cecum was found one small mass of yellow digested vegetation combined with yellowish, foul-smelling mushy substance that weighed 10½ oz.

Histopathology.   Clinician, Frank H. Wright, D.V.M.; pathologist, D. L. Graham, D.V.M.

1. Mass from stomach—Chronic granulomatous inflammatory tissue composed of scattered fibroblasts, capillaries, and numerous macrophages, many of which contain hematin pigment.
2. Clotted blood—Some clots contain large numbers of neutrophils and hematin-laden macrophages.
   Adipose tissue—Fat necrosis, accumulation of lipochromes (lipofuscin and ceroid) and infiltration of the interstitium with edema fluid, neutrophils, and macrophages.
3a. Lymph node—Hyperplasia and infiltration with neutrophils.
3b. Pancreas—Autolysis.
    Lymph node—See 3a.
4. Necrotic cellular debris.
5a & b. Appendix (5 sections)—The lumen contains numerous sections of nematodes (oxyurids) and the mucosa and submucosa are heavily infiltrated with lymphocytes; there is marked congestion of the submucosal and serosal venous channels. The muscular tunic is unremarkable.
6. Two pieces of adipose tissue—The adipose tissue is unremarkable save for the presence of several well-circumscribed masses of dark brown to black pigmented particulate debris, some of which have the appearance of plant material or mineralized fragments of dead metazoan parasites. Whatever the origin, the micronodules are well circumscribed and probably of no clinical significance.

7a & b. The specimens consist of myometrium and serosa; the endometrium is absent. There are, within the myometrium, several small focal aggregates of pigmented debris similar to those seen in the parauterine adipose tissue.

8. Kidney (2 pieces of cortex)—Moderate autolysis; there is intracellular accumulation of dark green-brown granules in the proximal tubule epithelium.

9. Spleen—Moderate autolysis; there is diffuse accumulation of hematin pigment with many focal aggregates within reticuloendothelial cells.

10. Stomach—Diffuse goblet cell hyperplasia accompanied by formation of occasional small mucus-filled cysts in distended gland crypts (mild catarrhal gastritis).

11, 12, 13. Small intestine—Autolysis of mucosa; numerous cross sections of nematodes in the lumen.

14. Large intestine scrapings—Plant debris and nematode sections.

15. Liver (3 sections)—There is intracanalicular and intracellular bile stasis and the sinusoids contain many hematin-laden macrophages.

16. Gallbladder—Autolysis; no microscopic lesions.

17. Clotted blood—Many monocytes contain dark brown to black birefringent pigment granules (hematin).

18. Lung—Severe subacute to chronic congestion and edema; there are several regions of acute hemorrhage as well as acute to subacute purulent pneumonia (probably secondary to congestion and edema).

19a & b. Brain—Well fixed, little autolysis; there are numerous monocytes, some laden with dark brown pigment granules, in the cerebral blood vessels.

20a & b. Nipples—No microscopic lesions.

21. Clotted blood intermixed with plant debris. *Special stains:* The hematin pigment observed in many of the specimens is light brown to dark brown, exhibits birefringence on polarization microscopy, and is iron-negative with the Prussian blue reaction (Perl's stain).

Diagnosis.   1. Malaria
         2. Pulmonary congestion, edema, and hemorrhage
         3. Subacute purulent pneumonia
         4. Intrahepatic bile stasis
         5. Cholemic nephrosis (probably not of clinical significance)
         6. Nematode (oxyurid) infection of the vermiform appendix
         7. Intestinal nematodes

Conclusions. The widespread and heavy accumulation of both free and intracellular hematin pigment is strongly suggestive of malaria; however, autolysis precludes specific identification of the parasite (attempts to identify the organism in suspicious areas with special stains and electron microscopy failed). Subacute to chronic pulmonary congestion, hemorrhage, and edema are among the likely results of the debility produced by malaria and in turn predisposed to secondary subacute purulent pneumonia. The degree to which the congestion, edema, and pneumonia contributed to Quince's death is difficult to assess with certainty. The gross description indicates that all lobes were affected to some degree and that pulmonary edema and pneumonia were, collectively, the cause of death but were in turn sequelae of malaria.

RAFIKI: Silverback of Group 8 approximately 55 to 60 years old. Estimated date of death April 22, 1974; body recovered April 23; autopsy
conducted by Dr. M. Leclerc of Ruhengeri Hospital.

Entirety of left lung adhered to chest wall, suggesting pleurisy of long
duration. Both lungs heavily edematous, the left more so than the right,
symptomatic of pneumonia. On one lobe of the right lung were a number
of ½" to 1" pale projections whose pathology was not determined. Liver,
gallbladder, kidney, spleen appeared healthy. No histological work done.
Anoplocephala gorillae and *Murshidia devians* found in small intestine. Skin
wound of long duration found in right lumbar region, possibly inflicted
in July 1969. Blood group O, rhesus negative.

Skeletal Examination.   Conducted by Jay Matternes. Skull: The only sutures
evident were the occipital-temporal sutures and the temporal-zygomatic suture on the zygomatic arch (a variable suture). The basal suture was invisible.
Degree of alveolar erosion very marked though virtually no wear on cheek
teeth; break in cusp of right lower canine; muzzle skewed slightly to the
right; exostosis observed in regions of attachment of temporal muscles.
Three mental foramina on both right and left. Single infraorbital foramina
bilaterally. Vertebrae: Lumbar 2 and 3 showed marked osteophytosis; a
marked asymmetry of 5th cervical caused first 4 to incline extremely to
the right.

THOR: 11-month-old female infant of Macho and Rifiki of Group 8.
Date of death, May 20, 1974; body recovered May 21; autopsy conducted
May 22 by Dr. F. Vounderick of the University of Butare.

External Examination.   Skin covering of os frontale through os parietale
torn open for length of 8 cm, exposing the encephalon. The latter nearly
intact except for 1 small puncture wound; sulci intact. Segment of upper
skull (5 cm by 5 cm) missing (split across the sutura sagittalis). Second
major wound on left lower ventral surface of abdominal wall and extending
to ileo-cecal junction. The skin broken for length of 12 cm, exposing colon,
cecum, and appendix. Weight = 10.13 lb (4.57 kg).

Internal Examination.   First noted that bladder had been pierced by bite
in addition to broken pubic symphysis. Spleen (20.6 gm), kidneys (right
= 20.1 gm; left = 19.1 gm), adrenals (900 mg), liver (116.9 gm), pancreas
(3.72 gm), heart (25.3 gm), thymus gland (3.2 gm), salivary gland (650
mg), lungs (47.3 gm), brain (376 gm) all examined and considered normal.
Adrenals surrounded by large deposits of fatty tissue considered normal.
Stomach and intestines very full of digested vegetative food contents with
no visible traces of milk; no adult parasites present; however, numerous
Strongylidae ova present. Blood group O, rhesus negative.

Histology Examination.   Not outstanding owing to autolysis of tissues. Noted
that the genitalia bore a striking resemblance to those of the more inferior

primates; the heavy density of ovules was undeveloped in the ovary and the uterus mucosa resembled that of *Galagos* and *Cercopithecus*.'

Skeletal Examination.   Conducted by Jay Matternes. Skull: Deciduous incisors completely erupted; deciduous canines largely in their crypts; first deciduous premolar erupted and tartar-stained; first deciduous molar just erupting with anterior buccal cusps tartar-stained. Occipital condyles implicit but not developed; anterior part of occipital (between basion and basal suture) is separate from rest of occipital. The tympanic and petrius bones of the temporal region are distinct from one another; premaxillary bone still separate from the maxillary.

Mandible: The symphysis of the chin completely fused; all deciduous premolars fully erupted, showing tartar stains; lower canines have just erupted and show tartar stains on cusps; first lower molars fully erupted showing tartar stains; second deciduous molar still in crypt. The mandibular foramen slopes backward bilaterally and the condyles slope sharply inward, which is the reverse condition of that found on most adult mandibles.

Long Bones: Ossification centers present inside of humeri cartilages. Neither humerus shows septal apertures. Radii well formed and, except for cartilaginous ends and indistinct muscle ridges, appear almost miniature replicas of adults'. Ulnae less well formed. Femora show large ossification centers with cartilaginous heads present and complete, separate from shafts; same true for tibiae. Fibulae well rounded in midshaft sections but lack indication of anterior crissa characteristic of adults'. The left fibula appears extraordinarily bowed.

Vertebrae: The atlas completely ossified except for anterior tuberculus; axis also completely ossified except for the dens, and joined to cervical 3 by cartilage; vertebral arches of remaining four cervical vertebrae separate from the corpi.

Innominates: All elements of the pelves completely unfused.

Scapulae: The sinuous borders almost miniatures of adult forms; acromion processes fairly well ossified; corocoid processes cartilaginous at scapular articulations but completely ossified at other ends.

Clavicles: Miniature duplicates of adult forms.

Myoglobin Examination.   Conducted by Dr. A. E. Romero-Herrera and Dr. H. Lehmann of the Department of Clinical Biochemistry, University of Cambridge.

"On aligning the tryptic peptides of the myoglobin from a gorilla [the above specimen] with the homologous human peptides, one amino acid difference was found. By dansyl-Dedman degradation this was shown to be at position 22, i.e., at a position other than those where man, chimpanzee and gibbon differ from one another." (A. E. Romero-Herrera, H. Lehmann and Dian Fossey, 1975)

WHINNY: Silverback of Group 4 approximately 50 to 55 years old. Estimated date of death May 1, 1968; body recovered May 3; autopsy conducted May 5, 1968 by Dr. Gourand of Ruhengeri Hospital.

Xiphoid-pubic Incision.   There existed abnormally thick and profuse peritoneal liquid plus numerous adhesions attaching the cecum and sigmoid to the parietal wall in the region of the iliac fossae. The mesenteric lining was free and of normal thickness but showed periodic lesions. Appendix normal. Liver, spleen, kidneys, stomach, and pancreas appeared normal.

Medial Sternum Incision.   The inferior right lung lobe appeared extremely pathological because the right pleura was strongly adhered to the top of the diaphragm. The pericardial cavity seemed normal.

All Organs Examined.   Excretory duct, testes, and the penis, which measured 6.5 cm.

Histological Examination.   Conducted by D. H. Wright, M.D., of Makerere University.

Material not ideal for histological examination owing to post-mortem autolysis. The only gross lesions identified were vast areas of consolidation in the lung of confluent bronchopneumonia. There were no abnormalities in any other organs.

Skeletal Findings.   Conducted by Jay Matternes. Extensive pathology on right side of skull indicative of a localized infection resulting from a wound, probably a bite. Infection appeared to be spreading to the occipital through the pneumatized area around the mastoid process, and if the lesions had entered the endocranium, meningitis would have resulted.

UNNAMED: Male juvenile captive from Zaire approximately 4 years old; date of death March 28, 1978, at Karisoke; autopsy conducted by Drs. Berger and Preciado of Ruhengeri Hospital on March 29, 1978.

External Examination.   There was a gangrenous bone infection of lower left leg that had spread from the ankle, where the trap wire was still embedded, into the flesh. The wounds were at least four months old and had caused the loss of all toes, resulting in a contracted, hairless stump and extreme atrophy of the entire extremity.

Internal Examination.   Infectious lesions were found in both lungs, which were white-gray in color, lacking elasticity and resembling the lungs of human tubercular patients.

Probable Cause of Death.   Gangrene coupled with pneumonia and possibly tuberculosis.

APPENDIX G

Introduction:
Karisoke Parasitology Research

Between November 1976 and April 1978, research assistant Ian Redmond accomplished the first long-term parasitology study ever undertaken at Karisoke. Working with an old microscope which he himself assembled, Ian painstakingly examined countless samples of gorilla feces to obtain specimens of endoparasites and made freehand drawings of his findings. Most of the samples were analyzed fresh at Karisoke, though some were preserved in 10 percent Formalin for later study at the British Museum (Natural History). There, Ian received enormous assistance with advice, facilities, and identifications from Mrs. E. Harris, Dr. David Gibson, and Mr. Charles Hussey of the Parasitic Worms Section of that institution. To them Ian wishes to express his deepest gratitude for their generous contributions of time and interest in his valuable pioneer field study. Specimens of ectoparasites were recovered from the bodies of deceased gorillas and identified in the Entomology Department of the British Museum (Natural History) by Dr. C. Lyal, to whom Ian remains extremely grateful.

The Karisoke Research Centre also wishes to express appreciation to the above for their kindly assistance to Ian Redmond, who, as a result of their efforts and his own, was able to write the following report based on his work. I will always remain grateful for all of Ian's contributions to Karisoke, where he was affectionately known as Toto ya Nyoka (The Worm Boy).

Dian Fossey

Summary of Parasitology Research, November 1976 to April 1978
by Ian Redmond

Endoparasites were studied by fecal analysis and, on one occasion, by dissecting a juvenile (Kweli) who died as a result of gun wound. Ectoparasites were never observed on healthy, free-living gorillas (although there were frequent opportunities to groom certain individuals) but were collected from deceased animals. See Appendix F.

Two kinds of ectoparasite were found, the gorilla louse *Pthirus gorillae* (see page 263) and a tiny, soft-bodied mite, *Pangorillalges gorillae.*

Fecal samples were collected in two ways—either when defecation was seen during behavioral observations or by collecting from identified night nests. Samples were analyzed fresh at Karisoke, though some were double

Table Summarizing Results of Parasitology Research Compiled by Ian Redmond

Hosts / Parasite	STUDY GROUPS			FRINGE GROUPS			TOTALS		
	Number of gorillas sampled regularly	Number infected	Percentage infection	Number of gorillas sampled occasionally	Number infected	Percentage infection	Total number sampled	Total number infected	Percentage infestation
Type A nematodes	32	29	90.7%	48	35	73%	80	64	80%
Type B nematodes	32	32	100%	48	35	73%	80	67	83.7%
Strongyloid nematode eggs	32	32	100%	19	17	89.5%	51	49	96.1%
Cestode, Anoplocephala gorillae	32	13	40.6%	52	30	57.8%	84	43	51.2%
Particles thought to be Protozoa	30	28	93.3%	4	4	100%	34	32	94.1%

checked by preserving 10 grams in 10 percent Formalin for later analysis in England.

The quantitative techniques which evolved over the first few weeks were as follows:

1. Samples were collected in polyethylene bags and weighed whole.

2. Direct smears were made by emulsifying 20 mg in 2 drops of 0.8 percent NaCl solution on a glass slide, then examined under a coverslip at 66× magnification.

3. 10 gm were separated from the bulk of the sample and washed through a sieve (1 mm × 1 mm mesh size) into a bucket, using filtered stream water. This was left to settle. The sediment was rinsed through a fine cloth sieve and then closely examined in a black tray, against which the white nematodes were visible. These were counted and examined, wet-mounted, under the microscope.

4. The remainder of the sample was washed through the large sieve, but this time the residue was searched for larger helminths such as cestode proglottids.

The results of this work are summarized below, and in the table following.

Nematodes (Roundworms)

Type A: A viviparous nematode to 4 mm in length, with a truncate head end. Adult females and larvae were recovered in varying numbers (0.1 to 10 per gram) but no males have yet been found. The specific characters of nematodes are often restricted to the shape of the male reproductive organs, so identification of Type A is not yet certain. It has obvious affinities, from the females, to the genus *Probstmayria,* but the shape of certain structures differs from the generic description. Either the genus must be redescribed to include Type A or a new genus must be created. The search for a male continues.

Type B: Similar to Type A but only to 2 mm in length, more slender and with a much more pointed head end. This is almost certainly *Probstmayria gorillae* Kreiss 1955, but as with Type A no males have been located to enable positive identification. The size and shape of females is exactly that of *P. gorillae.* Members of the genus *Probstmayria* are unusual for parasitic nematodes in that they exhibit continuity of generations within the same host. They live and reproduce within the gut of several diverse herbivores, including horses, pigs, apes, and tortoises. Larvae passed in the feces seem able to survive for several days in warm, damp soil or water, and from there may be accidentally ingested by a new host to begin an infection. Surprisingly, even in samples less than one hour old, none of the Type A or B nematodes were found alive. Possibly the ambient temperature at this altitude was too low for them to survive outside the host.

Strongyloid Eggs: These resemble eggs of the human hookworm, *Ancylostoma duodenale,* and were found in most samples examined by direct smear.

Identification of species from eggs alone is not possible, and no adults were found.

Impalaia sp.: Nematodes of this genus were collected from Kweli's small intestine during the post-mortem. Members of this genus are normally found in giraffes and antelopes (such as impala) and have never before been recorded from a primate host or from this geographical area. All the specimens found were immature, which tends to suggest that they had accidentally infected the wrong host and were unable to develop fully as a consequence. Without adults it is not possible to determine the exact species, but it is likely that they would normally parasitize bushbuck or duiker in the Virungas.

Types C, D, E, F, and H were found to be free-living or plant parasitic nematodes. Collection of feces without environmental contamination is not possible in the wild; nematodes can get into a sample from leaves or soil, or even be carried on the bodies of flies feeding or egg-laying on dung. Dr. L. F. Khalil of the Commonwealth Institute of Helminthology has identified the nonparasitic species as follows:

Type C: *Rhabditis* (*Cephaloboides*) sp., possibly *R. curvicaudata*.

Type D: Free-living soil nematodes with a spear, belonging to subfamily Dorylaiminae, possibly *Laimydorus* sp. and *Aporcelaimus* sp.

Types E and F: Rhabditid nematodes, possibly *Rhabditis* spp.

Type H: Unidentified.

Type G: Only seen on one occasion. It was alive and moving around the pharyngeal bulb of a dead adult female Type A. As the Type A was undamaged it is possible that Type G is a parasite of a parasite.

Cestodes (Tapeworms)

Anoplocephala gorillae Nybelin 1927 is the only cestode recorded from mountain gorillas. No complete worms were collected, but proglottids (tapeworm segments) were visible in the feces of infected animals. The adult cestode is attached to the gut wall by its hooked head (scolex) and the tape-like body (strobila) lies freely along the length of the intestine absorbing nutrients from the gut contents. As each segment matures it is released, full of eggs, into the feces. Cestode life cycles are often complex; that of *A. gorillae* is not known, but it may involve soil mites (often found around dung) as intermediate hosts.

Trematodes (Flukes)

No trematodes were collected by this author, but a single large specimen was collected from the left lung of Digit, during the post-mortem six days after his death. It is likely that it was not parasitizing the lung, but had moved there after Digit was killed. It measured 32 mm by 19 mm and had a median ventral sucker 14 mm by 6 mm.

Protozoans (Single-celled Animals)

When examining fresh feces by direct smear, one would expect to find ciliates (such as *Troglodytella,* which are found in chimpanzees) swimming around on the slide. No motile protozoa were seen during this study, even in samples less than one hour old on examination. Particles resembling protozoans, two to three times the size of the strongyloid eggs (see Fig. 2, April 1977 Report), were seen in most smears (32 out of the 34 animals examined for them).

Coprophagy and Parasites

In many herbivores, digestion of cellulose is aided by microbes living in the gut; young of the species establish this "gut flora" (or more correctly, fauna) by ingesting some of their parents' (or other adults') feces, or, in the ruminants, eructate from the fore-stomach. The exact role played by microbes in the gorilla's digestion has yet to be demonstrated, but coprophagy (eating feces) has been observed on many occasions. Usually an individual eats some of his or her own feces immediately on excretion, but sometimes an infant will eat feces of an older animal, or vice versa. This is clearly the most direct means for a young gorilla to establish an active gut fauna; it may also explain why Types A and B nematodes were always, like the "protozoans," dead on examination.

If infection and reinfection of the host is achieved by coprophagy, these parasites (perhaps "symbionts" would be a better term if the gorilla benefits by their presence) would continue for generation after generation in host after host without ever experiencing anything other than body temperature. This is providing that the feces are always eaten warm, which observations show to be the case.

Infants begin passing viviparous nematodes and "protozoan-like objects" in their first year or 18 months of life. By this time they have begun taking solid foods, including occasionally pieces of warm dung from their mother or peers. Once established in this way, populations of direct developers, such as *Probstmayria,* would continue in that host for life, perhaps being further supplemented if coprophagy becomes a habit.

Probstmayria vivipara larvae remain viable in pasture for days, and horses, their host, have ample opportunity to ingest them. Gorillas seldom feed in one location for more than a few hours, so opportunities for such accidental ingestion are minimal and would not explain the very high percentage of animals infected. Coprophagy would have the advantage of ensuring the parasite's transmission to a new host, but the drawback appears to be that larvae have lost any resistance they might have had to low temperatures. Thus, the cold water used in fecal analysis evidently finished them off.

This theory might be tested by microscopic examination of feces at a constant temperature of around 37°C—a simple idea but rather difficult in practice!

Parasites Recorded from Gorillas

Compiled by Ian Redmond from the records of the Parasitic Worm Section of the British Museum (Natural History), data supplied by Dr. L. F. Khalil of the Commonwealth Institute of Helminthology and Ian Redmond's own research.

Endoparasites

Nematodes (Roundworms)	References
Abbreviata caucasica (Linstow, 1902)	Khalil, pers. comm.; formerly in the genus *Physaloptera*.
Ancylostoma duodenale (Dubini, 1843)	Stiles and Speer, 1926; Stiles, Hassal, and Nolan, 1929.
Ascaris lumbricoides (Linné, 1758)	Stiles and Speer, 1926; Graber and Gevrey, 1981.
Chitwoodspirura wehri (Chabaud and Rousselot, 1956)	Yamashita, 1963; Graber and Gevrey, 1981.
Dipetalonema gorillae van den Berghe et al., 1957	van den Berghe and Chardome, 1949; van den Berghe, Chardome, and Peel, 1957 and 1964.
Dipetalonema leopoldi van den Berghe et al., 1957	van den Berghe, Chardome, and Peel, 1957 and 1964.
Dipetalonema perstans (Manson, 1891)	Yamashita, 1963.
Dipetalonema vanhoofi Peel and Chardome, 1946. *Note:* Some authors put this species in the genus *Tetrapetalonema* (*Esslingeria*).	van den Berghe, Chardome, and Peel, 1964; Rousselot, 1955 and 1956; Graber and Gevrey, 1981.
Dipetalonema streptocerca Macfie and Corson, 1922	van den Berghe, Chardome, and Peel, 1964.
Enterobius lerouxi Sandosham, 1950	Yamashita, 1963; Graber and Gevrey, 1981.
Hepaticola hepatica (Bancroft, 1893)	Paciepnik, 1976; formerly in the genus *Capillaria*.
Impalaia sp.	Redmond, in prep.
Libyostrongylus hebrenicutus Lane, 1923	Nagaty, 1938 (redescribed); Yamashita, 1963; *Note:* Skrjabin et al., 1952 and 1954, put this species in the genus *Paralibyostrongylus*.
Loa loa gorillae van den Berghe et al., 1964	Yamashita, 1963 (for *Loa loa*).
Microfilaria binucleata Peel and Chardome, 1946	van den Berghe, Chardome, and Peel, 1964 (redescribed).
Microfilaria gorillae Berghe and Chardome, 1949	Yamashita, 1963.
Murshidia devians Campana-Rouget, 1959	
Necator americanus (Stiles, 1902)	Stiles and Speer, 1926; Graber and Gevrey, 1981.
Necator congolensis Gedoelst, 1916	Lane, 1923; Stiles and Speer, 1926; Graber and Gevrey, 1981 (redescribed).

Nematodes (Roundworms), cont.	*References*
Necator gorillae Noda and Yamada, 1964	now considered a synonym of *N. congolensis*—see Graber and Gevrey, 1981.
Onchocerca volvulus Leuckart, 1893	van den Berghe, Chardome, and Peel, 1964.
Oesophagostomum stephanostomum Stossitch, 1904	Railliet and Henry, 1906; Lane, 1923; Travassos and Vogelsang, 1933—changed to genus *Ihleia*. Rousselot and Pellissier, 1952; Uehara et al., 1971; Yamashita, 1963.
Oesophagastomum apiostomum (Willach, 1891)	Paciepnik, 1976.
Probstmayria sp. inom.	Redmond, in prep.
Probstmayria gorillae Kreiss, 1955	Skrjabin et al., 1961; Yamashita, 1963; Redmond, in prep.
Protospirura muricola Gedoelst, 1916	Chabaud and Rousselot, 1956; Graber and Gevrey, 1981.
Strongylidae gen. and sp. inom. (eggs)	Redmond, in prep.
Strongylidea sp.	Cordero del Campillo, 1977.
Strongyloides sp. inom.	Stiles and Speer, 1926.
Strongyloides papillosus (Wedl, 1856)	Krynicka et al., 1979.
Strongyloides stercoralis (Bavay, 1876)	Krynicka et al., 1979.
Strongylus falcatus Linstow, 1907	
Ternidens deminutus (Railliet and Henry, 1905)	Skrjabin et al., 1952; Popova, 1958; Yamashita, 1963.
Trichurus sp.	Stiles and Speer, 1926.
Trichurus trichiura (Linné, 1771)	Krynicka et al., 1979; Graber and Gevrey, 1981; Amberson and Schwartz, 1952.

Acanthocephala (Thorny-headed Worms)

Prosthenorchis elegans (Diesing, 1851)	Moore, 1970.

Cestodes (Tapeworms)

Anoplocephala gorillae Nybelin, 1927	Sandground, 1927 and 1930; Lobez-Neyra, 1954; Redmond, in prep.
Bertiella studeri Blanchard, 1891	Graber and Gevrey, 1981.

Trematodes (Flukes)

Brodenia jonchi Berengner, Vallespinoza, and Fernandez, 1963	Gallego and Berengner, 1965; Cordero del Campillo et al., 1975.
Concinnum brumpti (Railliet, Henry, and Joyeux, 1912)	formerly in the genus *Eurytrema* in Stunkard, 1949; redescribed by Stunkard and Goss, 1950; Skrjabin, 1953; Yamashita, 1963;

Trematodes (Flukes), cont. | *References*

| | Cosgrove, 1966; Graber and Gevrey, 1981. |
| *Dicrocelium dendriticum* (Rud., 1819) | Paciepnik, 1976; Krynicka et al., 1979. |

Ectoparasites

Arachnida (Spiders, Mites, and Allies)

| *Pangorillalges gorillae* Gaud and Till, 1957 | formerly in the genus *Psoroptoides*, new genus created by Fain, 1962; Redmond, in prep. |
| Insecta (Insects) | |

| *Pthirus gorillae* Ewing, 1927 | Ke Chung Kim and Emerson, 1968. |

Notes on the Drawings

The following illustrations have been redrawn from originals prepared by Ian Redmond at Karisoke to illustrate monthly reports on the Parasitology Project. The explanatory notes accompanying them were also written in the field, but have been amended in the light of more recent findings.

All the original drawings were made freehand from fresh material, using the naked eye or an old binocular microscope. The latter was illuminated by either reflected daylight or paraffin pressure-lamp; chromatic aberration was a problem at higher magnifications (only 66×, 240×, and 360× were available). All specimens were wet-mounted since no sectioning, staining, or camera lucida facilities were available.

Drawings of Parasites Found in
Gorilla Dung during December 1976

Figure 1a: Generalized drawing of a Type A nematode 4th stage larva with scale shown to left (drawn after several specimens). It was thought that there were six lobes around the mouth.

Figure 1b: Enlargement of the central section of an adult Type A nematode showing development of larvae (this was best seen in specimens where the gut was disintegrating, which made other viscera more easily discernible). The gut is shown whole.

Figure 1c: Simplified sketch made when an adult Type A nematode was burst open (details of the uteri were omitted). The larger larva showed no sign of an enclosing membrane and floated free (having had its tail broken and pharynx everted, labeled eph, when the adult was burst); the smaller one remained half out of the rupture hole. They were thought to be 2nd and 3rd stage larvae.

Figure 2a: Drawing of a Type B nematode (adult female) with developing larva inside. Such larvae are most noticeable under the microscope by the curve, where they bend double; only one is shown for clarity (the scale is shown on left).

Figure 2b: Enlargement of anterior end of a Type B nematode, simplified to show the internal organs clearly. The dark spot to the right of the pharyngeal bulb appears as a clear brown oval under the microscope. It is a unicellular excretory gland.

Figure 3: Drawing of a Type C nematode, with scale shown to the left. Only three specimens have so far been found, of two different sizes.

All drawings in Figs. 1, 2, and 3 were done under 66× magnification. The main anatomical features are labeled as follows:

  m: mouth; situated at extreme anterior end surrounded by lobed lips
 ph: pharynx; a muscular organ, with a characteristic shape in each type, for pumping of food
 ex: excretory organ, a large unicellular gland known as a renette cell
  g: gut; a straight tube running the length of the body, ending in
 ao: anal opening; slightly forward of the tail
 lm: longitudinal muscles; these run the length of the body and are the only organs of locomotion found in nematodes, working against the hydrostatic pressure of the pseudocoelmoic fluid
 la: larva; in Types A and B, the larvae develop inside the adult's body
  v: vulva, the exterior opening of the vagina, leading from the uteri.

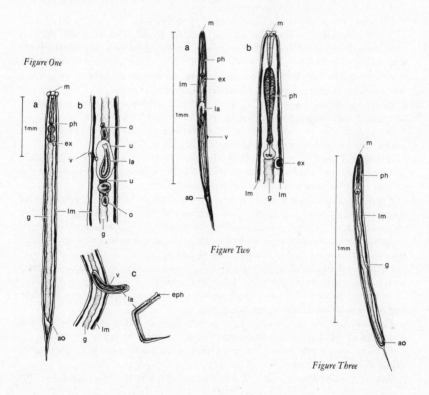

Figure One

Figure Two

Figure Three

Figure 4: All cestode proglottids thus far found were thought to be from the same species, *Anoplocephala gorillae.*

Figure 4a: The first proglottid found from Beethoven's dung was curled when discovered and has remained so.

Figure 4b: Most cestodes found resembled this form and are limp and pink tending toward a greenish-white color in older specimens. One of Uncle Bert's is shown here.

Both of these drawings were done with the naked eye. All the scales shown were estimated by using a ruler only.

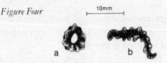

Figure Four   ⊢ 10mm ⊣

a   b

Drawings of Nematodes Found in Gorilla Dung during January 1977

Figures 1–4, attached to December 1976 Report, show nematode Types A, B, and C, and the cestode proglottids found.

Diagrams for all other types are attached to this report.

Figure 5: Based on a sketch of the first Type C2 found, with burst anal opening (bao—also found in several other specimens); drawn under 66× magnification (the scale is shown to the left).

Figure 6a: The first Type D found—with two dark ovals thought to be ova (o), but not seen in subsequent specimens; drawn under 66× mag.

Figure 6b: Anterior end of a Type D found in urine-soaked dust; no pharynx was visible, just the undulating line shown; drawn under 240× mag.

Figure 6c: Posterior end of a Type DT (otherwise identical with Type D) drawn under 240× mag. (the scale to the left only applies to Fig. 6a).

Figure 7a: Type E with ova (o) between loose cuticle (lc) and body wall of adult; in motile specimens the cuticle wrinkles up on the concave and stretches on the convex curve as shown; drawn under 66× mag.

Figure 7b: Two Type E with larvae—drawn after 48 hours in a petri dish of water (the right-hand one is the same specimen as that shown in Fig. 7a before the ova hatched); the larvae were inside the body wall and had apparently destroyed the adult viscera; several were seen to escape through the vulva (v). Drawn under 66× mag. (the scale applies to both Figs. 7a and b).

Figure 8a: Adult Type F with ova (o); the knurled effect at either end is only seen in some specimens; 66× mag. (scale to the left).

Figure 8b: Section of adult Type F showing embryonic development of larvae within the ova (o); drawn under 240× mag.

Figure 9: Anterior end of adult Type A with Type G coiled around the pharynx (the Type G was alive and motile); 240× mag. (scale to the left).

Figure 10a: Posterior end of the first Type H found; no viscera were discernible; drawn under 240× mag.

Figure 10b: The second Type H found (note: the first was not curved like this one and was slightly larger); the gut has a blotchy appearance and no pharynx could be seen in either specimen; drawn under 66× mag. (the scale on the right).

Annotations not mentioned above are the same as those in Figs. 1–4: m, mouth; ph, pharynx; g, gut; ao, anal opening; lm, longitudinal muscles; la, larvae; v, vulva.

The drawings are somewhat simplified for clarity. All scales shown are to serve as a guide only, since they were estimated by eye.

Note: Types C, D, E, F, and H were later identified as free-living or plant parasitic nematodes that had contaminated the samples. This illustrates the problems of collecting reliable data in field conditions.

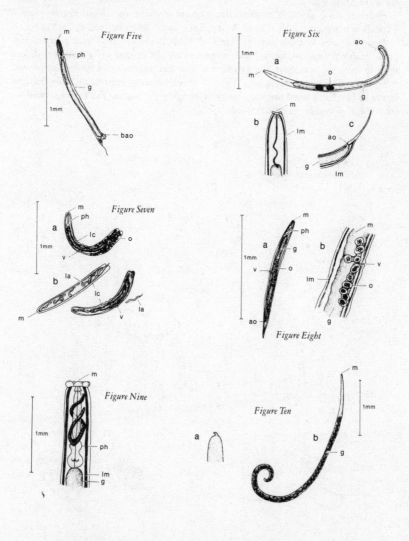

Drawings of Nematodes (Probstmayria *cf. sp.*) and Strongyloid Eggs from Gorilla Dung, March 1977

Finer details of the structures shown in Figs. 1–4 were largely obscured by chromatic aberration in the absence of either a blue light filter or a monochromatic light source.

Figures 1a and b: Generalized outline of the anterior tip of Type A nematodes showing lips and papillae (not the same specimen). 1a is modified by looking at several specimens under 240× mag.; 1b is just a line drawing of one individual at 360× mag. Dotted lines denote structures that were out of focus when looking at the solid lined structures.

Figure 1c: Line drawing to show the contrast in shape and detail visible when a Type B is compared with a Type A under the same magnification as Fig. 1a.

Figures 2a and b: Both anterior ends of Type B under 360× mag.; again, Fig. 2a is modified from numerous specimens and 2b shows one individual that differs slightly from the majority. The grooved cuticle was only noticed under this power.

Figure 3: Shows the vulva of a Type B gravid female, apparently after the first larvae in line had passed through, partially everting the vagina; the exterior swelling (vu) is not normally so pronounced.

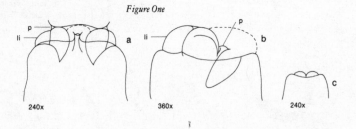

Figure One

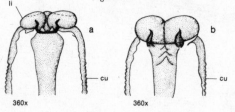

Figure Two

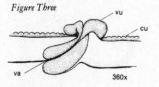

Figure Three

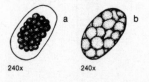

Figure Four

Figure 4: Shows the two types of egg found in direct smears; Fig. 4a is a thin-shelled oval with brown cells in the center (numerous are seen here owing to the time elapsed since excretion). The area between the shell and embryo appears clear under 240× magnification. This is typical of the appearance of many Strongyloid nematode eggs (e.g., human hookworm); identification is not possible without an adult specimen. Fig. 4b shows the second type, which is the same size, slightly more of a rounded oval, and completely filled with larger, gray cells, or refractive granules.

Annotations: p, papillae; li, lips; cu, cuticle; vu, vulva; va, vagina.

Drawings of Parasites Found in
Gorilla Dung during April 1977

Figure 1: Depicts two cestode proglottids found in a sample of Ziz's dung, which were found attached to one another; thus illustrating how they are joined to the strobila before being released in the host's gut. The drawing was made with the naked eye.

Figure 2: Shows a typical specimen of the large gray objects seen regularly in direct smears of Pantsy's and others' dung. These were thought possible to be dead protozoa, probably ciliates of the genus *Troglodytella*. Drawn under 240× magnification.

Figure One

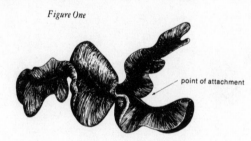

point of attachment

Figure Two

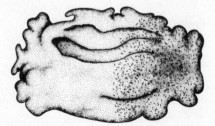

Drawings for Parasitology Report, May 1977

Figures 1a–e: Show five eggs of the human hookworm *Ancylostoma duodenale* in later stages of development, the curve of the larva being clearly visible in 1c and 1d. These were drawn from photomicrographs in the Ruhengeri Hospital and *not* from fresh material.

To investigate the development of eggs in dung after excretion, a number of samples were collected in the normal way, kept in polyethylene bags, and examined by direct smear at intervals for several days. The drawings show some of the results.

Figure 2a: Shows a single developing egg from a sample of Tiger's dung examined 23 hours after excretion. The embryo is undergoing gastrulation.

Figures 2b and c: Show two eggs from Uncle Bert's smear 22 hours after excretion. The embryo has not developed as far as in 2a in either of these.

Figures 3a–c: Depict three developing ova from Uncle Bert's dung (same sample) four days after excretion. In 3a the embryo is undergoing the rearrangement of cells that produces the elongated shape. This is already complete in 3b and 3c, where the folded body of the larva is discernible, completely filling the shell. The larva could occasionally be seen moving inside the shell.

a b c d e

Figure One

a b c

Figure Two

a b c

Figure Three

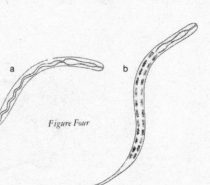

Figure Four

Figure 4a: Shows a motile larva, presumably newly hatched, found in Macho's dung four days after excretion.

Figure 4b: Shows a similar motile larva but with a less distinct gut and dark areas at intervals along each side of the gut. This was drawn from Tiger's dung four days after excretion.

All drawings of fresh material were made under 240× magnification.

Drawings for Parasitology Report, June 1977

Figure 1: Depicts a male Type D nematode (see also Appendix G, Fig. 6a, found in January 1977), drawn under 66× magnification.

Figure 2: Shows the anterior end of the Type D nematode drawn under 360× magnification.

Figure 3: Shows the posterior end of Type D nematode drawn under 360× magnification.

Note: This has since been identified as a free-living soil nematode and not a parasite. It was found in a sample from Simba that was collected off leaves literally seconds after excretion. This may imply that it had been ingested by Simba with food plants and passed through her digestive tract, or otherwise serve as an example of how easily samples could be contaminated by contact with the environment, no matter how brief the contact.

Figure One

Figure Two

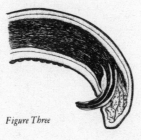

Figure Three

Drawings for Parasitology Report, March 1978

Eggs of *Anoplocephala gorillae*

Six eggs are illustrated to show the different appearance from various angles, drawn under 360× magnification. They were obtained by gently teasing apart a proglottid (previously fixed in 10 percent Formalin) on a microscope slide. Occasionally such eggs were seen in direct smears, but proglottids are usually passed whole, soon after release from the strobila, and eggs remain in the proglottid. Each proglottid contains thousands of eggs.

Bibliography

THE FOLLOWING LIST of publications pertaining to *Gorilla gorilla gorilla* is designed primarily for the scientific reader; many of the articles and journals are not readily available to the general public. The number of publications cited was limited by lack of complete bibliographic data or by space. The Bibliography is therefore not intended to be exhaustive.

Akeley, C. E. 1922a. Hunting gorillas in Central Africa. *World's Work,* June, 169; July, 307; August, 393; September, 525.

―――. 1922b. Is the gorilla almost a man? *World's Work,* September, 527.

―――. 1923a. Gorillas — real and mythical. *Nat. Hist.* 23:441.

―――. 1923b. *In Brightest Africa.* Garden City, N.Y.: Garden City Publishers.

Akeley, C. E., and Akeley, M. L. J. 1932. *Lions, Gorillas and Their Neighbors.* New York: Dodd, Mead.

Akeley, M. L. J. 1931. *Carl Akeley's Africa.* London: Gollancz.

―――. 1950. *Congo Eden.* New York: Dodd, Mead.

Akroyd, R. 1935. The British Museum (Natural History) expedition to the Birunga volcanoes, 1933-4. *Proc. Linn. Soc. London,* 147th sess., 17-21.

Alix, E., and Bouvier, A. 1877. Sur un nouvel anthropoide (*Gorilla mayêma*) provenant de la région du Congo. *Bull. Soc. Zool. Fr.* (Paris) 2:488-90.

Allen, J. G. 1931. Gorilla hunting in southern Nigeria. *Nigerian Field* 1:4-5.

Antonius, J. I., Ferrier, S. A., and Dillingham, L. A. 1971. Pulmonary embolus and testicular atrophy in a gorilla. *Folia Primatol.* 15:277-92.

Arnold, P. 1979. A preliminary report on the first mother-reared lowland gorillas (*Gorilla g. gorilla*) at the Jersey Wildlife Preservation Trust. *Dodo* (J. Jersey Wildl. Preserv. Trust), no. 16:60–65.

Aschemeier, C. R. 1921. On the gorilla and the chimpanzee. *J. Mammal.* 2:90–92.

———. 1922. Beds of the gorilla and chimpanzee. *J. Mammal.* 3:176–78.

Ashton, E. H., and Zuckerman, S. 1952. Age changes in the position of the occipital condyles in the chimpanzee and gorilla. *Am. J. Phys. Anthropol.* (n.s.) 10:277–88.

Aspinall, J. 1980. The husbandry of gorillas in captivity. *J. Reprod. Fertil.,* suppl., 28:71–77.

Awunti, J. 1978. The conservation of primates in the United Republic of Cameroon. In *Recent Advances in Primatology,* ed. D. J. Chivers and W. Lane-Petter, vol. 2 (Conservation), 75–79. London: Academic Press.

Babault, G. 1928. Note sur la biologie et l'habitat du gorille de Beringe. *Rev. Fr. Mammal.* 2:61–63.

Babladelis, G. 1975. Gorilla births in captivity. *Int. Zoo News* no. 130 (October).

Baldridge, C. 1978. The accuracy of measuring gorilla conadotropin using antibody to human LH in radioimmunoassay. Paper read at the 2nd Annual Meeting of the American Society of Primatologists, September 1978, at Emory University.

Baldwin, L. A., and Teleki, G. 1973. Field research on chimpanzees and gorillas: an historical, geographical, and bibliographical listing. *Primates* 14:315–30.

Barns, T. A. 1922. *The Wonderland of the Eastern Congo.* London: Putnam.

———. 1923. *Across the Great Craterland to the Congo,* 128–51. London: Benn.

———. 1926. *An African Eldorado: The Belgian Congo.* London: Methuen.

———. 1928. Hunting the morose gorilla. *Asia* (New York), February, 116, 154.

Bartholomew, G., and Birdsell, J. B. 1953. Ecology and the protohominids. *Am. Anthropol.* 55(4):481–98.

Bates, G. L. 1905. Notes on the mammals of southern Cameroons and the Benito. *Proc. Zool. Soc. London* 75:65–85.

Baumgärtel, M. W. 1958. The Muhavura gorillas. *Primates* 1:79–83.

———. 1959. The last British gorillas. *Geogr. Mag.* (London) 32:32–41.

———. 1960. *König im Gorillaland.* Stuttgart: Franckh.

———. 1961a. Death of two male gorillas and rescue of an infant gorilla. *Afr. Wild Life* 15:6–13.

———. 1961b. The gorilla killer. *Wild Life and Sport* 2(2):14–17.

———. 1976. *Up Among the Mountain Gorillas.* New York: Hawthorn Books.

———. 1977. *Unter Gorillas*. Berlin: Universitas Verlag.

Beck, B. B. 1982. Fertility in North American male lowland gorillas. *Am. J. Primatol. Suppl. 1*, 7–11.

Beebe, B. F. 1969. *African Apes*. New York: McKay.

Benchley, Belle J. 1932. Mbongo and Ngagi. *Touring Topics,* October, 15.

———. 1933. Mbongo and Ngagi. *Nature Mag.* 21:217–22.

———. 1942. *My Friends, the Apes*. Boston: Little, Brown.

———. 1949. Mountain gorillas in the Zoological Garden, 1931 to 1940. *Publ. San Diego Zool. Soc.,* 1–24.

Beringe, O. von. 1903. Bericht des Hauptmanns von Beringe über seine Expedition nach Ruanda. *Deutsches Kolonialblatt,* 234–35, 264–66, 296–98, 317–19.

Bernstein, I. S. 1969. A comparison of the nesting patterns among the three great apes. In *The Chimpanzee,* ed. G. J. Bourne, vol. 1, 393–402. Basel: Karger.

Bingham, H. C. 1928. Sex development in apes. *Comp. Psychol. Monogr.* 5(23):1–165.

———. 1932. Gorillas in a native habitat. *Carnegie Inst. Wash. Publ.* no. 426:1–66.

Bingham, L. R., and Hahn, T. C. 1974. Observations on the birth of a lowland gorilla in captivity. *1974 Int. Zoo Yearbook* 14:113–15.

Blancou, L. 1950. The lowland gorilla. *Anim. Kingdom* 58:162–69.

———. 1951. Notes sur les mammifères de l'équateur africain français: le gorille. *Mammalia* 15:143–51.

———. 1961. Destruction and protection of the wild life in French Equatorial and French West Africa. Pt. 5, Primates. *Afr. Wild Life* 15:29–34.

Blower, J. 1956. The mountain gorilla and its habitat in the Birunga Volcanoes, *Oryx* 3:237–97.

Bolwig, N. 1959. A study of the nests built by mountain gorilla and chimpanzee. *S. Afr. J. Sci.* 55(11):286–91.

Bourne, G. H., and Cohen, M. 1975. *The Gentle Giants: The Gorilla Story*. New York: Putnam.

Bradley, M. H. 1922. *On the Gorilla Trail*. New York: Appleton.

———. 1926. Among the gorillas. *Liberty* (weekly), 27 February, 31.

Broom, R. 1946. Notes on a late gorilla foetus. *Ann. Transvaal Mus.* 20:347–50.

Broughton, Lady. 1932. Stalking the mountain gorilla with the camera in its natural haunts. *Illustr. London News,* 5 November, 701, 710–13; 12 November, 756–57.

Brown, T., et al. 1970. A mechanistic approach to treatment of rheumatoid type arthritis naturally occurring in a gorilla. *Trans. Am. Clin. Climatol. Assoc.,* vol. 82.

Brown, T., Clark, H. W., and Bailey, J. S. 1975. Natural occurrence of rheumatoid arthritis in great apes — a new animal model. Paper presented at the Centennial Symposium on Science and Research, 13 November 1974. In *Proc. Zool. Soc. Phila.*, July.

———. 1980. Rheumatoid arthritis in the gorilla: a study of mycoplasma-host interaction in pathogenesis and treatment. In *Comparative Pathology of Zoo Animals*, 259–66. Washington, D.C.: Smithsonian Institution Press.

Bullick, S. H. 1978. Regeneration of *Musa* after feeding by gorilla. *Biotropica* 10:309.

Burbridge, B. 1928. *Gorilla*. New York: Century.

Burrows, G. 1898. *The Land of the Pigmies*. London: Pearson.

Burt, W. H. 1943. Territoriality and home range concepts as applied to mammals. *J. Mammal.* 24:346–52.

Burton, R. F. 1876. *Two Trips to Gorilla Land and the Cataracts of the Congo*. London: Low.

Burtt, B. D. 1934. A botanical reconnaissance in the Virunga volcanoes of Kigezi, Ruanda, Kivu. *Kew Bull.* 4:145–65.

Bush, M., Moore, J., and Neeley, L. M. 1971. Sedation for transportation of a lowland gorilla. *J. Am. Vet. Med. Assoc.* 159(5):546–48.

Cameron, V. L. 1877. *Across Africa*. New York: Harper.

Campbell, R. I. M. 1970. Mountain gorillas in Rwanda. *Oryx* 10:256–57.

Carmichael, L., Kraus, M. B., and Reed, T. H. 1962. The Washington National Zoological Park gorilla infant, Tamoko. *Int. Zoo Yearbook* 3(1961):88–93.

Caro, T. M. 1976. Observations on the ranging behaviour and daily activity of lone silverback mountain gorillas (*Gorilla gorilla beringei*). *Anim. Behav.* 24:889–97.

Carpenter, C. R. 1937. An observational study of two captive mountain gorillas (*Gorilla beringei*). *Human Biol.* 9:175–96.

Carter, F. S. 1973. Comparison of baby gorillas with human infants at birth and during the postnatal period. *Ann. Rep. Jersey Wildl. Preserv. Trust* 10:29–33.

———. 1974. Treatment of acute dehydration in a 293 day old lowland gorilla. *Ann. Rep. Jersey Wildl. Preserv. Trust* 11:60–62.

———. 1976. Significance of the crossed extension reflex in human and lowland gorilla neonates. *Ann. Rep. Jersey Wildl. Preserv. Trust* 13:85–86.

Casimir, M. J. 1975a. Feeding ecology and nutrition of an eastern gorilla group in the Mt. Kahuzi region (République du Zaïre). *Folia Primatol.* 24:81–136.

———. 1975b. Some data on the systematic position of the eastern gorilla population of the Mt. Kahuzi region (République du Zaïre). *Z. Morphol. Anthropol.* 66:188–201.

————. 1979. An analysis of gorilla nesting sites of the Mt. Kahuzi region (Zaïre). *Folia Primatol.* 32:290–308.

Casimir, M. J., and Butenandt, E. 1973. Migration and core area shifting in relation to some ecological factors in a mountain gorilla group (*Gorilla gorilla beringei*) in the Mt. Kahuzi region (République du Zaïre). *Z. Tierpsychol.* 33:514–22.

Chagula, W. K. 1961. The liver of the mountain gorilla (*Gorilla gorilla beringei*). *Am. J. Phys. Anthropol.* (n.s.) 19:309–15.

Chidester, J. A. 1980. Getting to know a gorilla. In the *Newsletter*, U.S. Department of State, April, pp. 13ff.

Clark, T. 1981. Born in captivity: detailed account of the first birth of a gorilla in New England. *Yankee*, October.

Clevenger, A. B., Marsh, W. L., and Peery, T. M. 1971. Clinical laboratory studies of the gorilla, chimpanzee, and orangutan. *Am. J. Clin. Pathol.* 55(4):479–88.

Clift, J. P., and Martin, R. D. 1978. Monitoring of pregnancy and postnatal behaviour in a lowland gorilla at London Zoo. *1978 Int. Zoo Yearbook* 18:165–73.

Coffey, P. F., and Pook, J. 1974. Breeding, hand-rearing and development of the third lowland gorilla at the Jersey Zoological Park. *Ann. Rep. Jersey Wildl. Preserv. Trust* 11:45–53.

Coffin, R. 1978. Sexual behavior in a group of captive young gorillas. *Bol. Estud. Med. Biol.* (Mexico) 30:65–69.

Coolidge, H. J., Jr. 1929. A revision of the genus *Gorilla. Mem. Mus. Comp. Zool.* (Harvard) 50:291–381.

————. 1930. Notes on the gorilla. In *The African Republic of Liberia and the Belgian Congo,* ed. R. P. Strong, vol. 2, 623–35. Cambridge: Harvard University Press.

————. 1936. Zoological results of the George Vanderbilt African expedition of 1934. Pt. 4, Notes on four gorillas from the Sanga River region. *Proc. Acad. Nat. Sci. Phila.* 88:479–501.

Cooper, D. 1975. Zoo's first baby gorilla dies at birth. *News from Phila. Zoo,* 16 June.

Corbet, G. B. 1967. Nomenclature of the "eastern lowland gorilla." *Nature* 215(5106):1171–72.

Cousins, D. 1972a. Body measurements and weights of wild and captive gorillas, *Gorilla gorilla. Zool. Garten* (N.F.) 41:261–77.

————. 1972b. Diseases and injuries in wild and captive gorillas (*Gorilla gorilla*). *Int. Zoo Yearbook* 12:211–18.

————. 1972c. Gorillas in captivity, past and present. *Zool. Garten* (N.F.) 42:251–81.

————. 1974a. Classification of captive gorillas, *Gorilla gorilla. 1974 Int. Zoo Yearbook* 14:155–59.

————. 1974b. A review of some complaints suffered by captive gorillas with notes on some causes of death in wild gorillas (*Gorilla gorilla*). *Zool. Garten* (N.F.) 44:201–19.

————. 1976a. The breeding of gorillas, *Gorilla gorilla,* in zoological collections. *Zool. Garten* (N.F.) 46:215–36.

————. 1976b. A review of the diets of captive gorillas (*Gorilla gorilla*). *Acta Zool. Pathol. Antverp.,* no. 66:91–100.

————. 1978a. Gorillas: a survey. *Oryx* 14:254–58, 374–76.

————. 1978b. Man's exploitation of the gorilla. *Biol. Conserv.* 13:287–97.

————. 1978c. The reaction of apes to water. *Int. Zoo News* 25(7):8–13.

————. 1980. On the Koolookamba: a legendary ape. *Acta Zool. Pathol. Antverp.,* no. 75:79–93.

Critchley, W. 1968. Report of the Takamanda gorilla survey. Unpublished manuscript.

Crook, J. H. 1970. The socio-ecology of primates. In *Social Behaviour in Birds and Mammals,* ed. J. H. Crook, 103–66. London: Academic Press.

Cunningham, A. 1921. A gorilla's life in civilization. *Zool. Soc. Bull.* (N.Y.) 24:118–24.

Dart, R. A. 1961. The Kisoro pattern of mountain gorilla preservation. *Current Anthropol.* 2(5):510–11.

Delano, F. E. 1963. Gabon gorilla hunt. *Sports Afield,* August, 42, 70.

Derochette, M. 1941. Les gorilles en Territoire de Shabunda. *Bull. Société Botan. Zool. Congolaises* 4(1):7–9.

Derscheid, J. M. 1928. Notes sur les gorilles des volcans du Kivu (Parc National Albert). *Ann. Soc. Roy. Zool. Belgique* 58(1927):149–59.

De Witte, G. F. 1937. Introduction. *Exploration du Parc National Albert, 1933–1935.* Fasc. 1. Inst. des Parcs Nat. du Congo (Brussels).

Didier, R. 1951. L'os pénien du gorille des montagnes. *Bull. Inst. Roy. Sci. Nat. Belgique* 27(35).

Dixson, A. F. 1981. *The Natural History of the Gorilla.* New York: Columbia University Press.

Dixson, A. F., Moore, H. D. M., and Holt, W. V. 1980. Testicular atrophy in captive gorillas (*Gorilla g. gorilla*). *J. Zool.* 191:315–22.

Dmitri, I. 1941. Ugh! ugh! ugh! *Sat. Eve. Post,* 4 January.

Donisthorpe, J. 1958. A pilot study of the mountain gorilla (*Gorilla gorilla beringei*) in South-West Uganda, February to September 1957. *S. Afr. J. Sci.* 54(8):195–217.

Du Chaillu, P. B. 1861. *Explorations and Adventures in Equatorial Africa.* New York: Harper.

————. 1867. *Stories of the Gorilla Country.* New York: Harper.

————. 1869. *Wildlife under the Equator.* New York: Harper.

Elliott, R. C. 1976. Observations on a small group of mountain gorillas (*Gorilla gorilla beringei*). *Folia Primatol.* 25:12–24.

Ellis, J. 1974. Lowland gorilla birth at Oklahoma City Zoo. *Keeper,* July–August, 9–10.

Ellis, R. A., and Montagna, W. 1962. The skin of the primates. Pt. 6, The skin of the gorilla (*Gorilla gorilla*). *Am. J. Phys. Anthropol.* 20:79–94.

Emlen, J. T., Jr. 1960. Current field studies of the mountain gorilla. *S. Afr. J. Sci.* 56(4):88–89.

———. 1962. The display of the gorilla. *Proc. Am. Philos. Soc.* 106:516–19.

Emlen, J. T., Jr., and Schaller, G. B. 1960a. Distribution and status of the mountain gorilla (*Gorilla gorilla beringei*) — 1959. *Zoologica* (N.Y. Zool. Soc.) 45:41–52.

———. 1960b. In the home of the mountain gorilla. *Anim. Kingdom* 63(3):98–108.

Endo, B. 1973. Stress analysis on the facial skeleton of gorilla by means of the wire strain gauge method. *Primates* 14:37–45.

Falkenstein, J. 1876. Ein Lebender Gorilla. *Z. Ethnol.* (Berlin) 8:60–61.

Farnsworth, LaM. 1976. A case of nasal hyperimia in a pregnant gorilla [Hogle Zoo, Salt Lake City]. *Newsletter,* Am. Assoc. Zool. Parks Aquar. (AAZPA), Wheeling, W. Va., vol. 17 (13 March).

Fischer, G. J. 1962. The formation of learning sets in young gorillas. *J. Comp. Physiol. Psychol.* 55:924.

Fischer, R. B., and Nadler, R. D. 1977. Status interactions of captive female lowland gorillas. *Folia Primatol.* 28:122–33.

———. 1978. Affiliative, playful, and homosexual interactions of adult female lowland gorillas. *Primates* 19:657–64.

Fisher, G., and Kitchener, S. 1965. Comparative learning in young gorillas and orangutans. *J. Genet. Psychol.* 107:337–48.

Fisher, L. E. 1972. The birth of a lowland gorilla at the Lincoln Park Zoo, Chicago. *Int. Zoo Yearbook* 12:106–8.

Fitzgerald, F. L., Barfield, M. A., and Grubbs, P. A. 1970. Food preferences in lowland gorillas. *Folia Primatol.* 12:209–11.

Fitzgibbons, J. F., and Simmons, L. 1974. Autopsy findings of a three month old zoo born female lowland gorilla. *J. Zoo Anim. Med.* 5:13–18.

Ford, H. A. 1852. On the characteristics of the *Troglodytes* gorilla. *Proc. Acad. Nat. Sci. Phila.* 6:30–33.

Fossey, D. 1968. J'observe les gorilles sur la chaine des Birunga. *Bull. Agricole Rwanda,* no. 3:163–64.

———. 1970a. Making friends with mountain gorillas. *Nat. Geogr.* 137:48–67.

————. 1970b. Mes amis les gorilles. *Bull. Agricole Rwanda,* no. 4:162–65.

————. 1971. More years with mountain gorillas. *Nat. Geogr.* 140:574–85.

————. 1972a. Living with mountain gorillas. In *The Marvels of Animal Behavior,* ed. T. B. Allen, 208–29. Washington, D.C.: National Geographic Society.

————. 1972b. Vocalizations of the mountain gorilla (*Gorilla gorilla beringei*). *Anim. Behav.* 20:36–53.

————. 1974. Observations on the home range of one group of mountain gorillas (*Gorilla gorilla beringei*). *Anim. Behav.* 22:568–81.

————. 1976a. Alternatives to female transfer trend among mountain gorillas. Unpublished paper.

————. 1976b. The behaviour of the mountain gorilla. Ph.D. diss., Cambridge University.

————. 1976c. Transcript of the great apes. *L.S.B. Leakey Foundation News* 6:4–5.

————. 1978a. His name was Digit. Int. Primate Protection League (IPPL) 5(2):1–7.

————. 1978b. Mountain gorilla research, 1969–1970. *Nat. Geogr. Soc. Res. Reps.,* 1969 Projects, 11:173–76.

————. 1979. Development of the mountain gorilla (*Gorilla gorilla beringei*) through the first thirty-six months. In *The Great Apes,* ed. D. A. Hamburg and E. R. McCown, 139–86. Menlo Park, Calif.: Benjamin-Cummings.

————. 1980. Mountain gorilla research, 1971–1972. *Nat. Geogr. Soc. Res. Reps.,* 1971 Projects, 12:237–55.

————. 1981. The imperiled mountain gorilla. *Nat. Geogr.* 159:501–23.

————. 1982a. An amiable giant: Fuertes's gorilla. *Living Bird Quarterly* 1(Summer):21–22.

————. 1982b. Berggorillas — von Austerben bedroht: Das geheimnis der Gorillas. *Stern Mag.,* no. 4 (21 January):24–40.

————. 1982c. Mountain gorilla research, 1974. *Nat. Geogr. Soc. Res. Reps.* 14:243–58.

————. 1982d. Reproduction among free-living mountain gorillas. *Am. J. Primatol. Suppl. 1,* 97–104.

Fossey, D., and Harcourt, A. H. 1977. Feeding ecology of free-ranging mountain gorilla (*Gorilla gorilla beringei*). In *Primate Ecology: Studies of Feeding and Ranging Behaviour in Lemurs, Monkeys and Apes,* ed. T. H. Clutton-Brock, 415–47. London: Academic Press.

Foster, J. W. 1982. Kin selection and gorilla reproduction. *Am. J. Primatol. Suppl. 1,* 27–35.

Freeman, H. E., and Alcock, J. 1973. Play behaviour of a mixed group of juvenile gorillas and orang-utans. *Int. Zoo Yearbook* 13:189–94.

Frueh, R. J. 1968. A captive-born gorilla (*Gorilla g. gorilla*) at St. Louis Zoo. *Int. Zoo Yearbook* 8:128–31.

Gaffikin, P. 1949. *Gorilla gorilla beringei* post-mortem report. *E. Afr. Med. J.* 26(8):1–4.

Galloway, A., Allbrook, D., and Wilson, A. M. 1959. The study of *Gorilla gorilla beringei* with a post-mortem report. *S. Afr. J. Sci.* 55(8):205–9.

Garner, R. L. 1896. *Gorillas and Chimpanzees.* London: Osgood McIlvaine.

———. 1914. Gorillas in their own jungle. *Zool. Soc. Bull.* (N.Y.) 17:1102–4.

Gatti, A. 1932a. Among the pygmies and gorillas. *Popular Mechanics,* September, 418–23.

———. 1932b. Gorilla. *Field 6 Stream,* October, 18–20, 66–67, 73.

———. 1932c. *The King of the Gorillas.* New York: Doubleday, Doran.

———. 1936. *Great Mother Forest.* London: Hodder 6 Stoughton.

Geddes, H. 1955. *Gorilla.* London: Melrose.

Gijzen, A., and Tijskens, J. 1971. Growth in weight of the lowland gorilla (*Gorilla g. gorilla*) and of the mountain gorilla. *Int. Zoo Yearbook* 11:183–93.

Golding, R. R. 1972. A gorilla and chimpanzee exhibit at the University of Ibadan Zoo. *Int. Zoo Yearbook* 12:71–76.

Good, O. I. A. 1947. Gorilla-land. *Nat. Hist.* 56(1):36–37, 44–46.

Goodall, A. G. 1977. Feeding and ranging behaviour of a mountain gorilla group (*Gorilla gorilla beringei*) in the Tshibinda-Kahuzi region (Zaïre). In *Primate Ecology: Studies of Feeding and Ranging Behaviour in Lemurs, Monkeys and Apes,* ed. T. H. Clutton-Brock, 449–79. London: Academic Press.

———. 1978. On habitat and home range in eastern gorillas in relation to conservation. In *Recent Advances in Primatology,* ed. D. J. Chivers and W. Lane-Petters, vol. 2 (Conservation), 81–83. London: Academic Press.

———. 1979. *The Wandering Gorillas.* London: Collins.

Goodall, A. G., and Groves, C. P. 1977. The conservation of eastern gorillas. In *Primate Conservation,* ed. Prince Rainier and G. H. Bourne, 599–637. New York: Academic Press.

Gould, K. G., and Kling, O. R. 1982. Fertility in the male gorilla (*Gorilla gorilla*): relationship to semen parameters and serum hormones. *Am. J. Primatol.* 2(3):311–16.

Greene, D. L. 1973. Gorilla dental sexual dimorphism and early hominid taxonomy. *Symp. 4th Int. Congr. Primatol.* 3:82–100. Basel: Karger.

Gregory, W. K. 1927. How near is the relationship of man to the chimpanzee-gorilla stock? *Quart. Rev. Biol.* 2:549–60.

———. 1950. *The Anatomy of the Gorilla.* New York: Columbia University Press.

Gregory, W. K., and Raven, H. C. 1937. *In Quest of Gorillas.* New Bedford, Mass.: Darwin Press.

Gromier, E. 1948. *La vie des animaux sauvages de la région des Grands Lacs.* Paris: Durel.

Groom, A. F. G. 1973. Squeezing out the mountain gorilla. *Oryx* 12:207–15.

Groves, C. P. 1967. Ecology and taxonomy of the gorilla. *Nature* 213(5079):890–93.

———. 1970a. *Gigantopithecus* and the mountain gorilla. *Nature* 226(5249):973–74.

———. 1970b. *Gorillas.* New York: Arco Publishing Co.

———. 1970c. Population systematics of the gorilla. *J. Zool.* 161:287–300.

———. 1971. Distribution and place of origin of the gorilla. *Man* (London), n.s., 6:44–51.

Groves, C. P., and Humphrey, N. K. 1973. Asymmetry in gorilla skulls: evidence of lateralized brain function? *Nature* 244(5410):53–54.

Groves, C. P., and Napier, J. R. 1966. Skulls and skeletons of *Gorilla* in British collections. *J. Zool.* 148:153–61.

Groves, C. P., and Stott, K. W., Jr. 1979. Systematic relationships of gorillas from Kahuzi, Tshiaberimu and Kayonza. *Folia Primatol.* 32:161–79.

Grzimek, B. 1953. Die gorillas ausserhalb Afrikas. *Zool. Garten* (N.F.) 20:173–85.

———. 1957a. Masse und Gewichte von Flachland-Gorillas. *Z. Säugetierk.* 21:192–94.

———. 1957b. Blinddarmentzündung als Todesursache bei Gorilla-Kleinkind. *Zool. Garten* (N.F.) 23:249.

Grzimek, B., Schaller, G. B., and Kirchshofer, R. (collab.). 1972. The gorilla. In *Grzimek's Animal Life Encyclopedia* 10:525–48. New York: Van Nostrand Reinhold.

Gyldenstolpe, N. 1928. Zoological results of the Swedish expedition to Central Africa 1921. Vertebrata 5, Mammals from the Birunga Volcanoes, north of Lake Kivu. *Ark. Zool.* (Uppsala) 20A:1–76.

Haas, G. 1958. Händigkeitsbeobachtungen bei Gorillas. *Säugetierk. Mitteil.* 6:59–62.

Haddow, A. J., and Ross, R. W. 1951. A critical review of Coolidge's measurements of gorilla skulls. *Proc. Zool. Soc. London* 121:43–54.

Hall-Craggs, E. C. B. 1961a. The blood vessels of the heart of *Gorilla gorilla beringei. Am. J. Phys. Anthropol.* (n.s.) 19:373–77.

———. 1961b. The skeleton of an adolescent gorilla (*Gorilla gorilla beringei*). *S. Afr. J. Sci.* 57(11):299–302.

———. 1962. The testis of *Gorilla gorilla beringei. Proc. Zool. Soc. London* 139:511–14.

Hanno. 1797. *The Voyage of Hanno*, tr. Thomas Falconer. London.

Harcourt, A. H. 1978a. Activity periods and patterns of social interaction: a neglected problem. *Behaviour* 66:121–35.

———. 1978b. Strategies of emigration and transfer by primates, with particular reference to gorillas. *Z. Tierpsychol.* 48:401–20.

———. 1979a. Contrasts between male relationships in wild gorilla groups. *Behav. Ecol. Sociobiol.* 5:39–49.

———. 1979b. Social relationships among adult female mountain gorillas. *Anim. Behav.* 27:251–64.

———. 1979c. Social relationships between adult male and female mountain gorillas in the wild. *Anim. Behav.* 27:325–42.

———. 1981. Can Uganda's gorillas survive? — a survey of the Bwindi Forest Reserve. *Biol. Conserv.* 19:269–82.

Harcourt, A. H., and Curry-Lindahl, K. 1979. Conservation of the mountain gorilla and its habitat in Rwanda. *Environ. Conserv.* 6:143–47.

Harcourt, A. H., et al. 1980. Reproduction in wild gorillas and some comparisons with chimpanzees. *J. Reprod. Fertil.,* suppl., 28:59–70.

Harcourt, A. H., and Fossey, D. 1981. The Virunga gorillas: decline of an "island" population. *Afr. J. Ecol.* 19:83–97.

Harcourt, A. H., Fossey, D., and Sabater Pi, J. 1981. Demography of *Gorilla gorilla. J. Zool.* 195:215–33.

Harcourt, A. H., and Groom, A. F. G. 1972. Gorilla census. *Oryx* 5:355–63.

Harcourt, A. H., and Stewart, K. J. 1977. Apes, sex, and societies. *New Scient.* 76:160–62.

———. 1978a. Coprophagy by wild mountain gorilla. *E. Afr. Wildl. J.* 16:223–25.

———. 1978. Sexual behaviour of wild mountain gorillas. In *Recent Advances in Primatology,* ed. D. J. Chivers and J. Herbert, vol. 1 (Behaviour), 611–12. London: Academic Press.

———. 1980. Gorilla-eaters of Gabon. *Oryx* 15:248–51.

———. 1981. Gorilla male relationships: can differences during immaturity lead to contrasting reproductive tactics in adulthood? *Anim. Behav.* 29:206–10.

Harcourt, A. H., Stewart, K. J., and Fossey, D. 1976. Male emigration and female transfer in wild mountain gorilla. *Nature* 263(5574):226–27.

———. 1981. Gorilla reproduction in the wild. In *Reproductive Biology of the Great Apes,* ed. E. E. Graham, 265–79. New York: Academic Press.

Hardin, C. J., Danford, D., and Skeldon, P. C. 1969. Notes on the successful breeding by incompatible gorillas (*Gorilla gorilla*) at Toledo Zoo. *Int. Zoo Yearbook* 9:84–88.

Hauser, F. 1960. Goma, des Basler Gorillakind; arztliche Berichte über "Goma." Documenta Geigy, *Bull.* no. 4 (Geigy Chemical Corp.).

Hedeen, S. F. 1980. Mother-infant interactions of a captive lowland gorilla. *Ohio J. Sci.* 80(4):137–39.

Heim, A. 1957. Auf den Spuren des Berg-gorillas. *Mitteil. Naturforsch. Gesell. Bern* (N.F.) 14:87–95.

Hess, J. P. 1973. Some observations on the sexual behaviour of captive lowland gorillas, *Gorilla g. gorilla* (Savage and Wyman). In *Comparative Ecology and Behaviour of Primates,* ed. R. P. Michael and J. H. Crook, 507–81. London: Academic Press.

Hess-Haeser, J. 1970. Achila and Quarta. *Museum Books,* February.

Hill, W. C. O., and Sabater Pi, J. 1971. Anomaly of the hallux in a lowland gorilla (*Gorilla gorilla gorilla* Savage and Wyman). *Folia Primatol.* 14:252–55.

Hobson, B. 1975. The diagnosis of pregnancy in the lowland gorilla and the Sumatran orang-utan. *Ann. Rep. Jersey Wildl. Preserv. Trust* 12:71–75.

Hofer, H. O. 1972. A comparative study on the oro-nasal region of the external face of the gorilla as a contribution to cranio-facial biology of primates. *Folia Primatol.* 18:416–32.

Hoier, R. 1955a. *À travers plaines et volcans au Parc National Albert.* 2nd ed. Inst. des Parcs Nat. du Congo Belge (Brussels).

———. 1955b. Gorille de volcan. *Zooleo* (Soc. Bot. Zool. Congolaises, Léopoldville), n.s., no. 30 (January):8–12.

Honegger, R. E., and Menichini, P. 1962. A census of captive gorillas with notes on diet and longevity. *Zool. Garten* (N.F.) 26:203–14.

Hornaday, W. T. 1885. *Two Years in the Jungle.* New York: Scribner.

———. 1915. Gorillas, past and present. *Zool. Soc. Bull.* (N.Y.) 18:1181–85.

Hosokawa, H., and Kamiya, T. 1961–62. Anatomical sketches of the visceral organs of the mountain gorilla (*Gorilla gorilla beringei*). *Primates* 3:1–28.

Hoyt, A. M. 1941. *Toto and I: A Gorilla in the Family.* Philadelphia: Lippincott.

Hughes, J., and Redshaw, M. 1973. The psychological development of two baby gorillas: a preliminary report. *Ann. Rep. Jersey Wildl. Preserv. Trust.* 10:34–36.

———. 1975. Cognitive manipulative and social skills in gorillas. Pt. 1, The first year. *Rep. Jersey Wildl. Preserv. Trust* 11:53–60.

Imanishi, K. 1958. Gorillas: a preliminary survey in 1958. *Primates* 1:73–78.

International Zoo Yearbooks for 1975–77. 1977–79. *See* gorilla births in captivity, vols. 17–19.

Jenks, A. L. 1911. Bulu knowledge of the gorilla and chimpanzee. *Am. Anthropol.* 13:56–64.

Johnstone-Scott, R. 1977. A training program designed to induce maternal behavior in a multiparous female lowland gorilla at the San Diego Wild Animal Park. *1977 Int. Zoo Yearbook* 17:185–88.

——. 1979. Notes on mother-rearing in the western lowland gorilla. *Int. Zoo News*, no. 161 (July–August):9–20.

Joines, S. 1976. The Gorilla conservation program at the San Diego Wild Animal Park. *ZooNooz* (San Diego) 49 (October).

Jonch, A. 1968. The white lowland gorilla at Barcelona Zoo. *Int. Zoo Yearbook* 8:196–97.

Jones, C., and Sabater Pi, J. 1971. Comparative ecology of *Gorilla gorilla* (Savage and Wyman) and *Pan troglodytes* (Blumenbach) in Rio Muni, West Africa. *Bibliotheca Primatol.* (Basel), no. 13:1–96.

Kagawa, M., and Kagawa, K. 1972. Breeding a lowland gorilla at Ritsurin Park Zoo, Takamatsu. *Int. Zoo Yearbook* 12:105–6.

Kawai, M., and Mizuhara, H. 1959. An ecological study of the wild mountain gorilla (*Gorilla gorilla beringei*). *Primates* 2:1–42.

Keiter, M., and Pichette, L. P. 1977. Surrogate infant prepares a lowland gorilla for motherhood. *1977 Int. Zoo Yearbook* 17:188–89.

——. 1979. Reproductive behavior in captive subadult lowland gorillas (*Gorilla g. gorilla*). *Zool. Garten* (N.F.) 49:215–37.

Keith, A. 1896. An introduction to the study of anthropoid apes. 1, The gorilla. *Nat. Sci.* (London) 9:26–37.

——. 1899. On the chimpanzees and their relationship to the gorilla. *Proc. Zool. Soc. London,* 296–312.

Kennedy, K. A. R., and Whittaker, J. C. 1978. The apes in stateroom 10. *Nat. Hist.* 85(9):48–53.

Kevles, B. A. 1980. *Thinking Gorillas,* ed. A. Troy. New York: Dutton.

King, G. J., and Rivers, J. P. W. 1976. The affluent anthropoid. *Ann. Rep. Jersey Wildl. Preserv. Trust* 13:86–95.

Kingsley, S. 1977. Early mother-infant behaviour in two species of great ape: *Gorilla gorilla gorilla* and *Pongo pygmaeus pygmaeus*. *Dodo* (J. Jersey Wildl. Preserv. Trust), no. 14:55–65.

Kirchshofer, R. 1970. Gorillazucht in Zoologischen Gärten und Forschungsstationeen. *Zool. Garten* (N.F.) 38:73–96.

Kirchshofer, R., et al. 1967. An account of the physical and behavioural development of the hand-reared gorilla infant, *Gorilla g. gorilla,* born at Frankfurt Zoo. *Int. Zoo Yearbook* 7:108–13.

——. 1968. A preliminary account of the physical and behavioural development during the first 10 weeks of the hand-reared gorilla twins born at Frankfurt Zoo. *Int. Zoo Yearbook* 8:121–28.

Knoblock, H., and Pasamanick, B. 1959. The development of adaptive behavior in a gorilla. *J. Comp. Physiol. Psychol.* 52:699–703.

Koch, W. 1937. Bericht über das Ergebnis der Obduktion des Gorilla "Bobby" des Zoologischen Gärtens zu Berlin: ein Beitrag zur vergleichenden Konstitutionspathologie. *Veröffentlich. Konst.-Wehrpathol.* 9:1–36.

Koppenfels, H. von. 1877. Meine Jagden auf Gorillas. *Gartenlaube* (Leipzig), 416–20.

Kraft, L. 1952. J'ai vu au Congo le gorille géant des montagnes. *Zooleo* (Soc. Bot. Zool. Congolaises, Léopoldville), n.s., no. 13 (February):187–93.

Krampe, F. 1960. Gorillas auf der Spur. *Kreis.* 4:104–12.

Lang, E. M. 1959. Goma, das Basler Gorillakind: die Geburt. Documenta Geigy, *Bull.* no. 1 (Geigy Chemical Corp.).

———. 1960a. The birth of a gorilla at Basle Zoo. *Int. Zoo Yearbook* 1(1959):3–7.

———. 1960b. Goma, das Basler Gorillakind, ein Jahr alt. *Zolli* (Bull. Zool. Garten, Basel) 5:8–13.

———. 1960c. Goma, das Basler Gorillakind: Gomas Fortschritte. Documenta Geigy, *Bull.* no. 3 (Geigy Chemical Corp.).

———. 1961a. Goma, das Gorillakind. Zurich: Müller.

———. 1961b. Goma, das Basler Gorillakind: Ruckblick und Ausblick. Documenta Geigy, *Bull.* no. 8 (Geigy Chemical Corp.).

———. 1961c. Jambo–unser zweites Gorillakind. *Zolli.* (Bull. Zool. Garten, Basel) 7:1–9.

———. 1962a. *Goma, the Baby Gorilla.* London: Gollancz.

———. 1962b. Jambo, the second gorilla born at Basle Zoo. *Int. Zoo Yearbook* 3(1961):84–88.

———. 1963. *Goma, the Gorilla Baby.* New York: Doubleday.

———. 1964. Jambo: first gorilla raised by its mother in captivity. *Nat. Geogr.* 125:446–53.

Lang, E. M., and Schenkel, R. 1961a. Goma, das Basler Gorillakind: die Entwicklung der sozialen Kontaktweisen. Documenta Geigy, *Bull.* no. 7 (Geigy Chemical Corp.).

———. 1961b. Goma, das Basler Gorillakind: die Reifung der Kontaktweisen im Umgang mit den Dingen. Documenta Geigy, *Bull.* no. 6 (Geigy Chemical Corp.).

Lasley, B. L., Czekala, N. M., and Presley, S. 1982. A practical approach to evaluation of fertility in the female gorilla. *Am. J. Primatol. Suppl. 1,* 45–50.

Lebrun, J. 1935. *Les essences forestières des régions montagneuses du Congo Oriental.* Inst. Nat. pour l'Étude Agronomique du Congo Belge, Sér. Sci. no. 1.

———. 1942. *La végétation du Nyiragongo.* Fasc. 3–5. Inst. des Parcs Nat. du Congo Belge (Brussels).

Lebrun, J., and Gilbert, G. 1954. *Une classification écologique des forêts du Congo.* Inst. Nat. pour l'Étude Agronomique du Congo Belge, Sér. Sci. no. 63.

Ledbetter, D. H., and Basen, J. A. 1982. Failure to demonstrate self-recognition in gorillas. *Am. J. Primatol.* 2(3):307–10.

Lequime, M. 1959. Sur la piste du gorilla. *La Vie des Bêtes* 14:7–8.

Liz Ferreira, A. M. de, Athayde, A., and Magalhaes, H. 1945. Gorilas do Maiombe Português. *Mem. Junta Miss. Geogr. Colon.,* Ser. Zool., Lisbon.

Lotshaw, R. 1971. Births of two lowland gorillas at Cincinnati Zoo. *Int. Zoo Yearbook* 11:84–87.

Lyon, L. 1975a. In the home of mountain gorillas. *Wildlife* 17 (January):16–23.

———. 1975b. The saving of the gorilla. *Africana* (Nairobi) 5(9):11–13, 23.

MacKinnon, J. 1976. Mountain gorillas and bonobos. *Oryx* 13:372–82.

Mahler, P. 1980. Molar size sequence in the great apes: gorilla, orangutan and chimpanzee. *J. Dental Res.,* April, 749–52.

Malbrant, R., and Maclatchy, A. 1949. *Faune de l'équateur Africain Français.* 2, *Mammifères.* In *Encycl. Biol.* 36:1–323.

Mallinson, J. 1974. Wildlife studies on the Zaire River expedition with special reference to the mountain gorillas of Kahuzi-Biega. *Ann. Rep. Jersey Wildl. Preserv. Trust* 11:16–23.

———. 1982. The establishment of a self-sustaining breeding population of gorillas in captivity with special reference to the work of the anthropoid ape advisory panel of the British Isles and Ireland. *Am. J. Primatol. Suppl.* 1, 105–19.

Mallinson, J., Coffey, P., and Usher-Smith, J. 1973. Maintenance, breeding, and hand-rearing of lowland gorilla at the Jersey Zoological Park. *Ann. Rep. Jersey Wildl. Preserv. Trust* 11:5–28.

———. 1976. Breeding and hand-rearing lowland gorillas at the Jersey Zoo. *1976 Int. Zoo Yearbook* 16:189–94.

Maple, T., and Hoff, M. 1982. *Gorilla behavior.* New York: Van Nostrand Reinhold.

March, E. W. 1957. Gorillas of eastern Nigeria. *Oryx* 4:30–34.

Marler, P. 1963. The mountain gorilla: ecology and behavior [review]. *Science* 140(3571):1081–82.

———. 1976. Social organization, communication and grades signals: the chimpanzee and the gorilla. In *Growing Points in Ethology,* ed. P. P. G. Bateson and R. A. Hinde, 239–80. Cambridge: Cambridge University Press.

Martin, R. D. 1975. Application of urinary hormone determinations in the management of gorillas. *Ann. Rep. Jersey Wildl. Preserv. Trust* 12:61–70.

————. 1976. Breeding great apes in captivity. *New Scient.* 72:100–2.

Martin, R. D., Kingsley, S. R., and Stavy, M. 1977. Prospects for coordinated research into breeding of great apes in zoological gardens. *Dodo* (J. Jersey Wildl. Preserv. Trust) no. 14:45–55.

Matschie, P. 1903. Einen Gorilla aus Deutsch-Ostafrica. *Sitzungsber. Gesell. Naturforsch. Freunde Berlin* no. 6 (9 June):253–59.

————. 1905. Merkwürdige Gorilla-Schädel aus Kamerun. *Sitzungsber. Gesell. Naturforsch. Freunde Berlin* no. 10 (12 December):279–83.

Maxwell, M. 1928. The home of the eastern gorilla. *J. Bombay Nat. Hist. Soc.* 32:436–49.

McKenney, F. D., Traum, J., and Bonestell, A. E. 1944. Acute coccidiomycosis in a mountain gorilla (*Gorilla beringei*), with anatomical notes. *J. Am. Vet. Med. Assoc.* 104(804):136–41.

Mecklenburg, A. F., Herzog zu. 1910. *In the Heart of Africa.* Tr. G. E. Maberly-Oppler. London: Cassell.

Merfield, F. G. 1956. *Gorillas Were My Neighbours.* London: Longmans, Green.

Merfield, F. G., and Miller, H. 1956. *Gorilla Hunter.* New York: Farrar, Straus & Cudahy.

Meyer, A. 1955. *Aperçu historique de l'exploration et de l'étude des régions volcaniques du Kivu.* Fasc. 1. Inst. des Parcs Nat. du Congo Belge (Brussels).

Miller, D. A. 1978. Evolution of primate chromosomes: man's closest relative may be the gorilla, not the chimpanzee. *Science* 198(4322):1116–24.

Milton, O. 1957. The last stronghold of the mountain gorilla in East Africa. *Anim. Kingdom* 60(2):58–61.

Morgan, B. J. T., and Leventhal, B. 1977. A model for blue-green algae and gorillas. *J. Appl. Probabil.* 14:675–88.

Moxham, B. J., and Berkovitz, B. K. B. 1974. The circumnatal dentitions of a gorilla (*Gorilla gorilla*) and chimpanzee (*Pan troglodytes*). *J. Zool.* 173:271–75.

Murnyak, D. F. 1981. Censusing the gorillas in Kahuzi-Biega National Park. *Biol. Conserv.* 21:163–76.

Murphy, M. F. 1978. *Gorillas Are Vanishing Intriguing Primates.* Published by the author.

Nadler, R. D. 1974. Periparturitional behaviour of a primiparous lowland gorilla. *Primates* 15:55–73.

————. 1975a. Cyclicity in tumescence of the perineal labia of female lowland gorillas. *Anat. Rec.* 181:791–97.

————. 1975b. Determinants of variability in maternal behavior of captive female gorillas. *Proc. Symp. 5th Congr. Int. Primatol. Soc.* (1974), 207–16. Tokyo: Japan Science Press.

————. 1975c. Second gorilla birth at Yerkes Primate Research Center. *1975 Int. Zoo Yearbook* 15:134–37.

————. 1975d. Sexual cyclicity in captive lowland gorillas. *Science* 189(4205):813–14.

————. 1976. Sexual behavior of captive lowland gorillas. *Arch. Sex. Behav.* 5(5):487–502.

————. 1978. Sexual behaviour of orang-utans in the laboratory. In *Recent Advances in Primatology*, ed. D. J. Chivers and J. Herbert, vol. 1 (Behaviour), 607–8. London: Academic Press.

————. 1982. Laboratory Research on sexual behavior and reproduction of gorillas and orang-utans. *Am. J. Primatol. Suppl. 1*, 57–66.

Nadler, R. D., and Green, S. 1975. Separation and reunion of a gorilla infant and mother. *1975 Int. Zoo Yearbook* 15:198–201.

Nadler, R. D., and Jones, M. L. 1975. Breeding of the gorilla in captivity. *Newsletter,* Am. Assoc. Zool. Parks Aquar. (AAZPA), Wheeling, W.Va., 16:12–17.

Nadler, R. D., et al. 1979. Plasma gonadotropins, prolacting, gonadal steroids and genital swelling during the menstrual cycle of lowland gorillas. *Endocrinology* 105:290–96.

Newman, K. 1959. Saza chief. *Afr. Wild Life* 13:137–42.

Noback, C. R. 1939. The changes in the vaginal smears and associated cyclic phenomena in the lowland gorilla (*Gorilla gorilla*). *Anat. Rec.* 73:209–25.

Noback, C. R., and Goss, L. 1959. Brain of a gorilla. 1, Surface anatomy and cranial nerve nuclei. *J. Comp. Neurol.* 3 (April):321–44.

Noell, A. M. 1979. *Gorilla Show.* Orlando, Fla.: Daniels Publishing Co.

Nott, J. F. 1886. The gorilla. In *Wild Animals Photographed and Described,* 526–38. London: Low.

O'Neil, W. M., et al. 1978. Acute pyelonephritis in an adult gorilla (*Gorilla gorilla*). *Lab. Anim. Sci.* 28(1).

O'Reilly, J. 1960. The amiable gorilla. *Sports Illus.* 12(25):68–76.

Osborn, R. M. 1957. Observations on the mountain gorilla, Mt. Muhavura, S.W. Uganda. Unpublished manuscript.

————. 1963. Observations on the behaviour of the mountain gorilla. *Proc. Symp. Zool. Soc. London* 10:29–37.

Osman Hill, W. C., and Harrison-Matthews, L. 1949. The male external genitalia of the gorilla, with remarks on the os penis of other Hominoidea. *Proc. Zool. Soc. London* 119:363–78.

Owen, R. 1849. Osteological contributions to the natural history of the chimpanzees (*Troglodytes,* Geoffroy) including the description of the skull of a large species (*Troglodytes gorilla,* Savage) discovered by Thomas S.

Savage, M.D. in the Gaboon country, West Africa. *Trans. Zool. Soc. London* 3:381–422.

———. 1859. On the gorilla (*Troglodytes gorilla* Sav.) *Proc. Zool. Soc. London*, 1–23.

———. 1862. Osteological contributions to the natural history of the chimpanzees (*Troglodytes*) and orangs (*Pithecus*). No. 4, Description of the cranium of a variety of the great chimpanzee (*Troglodytes gorilla*), with remarks on the capacity of the cranium and other characters shown by sections of the skull, in the orangs (*Pithecus*), chimpanzees (*Troglodytes*), and in different varieties of the human race. *Trans. Zool. Soc. London* 4(1851):75–88.

———. 1865. *Memoir on the gorilla* (*Troglodytes gorilla*, Savage). London: Taylor & Francis.

Parker, C. 1969. Responsiveness, manipulation, and implementation behavior in chimpanzees, gorillas, and orang-utans. *Proc. 2nd Int. Congr. Primatol.* 1:160–66. New York: Karger.

Patterson, F. 1978a. Conversations with a gorilla. *Nat. Geogr.* 154:438–66.

———. 1978b. The gestures of a gorilla: language, acquisition in another pongid. *Brain and Language* 5:72–97.

Patterson, T. L. 1979. Long-term memory for abstract concepts in the lowland gorilla (*Gorilla g. gorilla*). *Bull. Psychon. Soc.* 13(5):279–82.

Peden, C. 1960. Report on post-mortem performed on adult mountain gorilla. Mimeographed report by the Dept. of Veterinary Services and Animal Husbandry, Uganda.

Penner, L. R. 1981. Concerning threadworm (*Strongyloides stercoralis*) in great apes — lowland gorillas (*Gorilla gorilla*) and chimpanzees (*Pan troglodytes*). *J. Zoo Anim. Med.* 12:128–31.

Petit, L. 1920. Notes sur le gorille. *Bull. Soc. Zool. Fr.* (Paris) 45:308–13.

Philipps, T. 1923. Mfumbiro: the Birunga volcanoes of Kigezi-Ruanda-Kivu. *Geogr. J.* 61(4):233–58.

———. 1950. Letter concerning man's relation to the apes. *Man* (London) 50:168.

Pigafetta, F. 1881. A Report of the Kingdom of Congo . . . drawn out of the writings and discourses of the Portuguese, D. Lopes, by F. Pigafetta, in Rome [1591]. Tr. M. Hutchinson. London: Murray.

Pitman, C. R. S. 1931. *A Game Warden Among His Charges*. London: Nisbet.

———. 1935a. The gorillas of the Kayonsa Region, western Kigezi, s.w. Uganda. *Proc. Zool. Soc. London*, 477–94.

———. 1942. *A Game Warden Takes Stock*. London: Nisbet.

Platz, C. C., Jr., et al. 1980. Electroejaculation and semen analysis in a male lowland gorilla, *Gorilla gorilla gorilla*. *Primates* 21:130–32.

Plowden, G. 1972. *Gargantua: Circus Star of the Century.* New York: Bonanza Books.

Pretorius, P. J. 1947. *Jungle man.* London: Harrap.

Purseglove, J. W. 1950. Kigezi resettlement. *Uganda J.* 14(2):139–52.

Quick, R. 1976. Gorilla habitat display. *Int. Zoo News* 23(6):13–16.

Randall, F. E. 1943. The skeletal and dental development and variability of the gorilla. *Human Biol.* 15:235–337.

————. 1944. The skeletal and dental development and variability of the gorilla. *Human Biol.* 16:23–76.

Rankin, A. 1971. The gorilla is a paper tiger. *Reader's Digest,* April, 210–16.

Ratcliffe, H. L. 1940. New diets for the zoo. *Fauna* (Phila. Zool. Soc.) 2(3):62–65.

————. 1957. Diet keeps oldest gorilla healthy. *Sci. Digest,* February.

Raven, H. C. 1931. Gorilla: the greatest of all apes. *Nat. Hist.* 31(3):231–42.

————. 1936a. Genital swelling in a female gorilla. *J. Mammal.* 17:416.

————. 1936b. In quest of gorillas. 12, Hunting gorillas in West Africa. *Sci. Month.* 43:313–34.

Reade, W. W. 1863. *Savage Africa.* London: Smith (New York: Harper, 1864).

————. 1868. The habits of the gorilla. *Am. Naturalist* 1:177–80.

Redshaw, M. 1975. Cognitive, manipulative, and social skills in gorillas. Pt. 2, The second year. *Ann. Rep. Jersey Wildl. Preserv. Trust* 12:56–60.

————. 1978. Cognitive development in human and gorilla infants. *J. Human Evol.* 7:133–48.

Redshaw, M., and Locke, R. 1976. The development of a play and social behavior in two lowland gorilla infants. *Ann. Rep. Jersey Wildl. Preserv. Trust* 13:71–85.

Reed, T. H., and Gallagher, B. F. 1963. Gorilla birth at National Zoological Park, Washington. *Zool. Garten* (N.F.) 27:279–92.

Reichenow, E. 1920. Biologische Beobachtungen an Gorilla und Schimpanse. *Sitzungsber. Gesell. Naturforsch. Freunde Berlin,* no. 1:1–40.

Reisen, A. H., et al. 1953. Solutions of patterned string problems by young gorillas. *J. Comp. Psychol.* 46:19–22.

Retzius, G. 1913. Über die Spermien des Gorilla. *Anat. Anz.* 43:577–82.

Reynolds, V. 1965. Some behavioral comparisons between the chimpanzee and the mountain gorilla in the wild. *Am. Anthropol.* 67:691–706.

————. 1967a. *The Apes.* New York: Dutton.

————. 1967b. On the identity of the ape described by Tulp 1641. *Folia Primatol.* 5:80–87.

Riess, B. F., et al. 1949. The behavior of two captive specimens of the lowland gorilla, *Gorilla gorilla gorilla* (Savage & Wyman). *Zoologica* (N.Y. Zool. Soc.) 34:111–18.

Rijksen, H. D. 1975. De berggorilla in het Park der Vulkanen Rwanda. *Panda Nieuws* (11th year) no. 5 (May):41–45.

Riopelle, A. J. 1967. Snowflake, the world's first white gorilla. *Nat. Geogr.* 131:442–48.

Riopelle, A. J., Nos, R., and Jonch, A. 1971. Situational determinants of dominance in captive young gorillas. *Proc. 3rd Int. Congr. Primatol.* 3(1970):86–91. Basel: Karger.

Robbins, D., Compton, P., and Howard, S. 1978. Subproblem analysis of skill behavior in the gorilla; a transition from independent to cognitive behavior. *Primates* 19:231–36.

Robinson, P., and Benirschke, K. 1980. Congestive heart failure and nephritis in an adult gorilla. *J. Am. Vet. Med. Assoc.* 177(9):937–38.

Robyns, W. 1947–55. *Flore des spermatophytes du Parc National Albert.* Fasc. 1–3. Inst. des Parcs Nat. du Congo Belge (Brussels).

———. 1948a. *Les territoires biogéographiques du Parc National Albert.* Inst. des Parcs Nat. du Congo Belge (Brussels).

———. 1948b. *Les territoires phytogéographiques du Congo Belge et du Ruanda-Urundi.* Fasc. 1, Atlas général du Congo Belge et du Ruanda-Urundi. Inst. Roy. Col. Belge.

Rock, M. 1978. Gorilla mothers need some help from their friends. *Smithsonian* 9(July):4.

Romero-Horrera, A., Lehmann, H., and Fossey, D. 1975. The myoglobin of primates. 8, *Gorilla gorilla beringei* (eastern highland gorilla). *Biochimica Biophys. Acta* 393:383–88.

Rosen, S. I. 1972. Twin gorilla fetuses. *Folia Primatol.* 17:132–41.

Ross, R. 1954. Ecological studies on the rain forest of southern Nigeria. 3, Secondary succession in the Shasha Forest Reserve. *J. Ecol.* 42:259–82.

Rothschild, L. W. 1923. Exhibition of adult male mountain gorilla. *Proc. Zool. Soc. London,* 176–77.

Rumbaugh, D. M. 1967. "Alvila" — San Diego Zoo's captive-born gorilla. *Int. Zoo Yearbook* 7:98–107.

———. 1968. The behavior and growth of a lowland gorilla and gibbon. *ZooNooz* (San Diego) 39(7):8–17.

Sabater Pi, J. 1960. Beitrag zur Biologie des Flachlandgorillas. *Z. Säugetierk.* 25:133–41.

———. 1966a. Gorilla attacks against humans in Rio Muni, West Africa. *J. Mammal.* 47:123–24.

————. 1966b. Rapport préliminaire sur l'alimentation dans la nature des gorilles du Rio Muni (Ouest Africain). *Mammalia* 30:235–40.

————. 1967. An albino lowland gorilla from Rio Muni, West Africa, and notes on its adaptation to captivity. *Folia Primatol.* 7:155–60.

————. 1977. Contribution to the study of alimentation of lowland gorillas in the natural state, in Rio Muni, Republic of Equatorial Guinea (West Africa). *Primates* 18:183–204.

Sabater Pi, J., and Groves, C. 1972. The importance of higher primates in the diet of the Fang of Rio Muni. *Man* (London), n.s., 7:239–43.

Sabater Pi, J., and Lassaletta, L. de. 1958. Beitrag zur Kenntnis des Flachlandgorillas (*Gorilla gorilla* Savage u. Wyman). *Z. Säugetierk.* 23:108–14.

Saint-Hilaire, I. G. 1851. Note sur le gorille. *Ann. Sci. Nat.* 3rd ser. 16:154–58.

Sanchez, T. 1961. Gunning gorilla in Africa. *Safaris Unlimited,* January.

Sanford, L. J. 1862. The gorilla: being a sketch of its history, anatomy, general appearance and habits. *Am. J. Sci.* 33(2):48–64.

Savage, T. S., and Wyman, J. 1847. Notice of the external characters and habits of *Troglodytes gorilla,* a new species of orang from the Gaboon River, osteology of the same. *Boston J. Nat. Hist.* 5:417–43.

Schäfer, E. 1960. Über den Berggorilla (*Gorilla gorilla beringei*). *Z. Tierpsychol.* 17:376–81.

Schaller, G. B. 1960. The conservation of gorillas in the Virunga volcanoes. *Current Anthropol.* 1(4):331.

————. 1963. *The Mountain Gorilla: Ecology and Behavior.* Chicago: University of Chicago Press.

————. 1964. *The Year of the Gorilla.* Chicago: University of Chicago Press.

————. 1965a. The behavior of the mountain gorilla. In *Primate Behavior,* ed. I. DeVore, 324–67. New York: Holt, Rinehart and Winston.

————. 1965b. My life with wild gorillas. *True,* April, 39–41, 78–82.

Schaller, G. B., and Emlen, J. T. 1963. Observations on the ecology and social behavior of the mountain gorilla. In *African Ecology and Human Evolution,* ed. F. C. Howell and F. Bourlière. Chicago: Aldine.

Schenkel, R. 1960a. Goma, das Basler Gorillakind: Nesthocker oder Nestflüchter. *Documenta Geigy, Bull.* no. 2 (Geigy Chemical Corp.).

————. 1960b. Goma, das Basler Gorillakind: die Reifung artgemässen Fortbewegung und Körperhaltung. *Documenta Geigy, Bull.* no. 5 (Geigy Chemical Corp.).

Schouteden, H. 1927. Gorille de Walikale. *Rev. Zool. Afr.* 15:47.

————. 1947. De zoogdieren van Belgisch-Congo en van Ruanda-Urundi. Fasc. 1–3. *Ann. Mus. van Belgisch-Congo.*

Schultz, A. H. 1927. Studies on the growth of gorilla and of other higher primates with special reference to a fetus of gorilla, preserved in the Carnegie Museum. *Mem. Carnegie Mus.* 11(1):1–87.

————. 1930. Notes on the growth of anthropoid apes, with special reference to deciduous dentition. *Rep. Lab. Mus. Zool. Soc. Phila.* 58:34–45.

————. 1934. Some distinguishing characters of the mountain gorilla. *J. Mammal.* 15:51–61.

————. 1937. Proportions, variability and asymmetries of the long bones of the limbs and the clavicles in man and apes. *Human Biol.* 9:281–328.

————. 1938. Genital swelling in the female orang-utan. *J. Mammal.* 19:363–66.

————. 1939. Notes on diseases and healed fractures of wild apes and their bearing on the antiquity of pathological conditions in man. *Bull. Hist. Med.* 7:571–82.

————. 1942. Morphological observations on a gorilla and an orang of closely known ages. *Am. J. Phys. Anthropol.* 29:1–21.

————. 1950. Morphological observations on gorillas. In *The Anatomy of the Gorilla,* ed. W. K. Gregory, 227–51. New York: Columbia University Press.

Schwarz, E. 1927. Un gorille nouveau de la forêt de l'Ituri. *Rev. Zool. Afr.* 14:333–36.

Seager, S. W. J., et al. 1982. Semen collection and evaluation in *Gorilla gorilla gorilla. Am. J. Primatol. Suppl. 1,* 1–13.

Seal, U. S., et al. 1970. Airborne transport of an uncaged, immobilised 260 kg (572 lb) lowland gorilla. *Int. Zoo Yearbook* 10:134–36.

Sharp, N. A. 1927. Notes on the gorilla. *Proc. Zool. Soc. London,* 1006–9.

————. 1929. The Cameroon gorilla. *Nature* 123(3101):525.

Simons, E. L., and Pilbeam, D. 1971. A gorilla-sized ape from the Miocene of India. *Science* 173(3991):23–27.

Smith, R. R. 1973. Gorilla. *J. Inst. Anim. Technicians* 25(1). Paper read at the Congress of the Institute of Animal Technicians, Newcastle upon Tyne, 1972.

Snowden, J. D. 1933. A study in altitudinal zonation in South Kigezi and on Mounts Muhavura and Mahinga, Uganda. *J. Ecol.* 21:7–27.

Socha, W. W., et al. 1973. Blood groups of mountain gorillas. *J. Med. Primatol.* 2:364–68.

Spencer, D. 1977. A post mortem examination carried out on a stillborn [male] lowland gorilla (*Gorilla g. gorilla*) at the Jersey Zoological Park. *Dodo* (J. Jersey Wildl. Preserv. Trust), no. 14:96–98.

Stecker, R. M. 1958. Osteoarthritis in the gorilla: description of a skeleton with involvement of knee and spine. *Lab. Invest.* (Baltimore) 7(4):445–57.

Steiner, P. E. 1954. Anatomical observations in a *Gorilla gorilla. Am. J. Phys. Anthropol.* (n.s.) 12:145–79.

Steiner, P. E., Rasmussen, T. B., and Fisher, L. E. 1955. Neuropathy, cardiopathy, hemosiderosis and testicular atrophy in *Gorilla gorilla*. *Arch. Pathol.* 59:5–25.

Stewart, K. J. 1977. The birth of a wild mountain gorilla (*Gorilla gorilla beringei*). *Primates* 18:965–76.

Straus, W. L. 1930. The foot musculature of the highland gorilla (*Gorilla beringei*). *Quart. Rev. Biol.* 5:261–317.

———. 1942. The structure of the crown-pad of the gorilla and of the cheek-pad of the orang-utan. *J. Mammal.* 23:276–81.

———. 1950. The microscopic anatomy of the skin of the gorilla. In *The Anatomy of the Gorilla,* ed. W. K. Gregory, 213–21. New York: Columbia University Press.

Suarez, S. D., and Gallup, G. G., Jr. 1981. Self-recognition in chimpanzees and orangutans, but not gorillas. *J. Human Evol.* 10:175–88.

Theobald, J. 1973. Gorilla pediatric procedures. *Proc. Am. Assoc. Zoo. Vet.* 12–14.

Thomas, W. D. 1958. Observations on the breeding in captivity of a pair of lowland gorillas. *Zoologica* (N.Y. Zool. Soc.) 43:95–104.

Tijskens, J. 1971. The oestrous cycle and gestation period of the mountain gorilla. *Int. Zoo Yearbook* 11:181–83.

Tilford, B. L., and Nadler, R. D. 1978. Male parental behavior in a captive group of lowland gorillas (*Gorilla gorilla gorilla*). *Folia Primatol.* 29:218–28.

Tobias, P. V. 1961. The work of the gorilla research unit in Uganda. *S. Afr. J. Sci.* 57(11):297–98.

Tomson, F. N. (n.d.) Root canal therapy for a fractured canine tooth in a gorilla. *J. Zoo Anim. Med.* 9:101–2.

Trouessart, E. 1920. La pluralité des espèces des gorilles. *Bull. Mus. Nat. Hist. Nat.* (Paris) 26:102–8, 191–96.

Tullner, W. W., and Gray, C. W. 1968. Chorionic gonadotropin excretion during pregnancy in a gorilla. *Proc. Soc. Exp. Biol. Med.* 128:954–56.

Tuomey, T. J., and Tuomey, M. W. 1978. Bushman. *Chicago Tribune Mag.,* November.

Tuttle, R. H., and Basmajian, J. V. 1974. Electromyography of forearm musculature in *Gorilla* and problems related to knuckle-walking. In *Primate Locomotion,* ed. F. A. Jenkins, Jr., 293–347. New York: Academic Press.

———. 1975. Electromyography of *Pan gorilla:* an experimental approach to the problem of hominization. *Proc. Symp. 5th Congr. Int. Primatol. Soc.* (1974), 303–14. Tokyo: Japan Science Press.

Urbain, A. 1940. L'habitat et moeurs des gorilles. *Sciences* (Paris) 35:35.

Usher-Smith, J. H., et al. 1976. Comparative physical development in six hand-reared lowland gorillas (*Gorilla gorilla gorilla*) at the Jersey Zoological Park. *Ann. Rep. Jersey Wildl. Preserv. Trust* 13:63–70.

Van den Berghe, L. 1959. Naissance d'un gorille de montagne à la station de zoologie expérimentale de Tshibati. *Folia Sci. Afr. Centralis* 4:81–83.

Van den Berghe, L., and Chardome, M. 1949. Une microfilaire du gorille, *Microfilaria gorillae. Ann. Soc. Belge Med. Trop.* 29:495–99.

Van Straelen, V. 1960. Sanctity of gorilla fastnesses threatened. *Wild Life* 2(2):10–11.

Verhaeghe, M. 1958. *Le Volcan Mugogo.* Fasc. 3. Inst. des Parcs Nat. du Congo Belge (Brussels).

Verschuren, J. 1975. Wildlife in Zaire. *Oryx* 13:149–63.

Vogel, C. 1961. Zur systematischen Untergliederung der Gattung *Gorilla* anhand von Untersuchungen der Mandibel. *Z. Säugetierk.* 26:65–76.

Wallis, W. D. 1934. A gorilla skull with abnormal denture. *Am. Naturalist* 68:179–83.

Webb, J. C. Summary of additional data gathered in Cameroun on western lowland gorillas, *Gorilla g. gorilla* June–August 1974. Unpublished manuscript.

Weber, B. 1979. Gorilla problems in Rwanda. *Swara* (Nairobi) 2(4):28–32.

Werler, J. E. 1975. Gorilla habitat display at Houston Zoo. *1975 Int. Zoo Yearbook* 15:258–60.

Wildt, D. E., et al. 1982. Laparoscopic evaluation of the reproductive organs and abdominal cavity content of the lowland gorilla. *Am. J. Primatol.* 2(1):29–42.

William, Prince of Sweden. 1921. *Among Pygmies and Gorillas.* New York: Dutton.

Willoughby, D. P. 1950. The gorilla — largest living primate. *Sci. Month.* 70:48–57.

————. 1979. *All About Gorillas.* Cranbury, N.J.: Barnes.

Wilson, A. M. 1958. Notes on a gorilla eviscerated at Kabale, August 1958. Laboratory report on faeces and stomach contents. Unpublished report from the Animal Health Research Centre, Entebbe.

Wilson, M. E., et al. 1977. Characteristics of paternal behavior in captive orang-utans (*Pongo pygmaeus abelii*) and lowland gorillas (*Gorilla gorilla gorilla*). Paper presented at the Inaugural Meeting of the American Society of Primatologists at Seattle, Washington.

Wislocki, G. B. 1932. On the female reproductive tract of the gorilla, with a comparison of that of other primates. *Contrib. Embryol. Carnegie Inst. Wash.* 23(135):165–204.

———. 1942. Size, weight and histology of the testes in the gorilla. *J. Mammal.* 23:281–87.

Wolfe, K. A. 1974. Comparative behavioral ecologies of chimpanzees and gorillas. *Univ. Oregon Anthropol. Papers* 7:53–65.

Wood, B. A. 1979. Relationships between body size and long bone lengths in *Pan* and *Gorilla. Am. J. Phys. Anthropol.* (n.s.) 50(1).

Wordsworth, J. 1961. *Gorilla Mountain.* London: Lutterworth Press.

Yamagiwa, J. 1979. A sociological study of the mountain gorilla from a survey in the Kahuzi-Biega National Park (1978–1979). Unpublished manuscript.

Yerkes, R. M. 1927. The mind of a gorilla. Pts. 1, 2. *Genet. Psychol. Monogr.* 2(1, 2):1–191; 2(6):377–551.

———. 1928. The mind of a gorilla. Pt. 3, Memory. *Comp. Psychol. Monogr.* 5(2):1–92.

———. 1951. Gorilla census and study. *J. Mammal.* 32:429–36.

Younglai, E. V., Collins, D. C., and Graham, C. E. 1977. Progesterone metabolism in female gorilla. *J. Endocrinol.* 75:439–40.

Zahl, P. 1960. Face to face with gorillas in Central Africa. *Nat. Geogr.* 117:114–37.

Zucker, E. L., et al. 1977. Grooming behaviors of orang-utans and gorillas: description and comparison. Paper presented at the Animal Behavior Society at University Park, Pennsylvania.

Index